Darwin's Theory

An Introduction to Principles of Evolution

Edward L. Crisp

Printed by CreateSpace, An Amazon.com Company

DEDICATION

To Susan L. Sowards

CONTENTS

ACKNOWLEDGMENTS

I am very thankful and appreciative for the feedback, conversations, suggestions, and support that the following professional colleagues offered over the course of writing this book: Gary Casper, B.S., retired geologist, Office of Surface Mining Reclamation and Enforcement (OSMRE), U.S. Department of the Interior, Morehead, Kentucky; Charles Clonch, M.B.A., J.D., Portland, Oregon; Nancy Dew, M.S., professor emeritus of biology, West Virginia University at Parkersburg, Parkersburg, West Virginia; Marshall Griffin, Ed.D., professor emeritus of biology and chemistry, West Virginia University at Parkersburg, Parkersburg, West Virginia; Cecil Ison, M.A., retired archaeologist with the U.S. Forest Service, Morehead, Kentucky; Joyce Kronberg, Ed.D., professor emeritus of biology, West Virginia University at Parkersburg, Lowell, Ohio; Holly Martin, M.S., professor of biology, West Virginia University at Parkersburg, Parkersburg, West Virginia; Charles Mason, M.S., professor emeritus of geology, Morehead State University, Morehead, Kentucky; Robert Traylor, M.S., PG (Professional Geoscientist licensed in Geology by the State of Texas), CPG (Certified Petroleum Geologist by the American Association of Petroleum Geologists), CPG (Certified Professional Geologist by the American Association of Professional Geologists), retired petroleum geologist and geological advisor to the Texas Railroad Commission (which regulates the Texas petroleum industry), Houston, Texas; Susan Sowards, M.A., retired science teacher from West Virginia University at Parkersburg and Wood County Schools, Vienna, West Virginia; and Gary Waggoner, M.S., professor emeritus of biology, West Virginia University at Parkersburg, Vienna, West Virginia.

CHAPTER 1

THE EVOLUTION OF BIOLOGICAL EVOLUTIONARY THINKING

INTRODUCTION

The ancient Greeks had some thoughts on biologic evolution, but not much was passed on to later generations. From the time of Aristotle (384-322 BCE) until through the Middle Ages (476-1453 CE), there was not much thought on biologic evolution. In fact, Aristotle (and even Plato before him) believed that species were fixed (immutable) and could not change (Larson, 2004; Stott, 2013). Aristotle believed that all living organisms could be arranged on a scale of increasing complexity (later referred to as the *scala naturae* or scale of nature), with each kind of organism along this ladder of life more complex than the previous rung of life (Campbell, 1996). Also, traditional Judeo-Christian interpretations of the Bible (Genesis account in the Old Testament), particularly during the Middle Ages and before the 1700s, stressed the special (divine) creation of each kind of organism and the immutability of species (Campbell, 1996; Larson, 2004; Bowler, 2009).

With the dawning of the Enlightenment during the 1700s, ideas about nature began to change, becoming more materialistic and logic based. During the 1700s and into the 1800s, Enlightenment ideas and philosophies were spreading across Britain and the European continent, and even to America. Materialists of the Enlightenment preferred to remove any trace of design from nature (Bowler, 2009). As stated by Bowler, 2009, p. 81: "For these thinkers, everything we see arises from the ceaseless activity of nature proceeding without any direction or plan. In such a worldview, there could be no room for the concept of fixity of species; the living forms we see are produced by trial and error, and there is no guarantee that their structure will be perpetuated accurately by reproduction." So, during the late 1700s and early 1800s, several naturalists were proposing concepts of biologic evolution. However, to offset this challenge, the English ordained priest and Fellow of Cambridge University, William Paley (1743-1805) (Figure 1-1) revived the argument from design with his 1802 *Natural Theology or Evidence of the Existence and Attributes of the Deity, Collected from the Appearances of Nature* (Eddy and Knight, 2006; Bowler, 2009). Paley's publication was a repeat of an argument forcefully stated in 1691 by the Englishman John Ray (1627-1705) in his book *Wisdom of God as Manifested in the Works of Creation* (Bowler, 2009). Paley is known for comparing the design of a watch, which must have had a designer, to the design of adaptations of organisms by a designing God (Dawkins, 1996; Bowler, 2009). Most naturalists of the early nineteenth century accepted this view of nature and thus practiced natural theology, "…. a philosophy dedicated to discovering the Creator's plan by studying nature" (Campbell, 1996).

Figure 1-1. William Paley, the author of *Natural Theology* (1802). (Public Domain. Source: Wikimedia Commons.)

The Swedish naturalist (and natural theologian), Carolus Linnaeus (1707-1778) (Figure 1-2), devised a hierarchal classification system during the mid-1700s. His

Figure 1-2. Carolus Linnaeus, 1707-1778, devised a scheme for classifying organisms and he thought that species were immutable and remained as created by God. (Public Domain. Source: Wikimedia Commons.)

Systema Naturae, first published in 1735, but the most definitive edition (10th edition) in 1758, is still used today for naming and classifying organisms (Campbell, 1996; Prothero, 2004; Kardong, 2008; Wulf, 2008; Bowler, 2009; Erwin and Valentine, 2013). He further devised a system of Latinized binomial nomenclature for naming each organism on Earth with a unique two-part name consisting of the genus name followed by the species name (Clary and Wandersee, 2013) (for example: *Homo sapiens*; meaning "human" in Latin plus "thinking" in Latin or "thinking human" [Prothero, 2004]). Linnaeus also firmly abided by and promoted the view that species do not change over time and remain in the same form as created by God (Campbell, 1996; Kardong, 2008; Zimmer, 2010).

Charles Darwin's grandfather, Erasmus Darwin (1731-1802), and others of the late 1700s and early 1800s, challenged the design argument and suggested in their writings that organisms have slowly evolved. But the first naturalist to be taken seriously relative to the evolution of species was Jean-Baptiste Pierre Antoine de Monet, Chevalier de Lamarck (1744-1829), usually just referred to as Jean-Baptiste Lamarck (or simply Lamarck) (Figure 1-3). The following is stated by Charles Darwin himself, relative to the importance of Lamarck's work, in the third edition of *On the Origin of Species* in 1861 (p. xiii; see van Wyhe, 2002):

Figure 1-3. Portrait of Jean-Baptiste Lamarck, 1744-1829, by Jules Pizzetta, 1893. Lamarck was the first to be taken seriously relative to ideas about the evolution of organisms. (Public domain. Source: Wikimedia Commons.

> ...Lamarck was the first man whose conclusions on the subject excited much attention. This justly-celebrated naturalist first published his views in 1801; he much enlarged them in 1809 in his 'Philosophie Zoologique,' and subsequently, in 1815, in the Introduction to his 'Hist. Nat. des Animaux sans Vertébres.' In these works, he upholds the doctrine that all species, including man, are descended from other species. He first did the eminent service of arousing attention to the probability of all change in the organic, as well as in the inorganic world, being the result of law, and not of miraculous interposition...

Figure 1-4. Portrait of Charles Darwin in 1839 at the age of 31. The portrait was painted by George Richmond (Public Domain. Source: Wikimedia Commons.)

However, the real father of evolutionary theory entered the stage in the early 1800s, Charles Robert Darwin (1809-1882) (Figure 1-4). Although Darwin's ideas consisted of five different hypotheses relative to biologic evolution (see Mayr, 2001), he is best known today for two of those, 1) descent of organisms, with modification, from common ancestors (branching evolution) and 2) natural selection (Mayr, 2001, see p. 86-87). Descent with modification via natural selection (the primary mechanism of evolutionary change) was first presented as a joint paper read before the Linnaean Society of Great Britain and authored by Charles Darwin and Alfred Russell Wallace (1823-1913) on July 1, 1858 (although both Darwin and Wallace were absent from the proceedings) (Browne, 2002; Desmond and Moore, 1991; Kardong, 2008; van Wyhe, 2013). The following year, Darwin published *On the Origin of Species by Means of Natural Selection, or the Preservation of Favoured Races in the Struggle for Life* (often just referred to as *On the Origin of Species*) on November 24, 1859. All 1250 copies of the first edition were sold out on the first day it was released from publication (Browne, 2002; Kardong, 2008).

The basics of the Darwin and Wallace theory of evolution by natural selection are the following:

- Individuals within species produce many more offspring than can survive and become reproductively successful.
- Thus, there is a tendency for an intrinsic exponential increase in the number of individuals within a species.
- There is inherent variation in the traits of individuals within a species.
- There is competition for limited resources amongst the individuals within a species.
- Only a few of the individuals within a species survive to sexual maturity and become reproductively successful.
- Those individuals of a species with more favorable variations for survival in their environment will, on average, survive and become more reproductively successful than other individuals in the population, thus passing on their favorable inheritable traits to their offspring.

These basics, in their simplicity, result in natural selection. Darwin (1859) stated the concept of natural selection eloquently in *On the Origin of Species* as follows:

> Let it be borne in mind how infinitely complex and close-fitting are the mutual relations of all organic beings to each other and to their physical conditions of life. Can it, then, be thought improbable, seeing that variations useful to man have undoubtedly occurred, that other variations useful in some way to each being in the great and complex battle of life, should sometimes occur in the course of thousands of generations? If such do occur, can we doubt (remembering that many more individuals are born than can possibly survive) that individuals having any advantage, however slight, over others, would have the best chance of surviving and of procreating their kind? On the other hand, we may feel sure that any variation in the least degree injurious would be rigidly destroyed. This preservation of favourable variations and the rejection of injurious variations, I call Natural Selection.

THE CHALLENGE TO THE STATUS QUO OF IMMUTABILITY OF SPECIES

As we have seen above, by the late 1700s to early 1800s, the stability or immutability of species is beginning to be challenged by Enlightenment scholars. The time of the late 1700s to early 1800s is ripe for a challenge to the creationistic view of unchanging species and an independent, special, divine creation for each species. Modern creationists also believe that God has independently created each kind of organism. Of course, this was the view of Linnaeus who believed that his classification scheme reflected the order and pattern of God's construction of various types of similar, but independently created, organisms. Charles Darwin later used the Linnaeus classification scheme, based on the morphologic similarity of different organisms, as evidence of evolution.

Several naturalists of the late 1700s and early 1800s were starting to challenge the fixity of species and envisioned some form of evolutionary change, with species evolving to form new species. As we have already stated previously, Charles Darwin's grandfather, Erasmus Darwin (1731-1802), was one of the first to suggest a slow, gradual evolution of species. Dr. Darwin was a respected physician, botanist, and naturalist. Some of his first hypotheses about evolution were initiated in *Zoonomia, or, The Laws of Organic Life* (1794-1796) (University of California Museum of Paleontology; Erasmus Darwin; accessed October 30, 2013). He also hinted at his ideas on evolution in poetry, as illustrated in the following passage from the posthumously published *The Temple of Nature* (1802) (University of California Museum of Paleontology; Erasmus Darwin; accessed October 30, 2013):

> Organic life beneath the shoreless waves
> Was born and nurs'd in ocean's pearly caves;
> First forms minute, unseen by spheric glass,
> Move on the mud, or pierce the watery mass;
> These, as successive generations bloom,
> New powers acquire and larger limbs assume;
> Whence countless groups of vegetation spring,
> And breathing realms of fin and feet and wing.

Some of the thinking of Erasmus Darwin appears to be similar to that of Jean-Baptiste Lamarck (1744-1829). However, the French naturalist Lamarck was the first to develop a cohesive theory of evolution (first published in 1802 [Larson, 2004]), with organisms changing over time from primitive forms to more complex forms. In 1793, Lamarck became one of the founding professors of the National Museum of Natural History (Musee National d'Histoire Naturelle) in Paris (Larson, 2004). Lamarck was a botanist and an expert on invertebrate animals, thus his academic position gave him an excellent opportunity to promote the concept that species change over time (evolve). But, his mechanisms of inheritance of acquired characteristics, and an inherent driving force in organisms to adapt to their local environments, did not hold up over time (Zimmer and Emlen, 2013). Further, Lamarck believed that organisms sprang into existence by spontaneous generation as simple forms that adapted to their surroundings, but they contained a "life force" that pushed them from simple to ever more complex forms. Thus, Lamarck believed in a ladder approach for evolution, from simply upward to ever more complexity. Nonetheless, Lamarck did accept that species evolve as a result of natural processes and were not specially created by designing forces according to God's plan. This put Lamarck at odds with most naturalist of the day, particularly British naturalists that were steeped in Natural Theology, because it appeared that Lamarck was favoring random natural forces (Kardong, 2008; University of California at Berkeley, Early Concepts of Evolution: Jean-Baptiste Lamarck, accessed October 30, 2013).

Charles Darwin differed greatly from Lamarck in both his mechanism for evolutionary change, natural selection, and his pattern for evolutionary change. Darwin, of course, proposed that evolution was descent with modification from a common ancestor. This would result in a branching pattern of evolutionary divergence (Figure 1-5) rather than a ladder pattern of evolution as proposed by Lamarck.

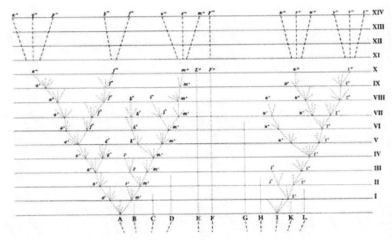

Figure 1-5. Darwin imagined a branching pattern for evolution as shown in his diagram above. This is the only figure that appeared in *On the Origin of Species*. (Public domain. Source: Wikimedia Commons.)

CHARLES DARWIN AND HIS THEORY OF EVOLUTION BY NATURAL SELECTION

Let us take a closer look at Darwin's life and the influences that may have changed his views about nature and the evolution of life on Earth. Charles Robert Darwin was born in Shrewsbury, England on February 12, 1809 (the same day that Abraham Lincoln was born) as the fifth child of Dr. Robert Waring Darwin and Susannah (Wedgewood) Darwin (the daughter of Josiah Wedgewood, the founder of the Wedgewood China dynasty). Dr. Robert Darwin was a successful physician and an astute financier (Browne, 1995). Thus, from the beginning, Charles Darwin was destined to become independently wealthy (Eldredge, 2005).

Darwin's mother died when he was eight years old (1817) and his two older sisters, Marianne and Caroline, stepped-up to run the household. He was enrolled in Mr. Case's school during part of 1817 to be tutored, and then attended Shrewsbury (Boarding) School from 1818 to 1825. At Shrewsbury, he studied the classics but was more interested in the natural sciences. At the age of sixteen, Darwin's father sent him off to Edinburgh University in Scotland to study medicine. Darwin attended for two years (1825-1827), but realized that he could not stomach the practice of medicine as it was done in those days (Eldredge, 2005). However, he did intensify his interest in the natural sciences while at Edinburgh (Browne,

1995; Eldredge, 2005). So, Darwin and his father concluded that he should attend Christ's College at Cambridge University to study for the clergy of the Anglican Church. He attended Cambridge from 1828-1831, taking a B.A. degree in 1831 (Desmond and Moore, 1991; Browne, 1995; Eldredge, 2005).

After graduating from Cambridge University in the spring of 1831, Darwin was led on a geological mapping expedition to northern Wales with Reverend Adam Sedgwick (Figure 1-6), Professor of Geology at Cambridge University (Desmond and Moore, 1991; Browne, 1995; Eldredge, 2005; Grant and Estes, 2009). Sedgwick would name the Cambrian Period (the first period of the Paleozoic Era) in 1835 based on a section of sandstones and shales in northern Wales (Larson, 2004; Grant and Estes, 2009; Erwin and Valentine, 2013). This geologic field trip for Darwin was set-up by his mentor at Cambridge University, Professor John Stevens Henslow (1796-1861) (Figure 1-7). Henslow was the former Professor of Mineralogy, but then Professor of Botany at Cambridge University (Desmond and Moore, 1991; Browne, 1995; Eldredge, 2005; Grant and Estes, 2009).

Figure 1-6. Reverend Adam Sedgwick, Professor of Geology at Cambridge University, as he appeared about 1831. (Public domain. Source: Wikimedia Commons.)

Figure 1-7. John Stevens Henslow (1796-1861), English clergyman, mineralogist, and botanist, is best known today as Charles Darwin's mentor at Cambridge University and recommended Darwin for the naturalist position onboard the HMS Beagle. (Public domain. Source: Wikimedia Commons.)

Darwin had taken a course in geology while at Edinburgh, but Professor Sedgwick introduced Darwin to detailed geologic field mapping and stressed to Darwin the importance of scientific reasoning relative to facts gathered from nature (Darwin, 1969; Browne, 1995; Berra, 2009; Grant and Estes, 2009). This geologic training turned out to be of great value to Darwin later, as an unpaid naturalist onboard the HMS *Beagle* (Grant and Estes, 2009). In fact, throughout his career, Darwin considered himself to be primarily a geologist (Eldredge, 2005) and on the HMS *Beagle* voyage around the world he studied volcanic eruptions, volcanic islands, how atolls form as coral reefs grow on slowly subsiding oceanic volcanoes, earthquakes in the Andes Mountains of South America, and evidence for the rising of the Andes Mountains

(Browne, 1995). He also collected and studied fossils found in Brazil and Argentina while on land excursions from the HMS *Beagle* voyage (Darwin, 1845; Stone, 1980; Browne, 1995).

Upon returning from Wales in the late summer of 1831, Darwin received a letter from his mentor at Cambridge University, John Henslow (Professor of Botany), informing him that he had an opportunity to be chosen as an unpaid naturalist on board the HMS *Beagle*. The captain of the *Beagle*, Robert FitzRoy, also wanted an intellectual companion for the voyage. The purpose of the voyage was to map the coast of South America and to circumnavigate the world. After meeting with Darwin, FitzRoy accepted Darwin as an unpaid naturalist for the journey. The *Beagle* sailed from England in late December of 1831 and returned five years later in October of 1836, circumnavigating the world (Figures 1-8 and 1-9) (Desmond and Moore, 1991; Browne, 1995; Eldredge, 2005).

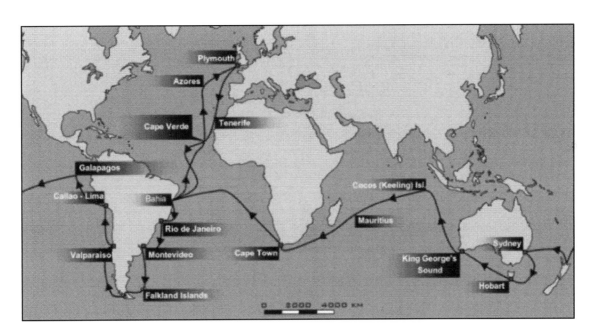

Figure 1-8. A general map that shows the route of the second voyage of the HMS *Beagle* during its trip around the world, from December 27, 1831, to October 2, 1836. Charles Darwin was an unpaid naturalist onboard the *Beagle*. (Permission is granted to copy, distribute and/or modify this document under the terms of the GNU Free Documentation License, Version 1.2 or any later version published by the Free Software Foundation; with no Invariant Sections, no Front-Cover Texts, and no Back-Cover Texts. A copy of the license is included in the section entitled GNU Free Documentation License. Source: http://commons.wikimedia.org/wiki/File:Voyage_of_the_Beagle.jpg.)

Figure 1-9. The HMS Beagle in the Straits of Magellan. Pencil sketch about 1900 by R. T. Pritchett (Public domain. Source:
http://commons.wikimedia.org/wiki/File:PSM_V57_D097_Hms_beagle_in_the_straits_of_magellan.png.)

Prior to the *Beagle* voyage, Darwin was not very focused on his future career with the clergy. He was not overly excited about becoming a country pastor. However, the *Beagle* voyage invigorated him and intensified his interests in the natural sciences. Up until then, he had not questioned the fixity of species nor the creation account for the origin of species (Browne, 1995; Grant and Estes, 2009). His ideas started changing quickly as the voyage continued and as he began observing and collecting plants, animals, and fossils. In particular, he made excursions onto the South American continent to observe the diversity of life in the Amazon rainforest and to collect fossils of extinct ground sloths (and other fossils) in Argentina. Darwin started thinking about how species change over time, with some going extinct and new species arising that are similar, but different than the extinct forms (Desmond and Moore, 1991; Browne, 1995; Grant and Estes, 2009).

The HMS *Beagle* finished mapping the coast of South America in late summer 1835 and then continued the journey around the world. One of the stops on the way was the Galapagos Islands off the coast of Ecuador. The *Beagle* explored the islands from September 17 to October 20, 1835 (Desmond and Moore, 1991; Browne, 1995; Grant and Estes, 2009). Darwin made observations and collections on several of the islands and was much struck by the differences in the various species of birds, iguanas, giant tortoises, etc. from island to island. Of course, Darwin's finches (Figure 1-10) of the Galapagos Islands are widely used today as an example of evolutionary adaptive radiation of a parent species to multiple daughter species, adapted to specific niches (Wicander and Monroe, 2010). However, although Darwin did collect finches on several of the islands, he did not see their full significance until later (after the voyage) as he was evaluating his collections and formulating his ideas on natural selection. Darwin did not even label which finches came from which islands, but luckily Captain FitzRoy, FitzRoy's steward, Harry Fuller, and even Darwin's servant, Syms Covington, made collections of finches and labeled which islands that the finches were collected on (Sulloway, 2005). After retruning to London, Darwin would eventually use his and their collections to hypothesize that the shape of the beak of the various

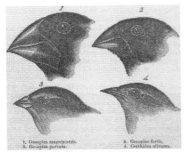

Figure 1-10. Darwin's finches in the publication Darwin, Charles, 1845, *Journal of researches into the natural history and geology of the countries visited during the voyage of H.M.S. Beagle round the world, under the Command of Capt. Fitz Roy, R.N. 2d edition.* (Drawn by John Gould. Image in public domain.)

finches was an adaptation to the different conditions and food sources on each of the islands and the partial isolation of the finches from each other on the several islands (Weiner, 1994; Sulloway, 2005). He further reasoned that a few individuals of a single species of finch were accidentally displaced (by a storm or some other mechanism) to the Galapagos Islands from the South American continent and adaptively radiated into the various species now found on the islands (Weiner, 1994; Sulloway, 2005; Grant and Grant, 2008; Grant and Grant; 2014).

Upon the return of the *Beagle* to London in October 1836, and after study of the finches by ornithologist John Gould who identified the finches as 13 unique species to the Galapagos Archipelago, Darwin publishes the first statement that hints at his ideas on natural selection. Darwin (1845) stated the following relative to the finches in the *Voyage of the Beagle Round the World (second ed.)*: "Seeing this gradation and diversity of structure in one small, intimately related group of birds, one might really fancy that from an original paucity of birds in this archipelago, one species had been taken and modified for different ends." Analysis of these finches led Darwin (and later other scientists) to hypothesize that they were derived from a few individuals of one ancestral species arriving from the mainland to populate and adaptively diversify across the islands (Weiner, 1994; Sulloway, 2005; Grant and Grant, 2008; Grant and Grant; 2014).

Back in London, after the voyage of the *Beagle*, Darwin decides not to become a pastor, but to start developing his ideas on the transmutation of species (evolution). In 1839, he married Emma Wedgewood and they lived in London for three years, before moving to the village of Downe in 1842, 16 miles to the south of London in the county of Kent. They called their home Down House (see Figures 1-11 through 1-15). Charles and Emma (Figure 1-16) would spend the rest of their life at Down House. They would have ten children; however, three did not live to adulthood. One of those three that did not make it to adulthood, not an uncommon occurrence in the 19th century, was 10-year-old Annie. She was the second born child of Charles and Emma and their first daughter. Annie died on April 23, 1851, after an extended illness of scarlet fever (of course, now curable) (Prothero, 2013). Annie was Charles' favorite and any faith in a kind and benevolent God was probably lost with her death (Prothero, 2013).

Darwin appears to have questioned the immutability of species while on the *Beagle* voyage, perhaps as early as 1832; however, it was 1837, while back in London and Cambridge, that Darwin began to start ordering his information and making various notebooks on his ideas about evolution. In 1842, he wrote a sketch of his theory, consisting of 35 pages, stressing the concept of natural selection. In 1844, he completed an enlargement of his ideas on natural selection in an essay of 230 pages. This more extensive essay was locked up in the family safe with instructions to his wife, Emma, to publish the manuscript upon his death (Browne, 2002).

Figure 1-11. The front of Down House showing a poster by English Heritage for Down House as the home of Charles Darwin, in the village of Downe, Kent, England (photo by E. L. Crisp, 2006).

Figure 1-12. The front of Down House in the village of Downe, Kent, England (photo by E. L. Crisp, 2006).

Figure 1-13.   The rear of Down House, and much of the back yard, in the village of Downe, Kent, England (photo by E. L. Crisp, 2006).

Figure 1-14. Darwin's greenhouse (photo by E. L. Crisp, 2006).

Figure 1-15. Inside Darwin's greenhouse (photo by E. L. Crisp, 2006).

Because he was aware of the impact his theory would have on Victorian England, Darwin was reluctant to publish his ideas. However, at the instigation of his close friends, Sir Charles Lyell (the geologist who published the first textbook on geology, entitled *Principles of Geology*, as three volumes from 1830-1833, which stressed the slow gradual geologic evolution of an extremely old Earth) and Joseph Hooker (botanist), Darwin began writing, in 1854, what he thought would be the major publication on his theory, entitled *Natural Selection*. However, Darwin never completed his writing of *Natural Selection* due to circumstances, namely the entrance of Alfred Russell Wallace on the scene (Desmond and Moore, 1991; Browne, 2002).

Figure 1-16. Portrait of Emma Wedgewood Darwin painted by George Richmond in 1840. This portrait was completed about one year after their wedding in January 1839. Emma, also Charles' first cousin, was about 31 at this time. (Public domain. (Source: Wikimedia commons.)

On June 18, 1858, Darwin received a letter from Alfred Russell Wallace asking Darwin if he would give an enclosed manuscript to Charles Lyell for possible publication. Wallace was a young naturalist, then on a collecting expedition in the Malay Archipelago. The title of the manuscript was *On the Tendency of Varieties to Depart Indefinitely from the Original Type*. Darwin was shocked! Wallace's short paper paralleled Darwin's ideas and although the word natural selection was not used in the manuscript, it was clear that Wallace's explanation for the mechanism of evolution was the same as Darwin's. An arrangement was made by Charles Lyell and Joseph Hooker for the joint reading of Wallace's paper and some excerpts from Darwin's work. The excerpts were an 1844 essay (read by Hooker in 1847) and from an 1857 letter from Darwin to Asa Gray of Harvard (about natural selection). The papers were read before the Linnaean Society by Hooker on July 1, 1858 and appeared in print on August 20, 1858 (Desmond and Moore, 1991; Browne, 2002).

Thus, spurred on by the work of Wallace, Darwin rushes to complete what he referred to as an abstract of his theory. *On the Origin of Species* was published on November 24, 1859. For this first edition, 1250 copies were printed and sold out within a few hours (Bowler, 2009). After publication of *On the Origin of Species*, Darwin continued to study, research, and publish. He completed a total of six editions of *On the Origin of Species*, the last in 1872. He wrote several other volumes dealing with geology, zoology, and botany. He also wrote several scientific journal articles. Darwin wrote about 2000 letters, that have been documented, to family, friends, colleagues, and other scientists. The scope of his scientific writing is almost unfathomable (Desmond and Moore, 1991; Browne, 2002).

Throughout the last 20 years or so of his life, Darwin (see Figure 1-17) was plagued by an unknown illness, which reduced his productivity (although he was still very productive). The strange illness left him fatigued, nauseated, and often with fits of vomiting. Darwin passed away at home on April 19, 1882 at the age of 73. He was buried in Westminster Abbey April 26, 1882 next to the grave of Sir John Herschel and about 20 feet from the grave of Sir Isaac Newton (Desmond and Moore, 1991; Browne, 2002).

The following two statements from Darwin's *On the Origin of Species* summarize his view of the evolution of life on Earth. The first relates to natural selection:

As many more individuals of each species are born than can possibly survive; and as, consequently, there is a frequently recurring struggle for existence, it follows that any being, if it vary however slightly in any manner profitable to itself, under the complex and sometimes varying conditions of life, will have a better chance of surviving, and thus be naturally selected. From the strong principle of inheritance, any selected variety will tend to propagate its new and modified form.

The second statement is the last sentence of *On the Origin of Species*: "There is grandeur in this view of life, with its several powers, having been originally breathed by the Creator into a few forms or into one; and that, whilst this planet has gone cycling on according to the fixed law of gravity, from so simple a beginning endless forms most beautiful and most wonderful have been, and are being, evolved."

(Note: The words "by the Creator" are not present in the first edition of *On the Origin of Species*, but are present in the second through sixth editions.)

Figure 1-17. Photograph of Charles Darwin in 1880, about two years before his death. (Public domain. Source: Wikimedia Commons.)

GREGOR MENDEL AND THE MECHANISM OF INHERITANCE

While writing *On the Origin of Species*, Darwin realized that one of the major problems with his theory of natural selection was explaining how traits are passed on from one generation to the next. The following direct quote from Darwin (1859) illustrates the problem: "The laws governing inheritance are for the most part unknown. No one can say why the same peculiarity in different individuals of the same species, or in different species, is sometimes inherited and sometimes not so."

The critics of Darwin's theory of natural selection quickly pointed this out and stated that even if beneficial heritable traits were to arise in a population (by mutation) they would be quickly blended out in the reproductive process from one generation to the next (and next, etc.). Even though Darwin's disciples of evolutionary theory, such as Thomas Henry Huxley in England (often referred to as Darwin's Bulldog), Asa Gray in America, Ernst Haeckel in Germany, and others, were, or became, strong ardent evolutionists; few, if any, completely accepted natural selection as the primary mechanism in evolutionary change (Ruse, 2009; Bowler, 2009). However, Darwin devoted three chapters of *On the Origin of Species* to discussing this topic and presenting evidence that characters do not blend out from one generation to the next. For example, a certain trait may disappear in one generation, only to reappear in a later generation. How could this be if traits were blended out? But, Darwin's critics were not convinced and still preached this concept of "blending out" of heritable traits from one generation to the next (Kardong, 2008). Darwin himself even accepted some blending of traits and also accepted some Lamarckian change (inheritance of acquired characteristics in the form of use and disuse of characters). The science of genetics would eventually come to the rescue of Darwin's theory.

So, critics of natural selection said that Darwin (and Wallace) could not account for the origin of variation and how variation is maintained in populations. Gregor Mendel (1822-1884) (Figure 1-18), an Austrian Monk (and called the Father of Genetics), solved this problem and founded the science of genetics.

Mendel did experiments with true breeding, self-fertilizing pea plants to establish that variable characters could be passed on to offspring and would not be blended out by combination with other characters. Mendel did his work in the 1860s and his results could have helped Darwin and Wallace in their arguments for natural selection. Unfortunately, Mendel published his results in an obscure journal that did not draw much attention. While he did send Darwin a copy of his paper, evidently Darwin did not read it. In fact, Mendel's results went almost unnoticed by the scientific community until his work was independently rediscovered in 1900 by several researchers (Henig, 2000; DeSalle and Yudell, 2005).

Figure 1-18. Photograph of Gregor Johann Mendel (1822-1884). (Unknown author. Public domain. Source: Wikimedia Commons.)

Mendel looked at characters like height, flower color, seed coat color, seed shape, etc. in his pea experiments. While pea plants normally self-fertilize (self-pollinate), Mendel mechanically (surgically) removed the anther (male flower part that contains the male sex gamete - pollen) and cross-fertilized peas with different characteristics. Typically, in self-fertilization, pollen from the anther falls on the stigma of the carpal (female flower part that produces the eggs). The pollen contains sperm nuclei, which fertilizes the egg, forming a zygote. The zygote (having received genetic material from the male part and female part) develops into an embryo. The embryo is coated with an envelope of nutrients and a coat to form a seed. The seed can then develop into a new individual (Campbell, 1996; Kardong, 2008). To prevent self-pollination, Mendel cut away the male anthers and dusted the female carpel with pollen from another plant of his choosing to control the cross. Seeds ripened in the pod; their characteristics were scored; and then these seeds were planted to grow and flower. The characteristics of these full plants were then scored (Kardong, 2008).

Mendel had thoroughly thought these experiments out and perhaps had been informally observing the products of plant hybridization for most of his life. After all, he was the son of a Moravian (Moravia is now part of the Czech Republic) farmer and had learned about the artificial pollination of fruit while a young boy working on his father's farm to raise apples and wine grapes. Mendel was also lucky, but he was still a genius (Jones, 2000). Jones (2000, p. 107) states the following relative to Mendel: "His luck was favored by a prepared mind: by his decision to study simple characters in a simple organism. His genius was to separate the products of inheritance from the mechanism of heredity." Mendel studied seven traits in true-breeding, self-fertilizing pea plants of the genus *Pisum* (Blumberg, 1997). However, this species of pea plant has several bivariable characters that are easy to identify in pea plants as being one or the other type; for example, white flowers or purple flowers, yellow pea seeds or green pea seeds, round pea seeds or wrinkled pea seeds, etc. Mendel listed fifteen characters that occur in two easily identified forms. He only chose to work with seven of these characters in his experiments, but does not explain how or why he chose exactly the seven characters that he did (Blumberg, 1997). We now know that the seven characters that Mendel chose are associated with genes on different chromosomes, except for two of the characters that are located at opposite ends of chromosome 4 *(Pisum* has seven chromosomes). If he had chosen some characters that resulted from genes linked closely on the same chromosome, he would have not obtained the same kind of results for his Law of Independent Assortment (DeSalle and Yudell, 2005). Mendel in his 1865 paper (actually published in 1866) *Experiments in Plant Hybridization* is very casual about throwing out data from pea plants that he considered having unreliable data. So, in the above two incidences, 1) not fully explaining why he chose the seven characters that he did (perhaps the others did not follow all the laws that he proposed, such as independent assortment) and 2) casually throwing data out (perhaps some consider this "tweaking" the data) (Blumberg, 1997).

Mendel's factors (he called them Elemente) for traits we now call genes. Factors that control different expressions of genes at a specific gene position, or locus, (such as purple flowers versus white flowers) we now refer to as alleles. From his studies of pea plants Mendel stated four laws of inheritance, restated as follows (summarized from Campbell, 1996; Kardong, 2008; and Wicander and Monroe, 2010):

- 1) Mendel concluded that, in a sexually reproducing individual organism, genes occur in pairs (one coming from the male parent and one coming from the female parent). Each member of a pair is called an allele (As we now know, there may be several alleles for a particular character trait in a population).

- 2) Mendel concluded that a pea seed received one allele from the pollen and one allele from the ovule (egg). In general, when sex cells (gametes) are formed, the two alleles of each pair separate from one another, and each sex cell receives only one allele of each pair. This is called the Law of Segregation.

- 3) When two alternative forms of the same gene (i.e., two different alleles) are present in an individual, often only one of the alleles is expressed. This is called the Law of Dominance. (Things are not always this simple, there are many cases of incomplete dominance and the crosses will have intermediate characteristics.)

- 4) Mendel concluded that, if two or more separate characteristics are considered in a cross, flower color, pea color, long stems versus short stems, etc., each trait is inherited without relation to other traits. Thus, all possible combinations of independently inherited characteristics will occur in the sex cells (gametes). This is called the Law of Independent Assortment. However, as we now know, this is only true for genes located on different chromosomes, or at least far apart if on the same chromosome, such that they may be separated during meiosis due to crossover.

For example, for flower color, Mendel noted that when true-breeding white flowered pea plants were crossed with true-breeding purple flowered pea plants, the resulting seeds of this F-1 hybrid cross when grown to maturity, all had purple flowers. However, when he let the F-1 pea plants self-fertilize to give an F-2 generation, the results showed phenotypes (external expression of genes) of 3 purple pea flowers and 1 white pea flower (ratio of 3:1 for phenotypes so 75% purple phenotypes and 25% white phenotypes in the F-2 generation). However, the genotypes (actual alleles present) will be 1 PP, 2 Pp, 1 pp (ratio of 1:2:1 for genotypes, so 25% or 1/4 are PP; 50% or 1/2 are Pp; 25% or 1/4 are pp) (P =

dominant allele and p = recessive allele). This is the result of a simple monohybrid cross in the F-2 generation of one gene with two possible alleles. PP is said to be homozygous dominant; Pp is heterozygous; and pp is homozygous recessive. So, to summarize this, when Mendel crossed homozygous purple flowered pea plants with homozygous white flowered pea plants all the offspring were purple in the first-generation cross (F-1 cross). Mendel then let the F-1 generation self-fertilize to give an F-2 generation. In the second generation (F-2) white flowers would again reappear, with a ratio of about 3 purple to 1 white (Mendel did a large number of trials). This was the pattern for all of the F-1 crosses and F-2 generations for all the characters that Mendel studied (see Figure 1-19 for a Punnett square that illustrates this). Of course, he went on to do additional crosses to show that traits do not blend and are passed on from parent to offspring as discrete particles, and are not blended out.

Mendel treated the results mathematically (statistically) and, based on the outcomes of his experiments, he concluded that factors (alleles) that control traits must appear in pairs. These pairs must combine and recombine according to laws of probability. Thus, he presented convincing evidence that traits are not blended out during heritance, but are passed on to offspring and the factors that control traits (alleles) are still present in the genome (all the genes of an organism) as discrete units, and may reappear after not expressing themselves in initial hybrid crosses. The reason that his work was not accepted in his day is most likely due to his rigorous mathematical and statistical treatment of the data. Mendel had a background in physics and mathematics (from the University of Vienna) that most of his contemporary botanists (and naturalists in general) did not have nor appreciate (Henig, 2000).

The importance of Mendel's work is that factors (genes) controlling characteristics (traits) are transmitted as discrete entities, and even though not always expressed, are not lost. Thus, much of the variation among the individuals of a species is accounted for by alternate expression of genes and variation can be maintained within a species without blending out.

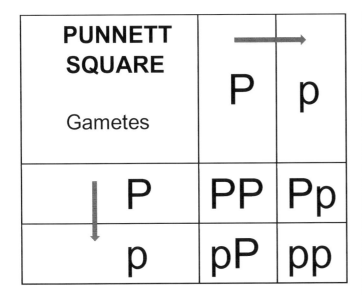

Figure 1-19. Example of a Punnett square for pea flower color of an F-2 generation (letting F-1 peas self-fertilize) in a simple monohybrid cross. This results in phenotypes of 3 purple pea flowers and 1 white pea flower (ratio of 3:1 for phenotypes; so 75% purple phenotypes and 25% white phenotypes). However, the genotypes will be 1 PP, 2 Pp, 1 pp (ratio of 1:2:1 for genotypes, so 25% or 1/4 are PP; 50% or 1/2 are Pp; 25% or 1/4 are pp). This is the result of a simple monohybrid cross in the F-2 generation of one gene with two possible alleles. PP is said to be homozygous dominant (purple); Pp is heterozygous (purple); and pp is homozygous recessive (white). (P = dominant allele which codes for purple flowers and p = recessive allele which codes for white flowers.)

Why didn't Charles Darwin discover Mendel's laws in his botanical experiments? He did many similar experiments as Mendel and even published a two-volume monograph in 1868 entitled *The Variation of Animals and Plants under Domestication*, but also performed countless experiments and observations that dealt with heredity and variation that were published elsewhere (Howard, 2009).

Darwin even performed some experiments with snapdragons (the genus *Antirrhinum*) that mimicked Mendel's experiments with peas. Darwin had a very similar set-up for these experiments, in which he crossed two true-breeding parent plants, one with a spiral arrangement of flowers on stems (the wild type) and the other with radially symmetrical (peloric) flowers on stems (a recessive character) (Kardong, 2008; Howard, 2009). In the F-1 hybrid generation the radial flower trait was not expressed, all the flowers were of the wild type, the spiral-flowered plants. He then crossed the F-1 hybrids to get an F-2 generation of flowers that gave 88 plants with the wild-type (spiral) flower arrangement and 37 plants that gave the radial flower arrangement. This gave a ratio of 2.38 to 1, which Mendel would have accepted as representing the 3:1 phenotypic ratio of dominant to recessive in the F-2 generation (Kardong, 2008; Howard, 2009). According to Howard (2009), "Darwin had other priorities and was in no way programmed to see the critical meaning in these numbers." Howard (2009) goes on to explain that Darwin's overriding purpose was to show that offspring produced by self-fertilization are less fit than progeny of cross-fertilization. He concentrated his studies on quantitative characters that agreed more with overall differential fitness, as Howard (2009) states "...attributes that, by their infinitesimal differences, determine life and death in the wild. So, Darwin counted seeds, weighed and measured them, planted them and looked at vitality. He measured growth and general thriftiness in his self-fertilized and cross-fertilized progeny."

Thus, Darwin stressed that hybrid vigor was important in increasing fitness, whereas Mendel diminished the importance of hybrid vigor in his studies (Blumberg, 1997). Blumberg (1997) states the following in MendelWeb Notes: "Mendel downplays the significance of hybrid vigor here (and throughout the paper), because he wants to stress that the hybrids usually look exactly like the dominant parental form; ..." Darwin was looking for infinitesimal small variations (rather than single or unit variations that interested Mendel) from parents to progeny in terms of fitness, changes that he felt were continuously occurring to create variation within natural populations. Darwin also felt that inheritance was not a problem in nature and that most traits that organisms possessed would be inherited. This may be because of his uneasiness with his understanding of inheritance, via blending inheritance and pangenesis, even bringing in some Lamarckism (inheritance of acquired characteristics). So, Darwin boxed himself in, so to speak, and "was unable to see the laws of inheritance in continuous variation, unable to see the real importance of discontinuous variation where the laws of inheritance could be discerned" (Howard, 2009).

Thus, if Darwin had been a little more open minded and had allowed himself to not be blinded by his ideas of continuous variation in traits and to have looked at what the numbers were telling him in his crosses, he might have stumbled onto the ideas of inheritance as Mendel had done. But alas, that did not happen. In fact, as late as 1872 (and well after Mendel's work had been published) with the publication of the 6th edition of *On the Origin of Species*, Darwin states the following relative to inheritance:

> The laws governing inheritance are for the most part unknown. No one can say why the same peculiarity in different individuals of the same species, or in different species, is sometimes inherited and sometimes not so; why the child often reverts in certain characters to its grandfather or grandmother or more remote ancestor; which a peculiarity is often transmitted from one sex to both sexes, or to one sex alone, more commonly but not exclusively to the like sex.

Oh well, I guess one revolutionary theory that became the central unifying concept in the biological sciences is perhaps enough for one person.

In conclusion, Mendel discovered the particulate nature of inheritance that helped explain Darwin's concept of natural selection. He showed that traits are not blended-out through inheritance. Darwin evidently did not read Mendel's work and therefore did not grasp the meaning of his work. However, Mendel did read Darwin's *On the Origin of Species* and supposedly sent Darwin a copy of his 1866 paper. The paper was found in Darwin's library, uncut (the reprint had been folded into page-sized segments, but had not been cut apart) and evidently had not been read by Darwin (Henig, 2000). Mendel's work was not rediscovered until 1900, and then scientists soon realized the importance of what he had discovered. Here was the basis for preservation of traits favored by natural selection. Here was the source of new traits as a result of mutations in genes. Here was the raw material of variation and evolution within a population. Here was the basis for inheritance of favored traits that are carried from one generation to the next via genes.

THE REDISCOVERY OF GENETICS AND THE MODERN EVOLUTIONARY SYNTHESIS (OR NEO-DARWINISM)

In 1866, Gregor Mendel sent out reprints of his paper on pea hybridization. He had requested 40 reprints and sent out most of those to various well-known botanists and professors of natural science (of course as previously stated, one went to Charles Darwin) in Europe; however, only eleven were later accounted for (Henig, 2000). Mendel only received one response to these mailings.

On March 30, 1868, Mendel was elected abbot of the St. Thomas Monastery and because of the demands of the position he had to give up his scientific research on plant breeding (Henig, 2000). He died in January of 1884, without receiving the credit for his great discovery of how determinants of biologic inheritance are passed on from one generation to the next (DeSalle and Yudell, 2005).

Thirty-five years after presenting his paper in Brunn, Gregor Mendel's work was independently rediscovered and duplicated by three scientists, Hugo de Vries (Dutch botanist, 1848-1935), Carl Correns (German botanist, 1864-1933), and Erich von Tschermak (Austrian agronomist, 1871-1962) during the spring of 1900 (Henig, 2000; DeSalle and Yudell, 2005). Mendel suddenly became famous and right away had a large following of scientists. Remember, in 1900, there was still no agreement on the primary mechanism of evolutionary change and scientists were still trying to understand inheritance. Henig (2000) states the following relative to the rediscovery of Mendel's work in 1900: "Within the year he was enshrined as an unappreciated genius born too soon, the guiding spirit behind the nascent (and still unnamed) science of genetics." The term genetics was proposed in 1905, for the study of heredity and variation, by William Bateson, a Cambridge University professor (Henig, 2000). Bateson became the leading British supporter of genetics (Bowler, 2009).

As the early 1900s progressed, the increased understanding of the cell and the understanding that genes occur on chromosomes in the nucleus of the cell led to a much greater understanding of genetics and how variation can be maintained in populations. However, the source of variation that resulted in new traits was still not resolved. But, one of the scientists credited with the rediscovery of genetics, Hugo de Vries (1848-1935) (Figure 1-20), a Dutch botanist, in 1886 noted that new varieties suddenly occurred within various plants within one generation. He referred to these sudden new heritable traits as resulting from mutations in the hereditary material (genes) of these new varieties of plants. His two-volume *The Mutation Theory* published 1901-1903 stated that species may originate suddenly due to a single advantages mutation, rather than gradually as Darwin had proposed (Bowler, 2009; Kardong, 2008; Wikipedia, Hugo de Vries, accessed 11/10/2013). But he did retain a place for natural selection; maladapted mutations would be removed from a population, whereas adaptive mutations would form new varieties or species. As

Bowler (2009, p. 269) states: "Sooner or later mutation would create a new form even better adapted to the prevailing conditions than the parent species, which would then become extinct." This type of sudden mutational change resulting in new species was an old idea (from even before Darwin, and certainly by many after Darwin), usual referred to as saltational evolution (Mayr, 1991). According to Kardong (2008), "The mutation theory drew many scientific disciples early in the twentieth century and was seen by some as an alternative to Darwin's ideas." Thus, many adherents of Mendelian genetics took the position that abrupt, inheritable mutations were sufficient in

Figure 1-20. Hugo de Vries about 1907. (Public domain. Source: Wikimedia Commons.)

themselves to generate new species and natural selection was not needed. Eventually, continued study and research during the early 1900s (particularly during the 1920s, 1930s. and 1940s) of natural populations and laboratory populations of different species, especially the work of Thomas Hunt Morgan and his students at Columbia University on laboratory populations of the fruit fly, *Drosophila melanogaster*, revealed the nature of the gene, gene positions on chromosomes, genetic mutations, and how this knowledge supported and complemented Darwin's theory of natural selection (Ruse, 2009). The following statement from Bowler (2009, p. 272) is relevant here:

> A few geneticists, including Morgan at one point, had toyed with the possibility that mutations might occur consistently in one direction, thus producing an orthogenetic trend by 'mutation pressure'. But the work on the fruit fly indicated no such trends; a whole range of apparently random mutations was observed, suggesting that Darwin had been right to suppose that variation by itself imposed no direction on evolution. Once the geneticists' initial enthusiasm for saltationism had waned, natural selection became the only plausible mechanism of evolution.

The genetic research undertaken in the 1920s and 1930s by Morgan and others showed that random mutations do occur in populations of organisms, but the supposed sudden changes (saltations) that resulted in new varieties or species of plants as observed by de Vries was determined to be due to polyploidy and new combinations of existing genes (Kardong, 2008; Wikipedia, Hugo de Vries, accessed 11/10/2013). Polyploidy is the duplication (doubling, tripling, quadrupling, etc.) of the entire genome of an organism from the parent(s) to the offspring. Of course, the offspring would then not have chromosomes that matched up with the parent and would not be able to cross back with the parent, but could self-fertilize, thus forming a new species. Polyploidy is common among plants, but rare in animals.

During the 1930s and 1940s, paleontologists, population biologists, geneticists, embryologists, comparative anatomists, zoologists, botanists, and others, developed ideas that merged to form the modern synthesis (sometimes called the Synthetic Theory of Evolution) or neo-Darwinian view of evolution (often just referred to as neo-Darwinism). As stated by Fairbanks (2012, p. 24), "Darwinian principles of natural selection informed by Mendelian principles of inheritance are now known as neo-Darwinism." This view incorporated the chromosome and genetic theories of inheritance into evolutionary thinking. The primary source of new traits in populations of organisms was agreed to result from mutational changes in genes on chromosomes. The scientists involved in the modern synthesis of evolution completely rejected Lamarck's idea of inheritance of acquired characteristics and they reaffirmed the importance of natural selection as the primary mechanism of evolutionary change. But since then, some scientists have challenged the emphasis in the modern synthesis of gradual evolutionary change. These scientists (although accepting that natural selection is the primary mechanism of evolutionary change) think that speciation occurs rapidly followed by long periods of very little change in species (stasis). This concept of rapid speciation interrupting long intervals of species stasis is referred to as punctuated equilibrium (Eldredge and Gould, 1972; Gould and Eldredge, 1977; Stanley, 1981; Gould, 2002).

Evolution by natural selection works on variation in populations, most of which is accounted for by the reshuffling of genes from generation to generation during sexual reproduction (for sexual reproducing organisms). The potential for variation is enormous, with thousands of genes, often each with several alleles, and with offspring receiving 1/2 of their genes from each parent. New traits arise by random mutations resulting in change in the chromosomes or genes.

By the early 1940s a comprehensive theory of biologic evolution resulted in the modern synthesis (neo-Darwinism). This synthesis integrated facts and inferences from several different fields of study, including population genetics, but also paleontology, biogeography, taxonomy and systematics, comparative anatomy, and comparative embryology (Campbell, 1996; Bowler, 2009). Some of the architects of the new synthetic theory of evolution included Russian born geneticist Theodosius Dobzhansky (1900-1975) (Who made the famous statement in 1973 that "Nothing in biology makes sense except in the light of evolution."); the taxonomist and ornithologist Ernst Mayr (1904-2005) of German descent; and the American paleontologist George Gaylord Simpson (1902-1984) (Campbell, 1996; Bowler, 2009). The modern synthesis stressed the evolution of populations, primarily by natural selection. The modern synthesis also supported the concept of gradually evolving populations due to differential reproductive success of certain individuals, those best adapted to specific environments. As a result, populations experience large changes due to the accumulation of small changes over extended periods of time (Campbell, 1996).

Even after the modern evolutionary synthesis (neo-Darwinism) was born, there was still one key component missing (Fairbanks, 2012). What is the molecular nature of the genetic material that results in inheritance of traits? By the 1940s, biologists knew that chromosomes consisted of two types of molecules, proteins and DNA (deoxyribonucleic acid). At first it was suspected that the hereditary material was associated with protein, but experiments with bacteria by Oswald Avery in 1944 and later experiments by Alfred Hershey and Martha Chase in 1952, demonstrated that DNA is the substance of inheritance (Campbell, 1996; Fairbanks, 2012). Later it was shown that DNA consists of long strands of monomers called nucleotides, with each nucleotide consisting of a nitrogenous base, a pentose sugar called deoxyribose, and a phosphate group. The nitrogenous base can be one of four different types: adenine (A), thymine (T), guanine (G), and cytosine (C) (Campbell, 1996). In 1953, the American James Watson and the Englishman Francis Crick solved the structure of DNA and showed the double helical nature of the molecule and how the nucleotide bases paired with each other, always A to T and C to G

(Watson and Crick, 1953a; Campbell, 1996). They further inferred how DNA may function in genetic inheritance and how duplication of DNA may occur (Watson and Crick, 1953b; Campbell, 1996).

Thus, as we have suggested previously, mutations result in a change in hereditary information. Mutations that take place in sex cells are heritable, whether they are chromosomal mutations, affecting a large segment of a chromosome or even the genome (all the DNA in the cell(s) of an organism), or point mutations of individual changes in particular nucleotide bases of DNA (Kardong, 2008). Mutations are random with respect to fitness. They may be beneficial, neutral, or harmful. If a species is well adapted to its environment, most mutations would not be particularly useful and perhaps would be harmful. But what was a harmful mutation can become a useful one if the environment changes. Information in cells is carried in genes (sequences of DNA) on chromosomes that direct the formation of proteins by selecting the appropriate amino acids and arranging them into a specific sequence. Neutral mutations may occur if the information carried in genes does not change the sequence of amino acids that result in a specific protein that is produced.

Some mutations are induced by mutagens, agents that bring about higher mutations rates, such as some chemicals, ultraviolet radiation, X-rays, extreme temperature changes, etc. But some mutations are spontaneous, occurring without any known mutagen, often as a result of mistakes in copying of DNA during the replication process in cells. If these mutations occur in the sex cells, they will be passed on to the next generation.

THE NATURE OF SCIENTIFIC INQUIRY

Before leaving this chapter, it is important that we discuss the nature of scientific inquiry and what constitutes sound science. After all, we owe our modern technological way of life to science and scientific inquiry. As stated by Haack, 2013: "Science is a good thing. As Francis Bacon foresaw centuries ago, when what we now call 'modern science' was in its infancy, the work of the sciences has brought both light, and ever-growing body of knowledge of the world and how it works, and fruit, the ability to predict and control the world in ways that have both extended and improved our lives." This is not to say that other areas of knowledge and inquiry are not important also. For example, knowledge of music, art, architecture, etc. and kinds of inquiry such as philosophical, historical, literary, etc. (Haack, 2013) are also valuable areas of study. However, here we are concerned with evolutionary biology and modern evolutionary biology is a sound science with sound scientific inquiry methodology.

Galileo Galilei (1564-1642) was the first person to seriously challenge the authoritative ideas of Aristotle and was one of the initiators of the modern form of scientific inquiry. Galileo was an Italian physicist, mathematician, and astronomer. He is often referred to as the father of modern science. He introduced the concept of time into ideas about motion and developed the concept of accelerated motion. Galileo was the first astronomer to use the telescope and made important contributions to astronomical ideas which eventually resulted in the acceptance of the Copernican view (Heliocentric view, that Earth and other planets orbit the Sun) of the Universe as opposed to the Aristotelian (Geocentric) view. He was the first to really practice the scientific method (or methods) in the investigation of nature. Galileo's work was another big step in the development of human thought, the separation of science from religion and philosophy (Arney, 2006; Wikipedia, Galileo Galilei, accessed January 2014).

However, Galileo was tried for heresy by the Roman Inquisition in 1633 and found guilty, but after he stated that it was never his intention to really believe the truth of the Copernican doctrine, he was "denounced only as 'vehemently suspected of heresy' and sentenced to punishment at the will of the court" (Hutchins, 1952, p. 126;). Because of this decision by the Roman Catholic Church, Galileo spent the latter years of his life basically under house arrest (Hutchins, 1952, p. 126; Sobel, 2000). In more recent times, the Catholic Church has "expressed regret for how the Galileo affair was handled" (Arney, 2006; Wikipedia, Galileo Galilei, accessed January 2014).

So, what is science? And, equally important, what is not science? A dictionary definition of science would read something like the following: 1) knowledge, 2) knowledge acquired by study, 3) systematized knowledge of any one department of the study of mind or matter, as, the science of physics. Obviously, not exactly how a scientist would define science. Science is really much more than the definition above; science involves a certain attitude about nature; science involves processes or methods; and science results in products. Let us summarize these attributes of science as follows:

- **Attitudes of Science:** Curious, objective, logical, and rational approach to the investigation of the world around us.

- **Scientific Methodology:** Sequential approach to the investigation of natural phenomena. For example, collecting data, evaluating data, formulating hypotheses, testing hypotheses by observation of nature or experimentation, devising appropriate and controlled experiments, and devising unbiased and objective measuring procedures to infer logical conclusions.

- **Products of Science:** facts, hypotheses, theories, principles, and laws of nature.

The following are definitions of terms used in science (see most introductory college science textbooks, such as Tillery (2007); or the National Academies Press (1999) *Science and Creationism: A View from the National Academy of Sciences, Second Edition* at http://www.nap.edu/catalog/6024.html for similar definitions):

- **Fact:** In science, an observation that has been empirically and repeatedly confirmed by observation.

- **Law:** A descriptive generalization about how some aspect of the natural world behaves under stated circumstances.

- **Hypothesis:** A testable statement about the natural world that can be used to build more complex inferences and explanations. A hypothesis may be based on inductive inference (from observable evidence) or may be merely an educated guess, but NOT JUST AN EDUCATED GUESS!!! THE HYPOTHESIS HAS TO HAVE THE POTENTIAL TO BE FALSIFIED OR REJECTED!!! If no data can be collected that will potentially falsify (or allow rejection of) the stated hypothesis, then the proposal is not testable and therefore is not a valid hypothesis (Falsification as used here is not quite in the same sense as used by Karl Popper, 2002 [first English edition 1959], in his *The Logic of Scientific Discovery* [see Haack, 2013, for a discussion of Popper's falsifiability]). A hypothesis, thus allows deductive predictions that are empirically tested by experimental observation or observation of nature.

- **Theory:** In science, a well-tested explanation of some aspect of the natural world that can incorporate facts, laws, inferences, and tested hypotheses. To be referred to as a theory, all data that has been collected to test the theory must support the theory, i.e., no evidence may be present that contradicts the explanatory statement(s) of the theory.

So, what is not science? Science is based on the objective analysis of data collected from the natural world. As such, speculation based on our own biases of how we view nature is not in the realm of science. Also, magic, mysticism, witchcraft, astrology, and other supernatural claims are outside the confines of scientific investigation. Religious beliefs also fall outside the boundaries of science. Science does not say that these beliefs are false, or true, but simply that these beliefs cannot be scientifically investigated. In other words, we cannot formulate hypotheses that are testable (have the potential to be falsified or rejected) relative to supernatural beliefs.

We sometimes hear the statement that there is no such thing as the scientific method. Certainly, not all scientists always follow exactly the same sequence of steps when investigating natural phenomena. Also, many scientific discoveries are a result of serendipity, i.e. making a lucky discovery when investigating another topic or idea. However, when investigating some aspect of nature, most scientists usually gather empirical evidence from nature (either via experiments or observing nature), then formulate explanations of natural phenomena, and ultimately test these predictive explanations by devising additional experiments or observations of nature. Then the process is repeated over again, and again, and again, etc. Susan Haack (2007, p. 94-95), in her book *Defending Science – Within Reason: Between Scientism and Cynicism*, states the following relative to the nature of scientific inquiry:

> ...there is less to 'scientific method' than meets the eye. Is scientific inquiry categorically different from other kinds? No. Scientific inquiry is continuous with everyday empirical inquiry – only more so. Is there a mode of inference or procedure of inquiry used by all and only scientists? No. There are only, on the one hand, modes of inference and procedures of inquiry used by all inquirers, and, on the other, special mathematical, statistical, or inferential techniques, and special instruments, models, etc., local to this or that area of

science. Does this undermine the epistemologically pretensions of science? No! The natural sciences are epistemologically distinguished, have achieved their remarkable successes, in part precisely because of the special devices and techniques by means of which they have amplified the methods of everyday empirical inquiry.

To further expand on this type of thinking relative to the "scientific method" and the nature of scientific inquiry, and empirical inquiry in general, Haack (2014, p. 66) further states the following:

> Any serious empirical inquirer, whatever his subject-matter, will make an informed guess at the possible explanation of the event or phenomenon that puzzles him, figure out the consequences of that guess, see how well those consequences stand up to the evidence he has and any further evidence he can lay his hands on, and then use his judgment whether to stick with the initial guess, modify it, drop it and start again, or just wait until he can figure out what further evidence might clarify the situation, and how to get it.
>
> Over centuries of work, however, scientists have gradually developed an array of tools and techniques to amplify and refine human cognitive powers and overcome human cognitive limitations: techniques of extraction, purification, etc.; instruments of observation from specific areas of science, are not used by all scientists. So, there is *no* 'scientific method' used by all scientists and only by scientists. But, far from suggesting that it is simply a mystery how the natural sciences can have 'made many true discoveries,' this approach suggests a plausible account of how they have gradually managed to refine, amplify, and extend unaided human cognitive powers. (Internal citation deleted.)

Scientific Methodology

The following is a listing that I used in all of my college science classes, with descriptions of the typical steps involved in scientific methodology based on employing empirical methods , induction, and the hypothetical-deductive method:

- A problem or idea about nature to investigate.

- Gathering of initial data (information) about the idea or problem. This may involve library research, observational data from nature, and preliminary experimentation.

- Formulation of a valid hypothesis (A valid hypothesis must be testable, in other words, it must be possible to prove whether it is false or is consistent with observed facts). A hypothesis is a very tentative explanation of some observed phenomenon of nature.

- Testing of the hypothesis by observation of nature and/or experimentation. In both cases, facts (data) are gathered and analyzed relative to the hypothesis. The hypothesis is a predictive model. The data will tell us whether the prediction(s) we have made relative to some natural phenomenon is (or phenomena are) consistent with the observed facts OR NOT. This is typically the point at which a scientific researcher would submit a manuscript of his work for publication in a scientific journal. The editor of the journal would solicit anonymous reviewers of the submitted manuscript. The reviewers (typically three or more that are prominent in the field of research relative to the submitted paper) may suggest that the manuscript be rejected, modified, accepted, or submitted to a different scientific journal. This process is referred to as peer review and is a very important part of scientific inquiry, preventing the publication of sloppy science, unsound science, and pseudoscience.

- Formulation of a theory. If, after repeated testing and wide acceptance by the scientific community, it is determined that none of the data is inconsistent with the stated hypothesis or hypotheses; the hypothesis or hypotheses may be elevated to the status of theory. A theory is an extensively tested and widely accepted explanation for some natural phenomenon. On the other hand, if any of the data is inconsistent with the hypothesis or hypotheses, then one or more of the hypotheses will be modified or rejected. Formulation of a theory usually happens slowly over many years or decades. More and more data is published that supports the hypothesis or hypotheses in question. Perhaps scientists investigating this particular area of study will convene a symposium at an annual scientific meeting of a major scientific organization, such as the Geological Society of America, with invited speakers to present summary papers relative to the topic. Over time, and perhaps with more symposia and published peer-reviewed papers, one or more of the hypotheses proposed to explain the phenomena being investigated is consistent with all of the evidence that has been gathered and brings together a large amount of data that is explained by the hypothesis or hypotheses. At this stage, the scientists working in this particular field may start referring to the explanation as a theory.

- Further testing of the theory. In fact, theories are continually being tested as more scientific data is gathered. Theories are also the spawning ground for additional hypotheses.

- Some hypotheses that are very descriptive of nature under certain sets of circumstances may eventually be elevated to the status of principles or laws of nature, but only after repeated testing and we cannot conceive of situations where the result would be different. A scientific principle is typically more specific than a scientific law, but the division is often arbitrary. Usually scientific laws are descriptions of natural phenomena of a more general nature and of more importance than a principle. Scientific laws and principles are descriptive and tell us how nature acts (not how nature ought to act). Newton's three Laws of Motion and his Law of Universal Gravitation are good examples.

- Some theories serve as broad scientific models of natural processes and involve many accumulated facts, inferences, tested hypotheses, principles, and laws. A good example is the Theory of Plate Tectonics, which explains internal movements within Earth and provides an explanation as to why earthquakes, volcanoes, mountain building, and other phenomena affect certain parts of Earth's surface. Plate Tectonic Theory, which was formulated from decades of research that came together in the late 1960s as an inclusive theory which explained a multitude of seemingly unrelated facts into a coherent, inclusive explanation of all the evidence. Plate Tectonic Theory resulted in a real revolution in the geosciences, with new explanations of formation of most Earth features and even the past geologic and biologic history of Earth. Another example of a paradigmatic theory is the modern Theory of Biologic Evolution, one of the strongest and most fact supported theories in science today, which acts as the central unifying concept in the biological sciences (including the medical and agricultural sciences).

CHAPTER 2

SEDIMENTARY ROCKS – ARCHIVES OF EARTH HISTORY: ORIGIN AND CLASSIFICATION

INTRODUCTION

Rocks are collections of minerals and/or organic material that record the evolutionary history of Earth, both the physical evolution and organic evolution of Earth. This chapter deals with sedimentary rocks. Sedimentary rocks (such as sandstone, shale, and limestone) are the primary archives of Earth history (Wicander and Monroe, 2013). Although geologists (scientists that study Earth materials, processes, and history) and paleontologists (scientists that study ancient life) primarily use sedimentary rocks to interpret Earth history and record evolutionary changes in both organisms and physical events, certainly igneous rocks and metamorphic rocks also reveal information about past life and physical processes. But typically, igneous and metamorphic environments of formation are high temperature and high-pressure ones that destroy the remains of once living organisms (but not always). This is usually not the case for sedimentary rocks.

Because I will be using terminology to identify certain sedimentary rocks and their relationships to fossils (remains of once living organism) in several chapters in this book, I thought it would be appropriate to include here a chapter on the origin and classification of sedimentary rocks. However, this material could be skipped for now and referred back to as necessary.

Minerals are homogeneous inorganic crystalline solids that have a definite chemical composition and characteristic physical properties. Rocks, on the other hand, are usually heterogeneous aggregates of minerals and/or organic material (coal, for example, is a sedimentary rock with a high content of organic material). Some rocks may be composed of primarily one mineral (most limestone, for example, is composed primarily of the mineral calcite). However, most rocks are mixtures of two or more minerals. Rocks are grouped into three major types based on their origin; these are igneous, sedimentary, and metamorphic. Igneous rocks crystallize from molten rock material (examples are granite and basalt) or form from the cementation of fragmented grains (pyroclastic material) blown from volcanic eruptions (examples are volcanic breccia and volcanic tuff). Sedimentary rocks are the end result of the lithification of sediment or crystallization of material dissolved in solution (examples are sandstone and limestone). Metamorphic rocks are the products of heat, pressure, and chemically active solutions that may alter any preexisting rock (examples are slate and gneiss).

Sedimentary rocks are formed from the weathered products of preexisting rocks. The weathered products consist of detrital particles transported from a source outside the depositional basin or dissolved material transported in aqueous solution due to chemical weathering of the source rocks. The weathered products are transported by water, wind, or ice to some depositional site (such as a river channel, a lake, an ocean or sea, a floodplain, a desert, or at the end of a melting glacier) where they are deposited as sediment and eventually lithified into hard rock by compaction and/or cementation of detrital grains or crystallization of material in solution.

There are two major groups of sedimentary rocks, detrital sedimentary rocks and chemical sedimentary rocks. Detrital sedimentary rocks are composed primarily of broken clasts (detritus) of other rocks and/or minerals. Particles of silicate minerals (such as quartz, feldspar, and clay minerals) and lithic (rock) fragments are the chief components of detrital sedimentary rocks. Grain size of the particles is the primary criteria used in the classification of detrital sedimentary rocks, with a secondary name given to the rock based on mineral composition. For example, quartz grains in the size range of 1/16th to 2 mm are referred to as quartz sand (such as typical ocean beach sand). A sedimentary rock composed of sand-sized quartz grains that are compacted and cemented together (lithified) is called quartz sandstone. However, a sedimentary rock composed of sand-sized clasts of quartz and with at least 25% sand-sized clasts of feldspar is called an arkose. If a detrital sedimentary rock is composed of sand-sized grains that consist of quartz, feldspar, and rock fragments and, in addition, has abundant mud-sized grains (<1/16th mm) between the larger grains, it is referred to as a graywacke or subgraywacke (depending on the amount and type of mud matrix).

Chemical sedimentary rocks form by the precipitation (crystallization) of mineral matter dissolved in aqueous solution, either inorganically or biochemically. Many of these rocks have an interlocking crystalline texture and examples are rock salt and rock gypsum that crystallize directly from concentrated seawater or brine. Chemical sedimentary rocks may also have a clastic texture if the precipitated material consists of whole or broken particles that have been transported before final deposition. For example, many marine organisms secrete calcareous exoskeletons (shells) that are often transported and broken by current action. Once the broken shell fragments are compacted and cemented, the resulting rock displays a clastic texture consisting of shells and broken shell fragments. This type of texture is referred to as a bioclastic texture.

THE ORIGIN OF SEDIMENT

The material that ultimately results in the formation of sedimentary rocks is called sediment. Typically, sediment is thought of as loose clastic particles of some sort of material that has been fragmented. When talking of geologic materials, sediment usually refers to particles (grains) of minerals or rocks that have been fragmented by weathering processes, with the sediment eroded (removed from the site of weathering), transported, and deposited at some location in a depositional basin (any area receiving sediment). Examples of this type of sediment are gravel, sand, silt, and clay (note: mud as a particle size indicator usually refers to a mixture of silt- and clay-sized material). In addition to detrital sediment, material that has been chemically dissolved from rocks by aqueous (watery) solutions during the weathering process can also be carried away (transported) to a depositional basin where the material in solution may precipitate (either inorganically or biochemically) to form sediment. Salt (NaCl, the mineral halite) crystals, gypsum ($CaSO_4 \cdot 2H_2O$) crystals, and shells or broken shell fragments (of calcite or aragonite [$CaCO_3$] or silica [SiO_2]) are examples of this type of sediment. Plants and plant fragments are also considered as sediment of biochemical origin and are important in the formation of the sedimentary rock bituminous coal.

Sediment is derived from a source (place of origin) and is produced by the mechanical (physical) and chemical weathering of rocks or the activities of organisms. Erosion removes this material by mass wasting and/or the action of water, wind, or ice. Further transportation of the eroded material takes place by water, wind, or ice to some final depositional site, referred to as a depositional basin.

LITHIFICATION OF SEDIMENT

Sediment is changed into rock, a process called lithification, by a combination of processes. As sediment is buried by additional sediment, the grains are compacted and much of the water that was originally present (for sediment deposited in an aqueous environment) between the grains is squeezed out. Water still remaining in the pore space of the sediment, or ground water that may percolate through the pores of the sediment, becomes saturated with certain minerals, which precipitate in the pore space and cement the grains together. For most siliciclastic detrital sedimentary rocks and many clastic chemical sedimentary rocks, compaction and cementation are the major lithification processes. However, some chemical sedimentary rocks are precipitated (by crystallization) directly from solution and may develop an interlocking crystalline texture at the time of formation. Rock salt and rock gypsum are good examples of this type of lithification.

CLASSIFICATION OF DETRITAL SEDIMENTARY ROCKS

Detrital sedimentary rocks are primarily composed of the clasts (grains) of silicate minerals and rocks. The basis for classification of these sedimentary rocks is grain size. Grain size is one component of the texture of a rock; the texture being the size, shape, and arrangement of the grains within a rock. The grain size of detrital sedimentary rocks ranges from boulder size to minute particles less than 1/256th of 1 mm or 0.0039 mm (3.9 micrometers). Thus, grain size is the most important factor in the naming of detrital sedimentary rocks. A secondary name for a particular detrital sedimentary rock is often given based on mineral composition. Table 2-1 is a simplified version of the grain size classes for sediment and sedimentary rocks used most often by geologists and engineers. Table 2-2 is a classification of the common detrital sedimentary rocks.

Sediment Size Class	Size Range (mm)	Associated Rock Name
Gravel	Greater than 2 mm	Conglomerate (if grains rounded) or Breccia (if grains angular)
Sand	1/16 to 2 mm	Sandstone
Silt	1/256 to 1/16 mm	Siltstone
Clay	Less than 1/256 mm	Shale (if fissile) or Claystone (if nonfissile, blocky) Note: Mudstone is the name given to a rock that is nonfissile and consists of a mixture of silt- and clay-sized grains.

Table 2-1. Grain size classes and associated detrital sedimentary rock names.

CLASSIFICATION OF DETRITAL SEDIMENTARY ROCKS

Grain Size Class	Composition	Comments	Rock Name		
GRAVEL (>2 mm)	Variable, Quartz, Feldspar, Silt, Clay, Rock Fragments	Rounded Grains	CONGLOMERATE		
		Angular Grains	BRECCIA		
SAND (1/16 mm to 2 mm or 0.0625 mm to 2 mm)	Quartz (> 95%)	Typically Rounded to Subrounded Quartz Grains with < 5% Mud Matrix	QUARTZ SANDSTONE		SANDSTONE
	Quartz with > 25% Feldspar	Angular to Subangular Grains of Quartz and Feldspar, Poor Sorting	ARKOSE		
	Quartz, Feldspar, Volcanic and/or Metamorphic Rock Fragments, and Abundant Clay Matrix (typically > 20%)	Angular to Subangular Grains, Poor to Very Poor Sorting	GRAYWACKE	WACKE SANDSTONE	
	Quartz, some Feldspar, some Rock Fragments, Abundant Mud (Silt and Clay) Matrix (typically > 15% but < 20%)	Subrounded to Subangular Grains, Poor Sorting	SUBGRAYWACKE		
MUD (< 1/16 mm) — SILT (1/256 mm to 1/16 mm or 0.0039 mm to 0.0625 mm)	Quartz, Feldspar, and Mica	Nonfissile, Blocky	SILTSTONE		MUDROCKS
CLAY (< 1/256 mm or 0.0039 mm)	Quartz, Feldspar, Mica, and Clay Minerals	Fissile	SHALE		
		Nonfissile, Blocky, clay minerals	CLAYSTONE		
MUD (mix of silt and clay)	Quartz, Feldspar, Mica, and Clay Minerals	Nonfissile, Blocky	MUDSTONE		

Table 2-2. Classification of detrital sedimentary rock.

Conglomerates and Breccias

Conglomerates and breccias are the coarsest-grained detrital sedimentary rocks and contain a significant portion of clastic grains that are larger than 2 mm in diameter. The larger clasts in a conglomerate may vary from granule-size (2 to 4 mm) or to boulders (greater than 256 mm). Many geologists would classify a rock as a conglomerate or breccia if the rock contained 15% or more grains that are larger than 2 mm. Conglomerates (Figure 2-1) consist of rounded to subrounded grains, typically in a finer matrix (groundmass) of sand-sized and/or silt-sized grains. Breccias (Figure 2-2), usually deposited closer to their source areas than conglomerates, consist of angular clasts in a matrix of sand-sized, silt-sized, or clay-sized grains. Clay-sized matrix material is usually more abundant in breccias than in conglomerates.

Figure 2-1. A typical conglomerate. This sample is from the late Jurassic Morrison Formation of Utah and was formed from gravel and sand that accumulated in an ancient stream. The scale is in centimeters. (Photo by E. L. Crisp)

Figure 2-2. A breccia. Notice the large angular grains in a fine mud matrix. The large angular grains in this sample are fragments of chert (microcrystalline quartz), thus this rock may be called a chert breccia. The coin for scale is a quarter. (Photo by E. L. Crisp)

Sandstones

Detrital sedimentary rocks that consist of sand-sized particles in the range of 1/16th mm to 2 mm are called sandstones. Sandstones are further classified according to their composition of grains, cement, and matrix. There are four common sandstone types. Quartz sandstones (Figures 2-3 and 2-4) contain 95% or greater quartz grains in a matrix of less than 5% mud (silt- and clay-sized grains) and are commonly cemented with silica (microcrystalline quartz, SiO_2), calcium carbonate (calcite, $CaCO_3$), iron oxide (hematite, Fe_2O_3), or iron hydroxide (limonite, $FeO(OH) \cdot nH_2O$). A sandstone that consists primarily of quartz grains, but contains 25% or greater feldspar grains is classified as an arkose (Figure 2-5).

Figure 2-3. A quartz sandstone cemented with silica (microcrystalline quartz). This is a coarse quartz sandstone with grain diameters between 1 to 2 mm. Scale numbers are in centimeters. (Photo by E. L. Crisp)

Figure 2-4. A quartz sandstone cemented with hematite (iron oxide). Scale numbers are in centimeters. The darker bands have a reddish color. (Photo by E. L. Crisp)

Figure 2-5. Photo of an arkose sandstone. Notice the relatively large potassium feldspar grains and the poor sorting. This sandstone is also cemented with hematite (iron oxide) and has a reddish color. (Photo by E. L. Crisp)

Sandstones that consist primarily of quartz and chert (microcrystalline quartz) grains, but with mud (silt- and clay-sized material) matrix of greater than 15% of the bulk rock volume, are referred to as subgraywackes (Figure 2-6). Graywackes, on the other hand, are very "dirty-looking sandstones" consisting of grains of quartz, feldspar, and chemically unstable rock fragments in a dark gray to greenish gray matrix (typically exceeding 20% of the bulk rock volume) of mud-sized material (usually containing a high content of the clay mineral chlorite). We often refer to these impure sandstones that have greater than 15% mud matrix as simply wackes (for example quartz wacke or lithic wacke).

Figure 2-6. A subgraywacke sandstone. Although this sandstone has greater than 75% quartz grains, it also has greater than 15% mud matrix. Also note the grains of biotite mica (black) and muscovite mica (light colored). (Photo by E. L. Crisp)

Sandstone types are representative of the depositional environmental conditions (particularly energy of the transporting medium) at the time of sediment deposition and are a great aid to the historical geologist in the interpretation of depositional paleoenvironments (ancient + environments). Quartz sandstones represent stable conditions of sedimentation with mild subsidence (sinking due to sediment loading) and considerable transport and winnowing action of the grains prior to final deposition and burial. They have well-sorted and rounded grains and represent texturally mature and compositionally mature (chemically stable) sandstones, most typically indicative of a beach or shallow marine environment, but may also be found in other environments of deposition, such as desert dune deposits or river bar deposits.

Other sandstone types are representative of depositional environments with less consistent energy conditions (more variable current and winnowing action) than quartz sandstone. Arkose sandstones characteristically occur as thick sedimentary bodies of limited areal extent and usually represent burial in rapidly subsiding basins or fault troughs, resulting from the rapid erosion of granite source areas. They are usually moderately to poorly sorted, have angular to sub-rounded grains, and are typically pinkish in color due to the high content of potassium feldspar that makes up greater than 25% of the grains. They are often deposited in continental settings, such as braided streams and alluvial fans, and may be reddish in color due to a high content of hematite cement (due to the common oxidation of iron present in the surface or ground waters in continental depositional settings). Subgraywackes represent deposition under moderate conditions of subsidence in sedimentary basins where the rate of deposition is rapid enough to prevent winnowing of mud during transportation and final deposition, such as in deltaic and fluvial (river) depositional environments. They typically show moderate sorting and rounding, but with silt and clay content exceeding 15% of the bulk rock volume. Graywackes occur as thick lenticular masses usually associated with linear, rapidly subsiding troughs (for example, the lower continental slope, continental rise, and proximal abyssal plain regions bordering continental shelves). They represent a "poured-in" type of rapid sedimentation with a lack of winnowing action and are often associated with turbidity current deposition (due to slumping of sediment off of the continental shelf edge and the rapid mixing of sediment and water to form a turbid mixture that flows down the continental slope, until the gradient decreases in the lower part of the slope that allows deposition to take place). These sandstones, that have very poor sorting and contain grains that are often angular to subangular, are typically derived from the rapid erosion of volcanic and metamorphic tectonic source areas and contain chemically unstable mineral grains that are rapidly deposited and buried (such as feldspars, micas, amphiboles, pyroxenes, and olivine that are often chemically degraded to clay after deposition).

Mudrocks

Mudrocks are by far the most abundant sedimentary rocks, making up about 70 to 80% of all sedimentary rocks. Mudrocks are detrital sedimentary rocks consisting primarily of silt-sized to clay-sized particles. A mudrock with predominantly silt-sized detrital grains is called a siltstone. Most of the grains within a siltstone cannot be seen with the naked eye, but some of the larger silt-sized particles may be barely visible. Siltstones commonly have a slightly gritty feel as compared to mudrocks with smaller grain size, such as claystones, mudstones, and most shales. Claystone (Figure 2-7) is a rock consisting predominantly of particles (primarily clay minerals and micas, but may also have quartz and feldspar grains) of clay size (less than 1/256 mm, or less than 4 micrometers). Claystones have a very smooth texture and sometimes a soapy feel if most of the minerals present are clay minerals. They have a blocky appearance when broken. Mudstones are also fine-grained detrital sedimentary rocks with a blocky appearance in hand specimens, but consist of a mixture of silt-sized and clay-sized material. It is often difficult to distinguish between claystone and mudstone, but mudstone is typically not quite as smooth feeling as claystone and does not have a soapy feel. Many mudrocks exhibit fissility, the tendency of the rocks to split into fine, parallel layers upon weathering or breakage. For our purposes here, we will define fissility as fine layering of less than 5 mm within a mudrock. A mudrock that exhibits fissility (has a fissile structure) is referred to as shale (see Figure 2-8).

Figure 2-7. Claystone. Note the very fine texture and the blocky nature (not fissile). (Photo by E. L. Crisp)

Figure 1. A black shale with fern fossils. The black color is due to organic carbon in the rock. The coin for scale is a quarter. (Photo by E. L. Crisp)

CLASSIFICATION OF CHEMICAL SEDIMENTARY ROCKS

Chemical sedimentary rocks are those that are predominantly composed of chemical and biochemical precipitates. Carbonates (limestones, composed of the minerals calcite or aragonite, [$CaCO_3$]) and dolostones, composed of the mineral dolomite, [$CaMg(CO_3)_2$]) are the most common chemically formed sedimentary rocks. Carbonates account for about 15% of all sedimentary rocks, with limestones predominating over dolostones. Other chemically derived sedimentary rocks, which are common in the rock record, are rock gypsum (composed of the mineral gypsum, [$CaSO_4 \cdot 2H_2O$]), rock salt (composed of the mineral halite, [$NaCl$]), and chert (composed of microcrystalline quartz [SiO_2]). Bituminous coal is also classified as a chemical sedimentary rock because it consists primarily of organic carbon derived from plant remains.

Chemical sedimentary rocks are basically classified according to their composition (see Table 2-3), with secondary names based on textural parameters.

For example, any sedimentary rock consisting of more than 50% calcium carbonate is called a limestone. There are several types of limestone depending on origin and textural characteristics.

Carbonates (Limestones and Dolostones)

Any sedimentary rock with greater than 50% carbonate minerals is classified either as a limestone or dolostone, depending on whether the rock is composed predominantly of calcium carbonate ($CaCO_3$, as calcite or aragonite) or calcium magnesium carbonate [$CaMg(CO_3)_2$, dolomite]. Most limestones are a combination of clastic particles and calcium carbonate cement or calcium carbonate mud matrix (fine mud-sized particles of calcium carbonate, sometimes referred to as a lime mud). The calcium carbonate mineral in all ancient limestones is calcite; however, most modern carbonate deposition is in the form of aragonite. Aragonite (belonging to a different crystal class than calcite) is relatively unstable at burial conditions near Earth's surface and changes to the crystalline structure of calcite with time. Most dolostones were originally deposited as calcium carbonate, but the carbonate type has been converted to dolomite by percolating ground waters rich in magnesium, as illustrated by the following equation:

$$Mg^{+2} + 2CaCO_3 \text{---------}> CaMg(CO_3)_2 + Ca^{+2}.$$

Limestone (Figures 2-9 through 2-12) and dolostone (Figure 2-13) are relatively soft and can be easily scratched by a steel nail. Limestone can easily be distinguished from other sedimentary rocks because it will effervesce (fizz) readily in dilute (10%) hydrochloric acid (HCl). Dolostone may resemble some limestones in appearance, but will not effervesce readily in HCl unless it is powdered (this can be done by scratching the surface with a steel nail).

CLASSIFICATION OF CHEMICAL SEDIMENTARY ROCKS		
Composition	**Texture and Comments**	**Rock Name**
Greater than 50% Calcite or Aragonite (Calcium Carbonate, $CaCO_3$)	Coarse (>2 mm) to Fine (0.0039 mm) Crystalline	Crystalline Limestone
	Microcrystalline (Lithified Carbonate Mud	Micrite (a limestone)
	Clastic, fossils or fossil fragments, densely cemented or with micrite matrix	Fossiliferous Limestone
	Clastic, shells or shell fragments, coarse sand-sized to greater than 2 mm, loosely cemented	Coquina (a limestone)
	Clastic, skeletal grains, silt to clay size, loosely cemented, friable (grains separate easily)	Chalk (a limestone)
	Clastic, oolite grains, sand-sized	Oolitic Limestone
Dolomite [$CaMg(CO_3)_2$]	Fine to Coarse Crystalline	Dolostone
Halite (NaCl)	Fine to Coarse Crystalline	Rock Salt
Gypsum ($CaSO_4 \cdot 2H_2O$)	Fine to Coarse Crystalline	Rock Gypsum
Quartz (SiO_2)	Microcrystalline, bedded variety or as nodules in limestone, either organic or inorganic	Chert (usually tan or gray, black or brown varieties called flint, red variety called jasper, banded varieties called agate)
Carbonized Macerated Plant Fragments (Carbon, with impurities)	Very Fine, black and nonporous, brittle, dull to shiny	Bituminous Coal

Table 2-3. Classification of chemical sedimentary rocks.

Figure 2-9. Oolitic limestone. Note the spherical oolites that make up the rock. The oolites (or ooids or ooliths) have a diameter of between 0.5 to 1 mm and are formed by inorganic precipitation in current agitated, relatively shallow water environments (usually less than 5 meters, see Prothero and Schwab, 1996, p. 239) that are supersaturated with calcium carbonate ($CaCO_3$). (Photo by E. L. Crisp)

Figure 2-10. Micritic limestone. Basically, lime mud that has lithified. (Photo by E. L. Crisp)

Figure 2-11. Coquina, a bioclastic limestone consisting of relatively large loosely cemented shell fragments. (Photo by E. L. Crisp)

Figure 2-12. Fossiliferous Limestone. This bioclastic limestone consists of sand-sized fragments of fossils and some larger whole fossils. Some geologists would refer to this rock as a biosparite or a bioclastic grainstone (see Prothero and Schwab, 1996, p. 237-244). Fossiliferous Limestone is a common building and monument stone across the U. S. This particular sample was photographed on a building in Marietta, Ohio. (Photo by E. L. Crisp)

Figure 2-13. Dolostone. This specimen has a fine "sugary" crystalline texture. (Photo by E. L. Crisp)

Evaporites

Evaporite sedimentary rocks, rock salt (consisting mostly of the mineral halite) and rock gypsum (consisting mostly of the mineral gypsum) (Figure 2-14), form by crystallization from highly saline bodies of water. Typically, these rocks are deposited in restricted basins (inland seas or lakes) where evaporation exceeds rainfall. In these evaporating basins, halite or gypsum crystals precipitate from solution and fall to the bottom to form a growing, layered mass of interlocking crystals. Upon increased burial and due to increased confining pressure, gypsum ($CaSO_4.2H_2O$) may alter to anhydrite ($CaSO_4$) with the release of water.

Figure 2-14. Rock gypsum. This specimen has a fine crystalline texture. (Photo by E. L Crisp)

Chert

Chert (Figure 2-15) is a sedimentary rock composed of microcrystalline quartz Bedded varieties of chert are the accumulations of the siliceous (composed of SiO_2) skeletons of microscopic protozoans and algae or, in some cases, the larger spicules of sponges. Bedded chert usually forms in deep marine (oceanic) environments. Chert is also common as lensoid or nodular deposits in limestone. It is unclear whether this form of chert is of organic or inorganic origin. Chert has many of the characteristics of other types of quartz, such as hardness, conchoidal fracture, and no cleavage. Chert may occur in various colors with the different colors given a more specific name. Red varieties of chert are called jasper, the black and brown varieties are called flint, the banded varieties are called agate, but the tan to gray varieties are typically just called chert.

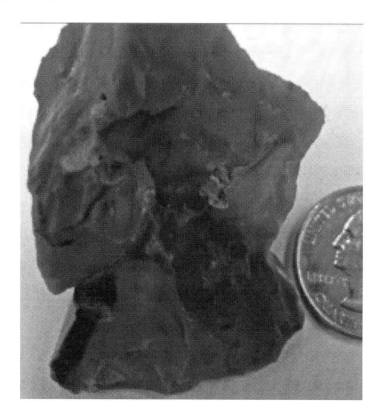

Figure 2-15. Chert, composed of microcrystalline quartz. Note the conchoidal fracture in the lower left. The specimen is also stained with hematite (iron oxide). (Photo by E. L. Crisp)

Sedimentary Coal

Peat, lignite, and bituminous coal (Figure 2-16) are classified as biochemical sediment (peat) or biochemical sedimentary rocks (lignite and bituminous coal) that formed in swampy, low oxygen depositional environments (Figure 2-17) and all consist of 50% or greater carbon derived from plant remains mixed with various amounts of impurities, such as silt and clay. Peat is a soft, brown, porous, and unconsolidated sediment with visibly identifiable plant remains. With further compaction, peat is compressed into a more compact brown coal called lignite. With further burial compaction and chemical alteration of lignite, volatiles (such as water vapor, methane gas, etc.) are squeezed out, thus fixing more carbon in the coal to form bituminous coal (Figure 2-18). Bituminous coal is black, relatively hard (compared to peat and lignite), brittle, dull to shiny, blocky when broken, and relatively free of identifiable plant remains.

Figure 2-16. Bituminous coal. Composed of carbonaceous material of plant origin. (Photo by E. L. Crisp)

Figure 2-17. A diorama at the Carnegie Museum of Natural History in Pittsburgh of a Pennsylvanian Period coal swamp. The large amphibian (about 6 feet long) is *Eryops*. (Photo by E. L. Crisp)

Figure 2-18. Photograph of a Pennsylvanian aged coal seam (about 2 feet thick) in the Allegheny Formation along U. S. Route 50 in Wood County, West Virginia. Susan Sowards is standing near the lower portion of the outcrop for scale and is pointing at the coal seam. (Photo by E. L. Crisp)

CHAPTER 3

GEOLOGIC TIME AND PRINCIPLES OF RELATIVE AND ABSOLUTE AGE DATING OF THE ROCK RECORD

INTRODUCTION

When Charles Darwin developed his theory of biologic evolution he knew that vast amounts of time were necessary for his theory to have validity. He envisioned hundreds of millions of years in order for organisms to diversify to the extent that he observed and for Earth to form the rock and fossil record that he and other scientists were observing. Of course, many naturalists of Darwin's time were creationists (i.e. believed in special creation, each species of organism was individually created by God). Many fundamentalist Christians of today also believe in special creation. Many of these fundamentalist Christians argue that Earth is less than 10,000 years old (Young Earth Creationists) and that evolution via natural selection, as argued by Darwin, could not take place in such a short amount of time to produce the diversity and complexity of life that we observe on Earth today. There are also creationists today that accept the old age for Earth, but still believe that God created each species separately. Many of these Old Earth Creationists (see Scott, 2004, p. 61-63) argue that the rock record shows the sudden appearance of fossil species without precursor or intermediate forms; most (but not all) Intelligent Design (ID) Creationists fall into this category (see *Darwin's Doubt – The Explosive Origin of Animal Life and the Case for Intelligent Design* by Stephen C. Meyer, 2013). Most of these Young Earth Creationists, Old Earth Creationists, and ID Creationists do not understand the nature of the rock and fossil record and the geologic dating process and are not qualified geologists or paleontologists. This chapter will discuss a brief history of geologic time concepts and the processes involved in dating the rock and fossil record.

Time is what sets geology apart from other sciences (except Astronomy). Geology deals with deep time; immense spans of time that is hard for most people to comprehend. Our Earth has been dated by geochemists to be 4.6 billion years old (4.6 Ga [Giga-annum]), based on radiometric dating of most meteorites (representing solidified, primarily unevolved, solar system objects from the asteroid belt and which accreted from the solar nebula approximately 4.6 billion years ago) (Ogg, Ogg, and Gradstein, 2008; Wicander and Monroe, 2010; or review any recent introductory geology textbook). According to Ogg, Ogg, and Gradstein (2008, p. 26), "The oldest dated rock on Earth is 4.03 Ga from the Acasta gneiss complex of the Slave Craton, northwestern Canada (Stern and Bleeker, 1998)."

Geologists measure the passage of geologic time in two ways, relative dating and absolute dating of rocks and geologic events. Relative ages of rocks are concerned with the relative order (chronologic order) in which geologic events occurred in the past. Relative dating of geologic history refers to placing the sequence of Earth history events in the proper chronologic order and does not give the age of Earth history events in years. Using principles of relative age dating, geologists have divided geologic time into subdivisions. The subdivisions of geologic time (Figure 3-1) are based on changes that have occurred to the surface of Earth during past times and to changes in life forms that have existed in the past (based on fossils found in the rocks). Absolute dating, on the other hand, refers to determining the age of Earth history events in years before present. Often this is expressed as Ma (Mega-annum [million years]) or Ga (Giga-annum [billion years]). This chapter will introduce the principles that geologist use to determine the relative and absolute ages of Earth's rocks. Along the way we will discuss some of the early history of dating rocks of Earth.

As you can see from the geologic time scale (Figure 3-1), the largest subdivisions of geologic time are Eons. Below the level of Eons are, in decreasing order of rank, Eras, Periods, and Epochs (There are smaller subdivisions, but we will not concern ourselves with those now). Geologists determine the subdivisions of these episodes of time based on changes that have occurred on Earth (both in terms of physical and biologic events) in the past.

EON	ERA	PERIOD		EPOCH		Ma
Phanerozoic	Cenozoic	Quaternary		Holocene		0.011
				Pleistocene	Late	0.8
					Early	2.4
		Tertiary	Neogene	Pliocene	Late	3.6
					Early	5.3
				Miocene	Late	11.2
					Middle	16.4
					Early	23.0
			Paleogene	Oligocene	Late	28.5
					Early	34.0
				Eocene	Late	41.3
					Middle	49.0
					Early	55.8
				Paleocene	Late	61.0
					Early	65.5
	Mesozoic	Cretaceous		Late		99.6
				Early		145
		Jurassic		Late		161
				Middle		176
				Early		200
		Triassic		Late		228
				Middle		245
				Early		251
	Paleozoic	Permian		Late		260
				Middle		271
				Early		299
		Pennsylvanian		Late		306
				Middle		311
				Early		318
		Mississippian		Late		326
				Middle		345
				Early		359
		Devonian		Late		385
				Middle		397
				Early		416
		Silurian		Late		419
				Early		423
		Ordovician		Late		428
				Middle		444
				Early		488
		Cambrian		Late		501
				Middle		513
				Early		542
Precambrian	Proterozoic	Late	Neoproterozoic (Z)			1000
		Middle	Mesoproterozoic (Y)			1600
		Early	Paleoproterozoic (X)			2500
	Archean	Late				3200
		Early				4000
	Hadean					

Figure 3-1. The geologic time scale. (This image is in the public domain in the United States because it contains data compiled by the U.S. Geological Survey.) (Also see the Geological Society of America Geologic Time Scale at the following website: http://www.geosociety.org/science/timescale/)

BRIEF HISTORY OF GEOLOGY AND DISCUSSION OF GEOLOGIC TIME CONCEPTS

There was very little advancement in geology until the middle of the eighteenth century. Except for the contributions of a few brave souls, such as Copernicus,

Galileo, and Newton, this dark time (prior to mid-1700s) for scientific and original thought was mostly due to a strict interpretation of the Book of Genesis in the Old Testament. Geologic time was considered to be but a few thousand years, and some people today still adhere to a young Earth based on a literal interpretation of the Bible. Fossils were regarded as creatures engulfed by the Biblical Flood, freaks of nature, inventions of the devil, or figured stones (Krumbin and Sloss, 1963).

Most naturalists of the late 1700s and early 1800s were Young Earth Creationists. Many biblical scholars throughout the Middle Ages had studied the scriptures of the Old Testament (basically the Hebrew Bible translated into Greek, which essentially became the Old Testament) and determined that Earth was about 6000 years old (Repcheck, 2003). In fact, in the year 1650, James Ussher (1581-1665), Archbishop of Armagh, Ireland (and one of the last well-known biblical chronologists) (Repcheck, 2003), calculated, using genealogies described in the Old Testament, that Earth was created on October 22, 4004 B. C. (Wicander and Monroe, 2010). Thus, indicating that Earth is only about 6000 years old. (INTERESTING NOTE: Leonardi da Vinci (1452-1519) estimated that it took 200,000 years just to deposit the sediments in the Po River Valley in Italy.) Archbishop Ussher's calculations and dates were actually written into the margins of the Old Testament by the editors of the King James Bible for several generations (Repcheck, 2003; Larson, 2004).

During the late 1700s Abraham Gottlob Werner (1749-1817) (Figure 3-2), a German professor of mineralogy at the Frieberg Mining Academy in Germany, assimilated the chaotic geologic data of his time and proposed the first widely accepted classification of rocks and a crude relative time scale. Werner's classification was actually a modification of an earlier (1756) classification by a German mineralogist and mining engineer, Johann Lehmann (Krumbein and Sloss, 1963). Werner's classification is summarized as follows (from Krumbein and Sloss, 1963), from youngest to oldest:

Figure 3-2. Abraham Gottlob Werner. (Wikimedia Commons. This image is in the public domain.)

Volcanic Series: Werner thought was minor and due to local effects of combustion of underground coal beds.

Alluvial (or Tertiary) Series: poorly consolidated sands, gravels, and clays that formed after the withdrawal of the ocean from the continents.

Stratified (or Secondary) Series: included the majority of stratified, fossiliferous rocks.

Transition Series: thoroughly indurated limestones, dikes and sills, and thick graywackes (sandstones with rock fragments and considerable clay matrix). Werner thought these rocks were the first orderly deposits formed from a worldwide ocean.

Primitive (or Primary) Series: crystalline rocks, now known as igneous and high-rank metamorphic rocks. Werner thought these were chemical precipitates from a worldwide ocean before the emergence of land.

Werner's concept, that most rocks were formed within the ocean and that igneous and metamorphic rocks were chemical precipitates from the ocean, became known as Neptunism (Neptune being the god of the sea). Werner had a very strong personality, such that his ideas were strongly impressed upon his students. Some say that Werner retarded the science of geology, however others say, because of the heated controversies resulting from his ideas, gathering of field data was stimulated and others interested in geology became more active (Krumbein and Sloss, 1963).

Figure 3-3. James Hutton (1726-1797), Father of Geology. Portrait by Sir Henry Raeburn, 1776. (Wikimedia commons. Image is in the public domain.)

James Hutton (1726-1797) (Figure 3-3), a Scottish medical doctor, was one of the first to seriously challenge the ideas of Werner. Hutton never practiced medicine, but was very interested in the processes that formed and shaped Earth. Hutton was a contemporary of Werner, but displayed a much more scientific approach than Werner. Through lab and fieldwork, he postulated that igneous rocks cooled from a molten state, not as precipitates from the ocean. He imagined Earth as a great heat engine and believed subterranean heat supplied the heat for molten rock material and volcanism (Repcheck, 2003).

Hutton (1788) published a paper entitled *Theory of the Earth*, which laid the ground work for modern geology: "...the full account of his theory as read at the 7 March 1785 and 4 April 1785 meetings did not appear in print until 1788. It was titled *Theory of the Earth; or an Investigation of the Laws Observable in the Composition, Dissolution, and Restoration of Land upon the Globe* and appeared in Transactions of the Royal Society of Edinburgh, vol. I, Part II, pp. 209–304, plates I and II, published 1788" (Wikipedia, James Hutton, last accessed 01/10/2018). However, Hutton was a less inspiring personality than Werner, thus his ideas were overshadowed by Werner's for many years. Hutton also wrote in a difficult style, so his ideas on the theory of Earth were not widely read or accepted until John Playfair, and associate of Hutton, realized the fault and published in 1802 a book entitled *Illustrations of the Huttonian Theory of the Earth* (Gillispie, 1951; Krumbein and Sloss, 1963; Repcheck, 2003; Secord, 1997; Wikipedia, James Hutton, last accessed 01/10/2018).

Hutton's ideas on subterranean heat causing rock material to be in a molten state until brought near to the surface or onto the surface became known as Plutonism (for Pluto, god of the underworld). Heated controversies between the Neptunists and the Plutonists lasted about half a century (Krumbein and Sloss, 1963; Repcheck, 2003; Wikipedia, James Hutton, last accessed 01/10/2018). Hutton imagined Earth undergoing countless cycles of 1) uplift (due to activity in the hot depths below the surface), 2) weathering and erosion of surface rocks, and 3) deposition of sediment to be lithified into hard rock, then 4) deep burial as more rocks formed above older rocks, and finally 5) the deeply buried rocks melted and 6) were uplifted towards the surface again to repeat the cycle. This process has become known as the rock cycle (with some modification to make a distinction between sedimentary, igneous, and metamorphic rocks within the cycle) (Monroe and Wicander, 2011).

One of Hutton's greatest contributions to geology was his concept of uniformitarianism. This concept, meaning, "the present is the key to the past", states that by studying geologic processes in operation today we can safely assume that such processes operated in the past and thus we can interpret rocks as a response to geologic processes. With modification, this concept is still the basis for modern geologic thought. We now realize that, although the processes themselves probably have not changed with time, the rates of some geologic processes may have varied drastically from time to time. In other words, the basic laws and principles of physics, chemistry, and biology have not changed over time. Some geologists today prefer the term actualism for the uniformitarian concept. Another great contribution of Hutton to modern geologic thought is the concept of deep time. Based on uniformitarian geologic processes, Hutton proposed that Earth must be many millions of years old (Krumbein and Sloss, 1963; Repcheck, 2003; Wicander and Monroe, 2010; Tarbuck, Lutgins, and Tasa, 2012; Wikipedia, James Hutton, last accessed 01/10/2018). Hutton, and later Charles Lyell (who continued Hutton's uniformitarian approach to the study of geology), believed that most geologic processes (for example weathering of rocks, deposition of sediment, uplift of mountains, etc.) acted slowly and gradually over long episodes of time. However, modern geologists realize that volcanoes are explosive and violent collisions occur when meteorites impact Earth, but even here the basic laws of nature are not violated.

Hutton also was the first to clearly state and use the principle of superposition, which says that older rock beds (strata) are at the bottom of an undeformed sedimentary sequence of strata. However, Nicolas Steno (1669) has been given credit as the first to actually suggest that sedimentary strata are deposited layer upon layer, that younger beds are laid down on top of older beds, and that strata are initially deposited in nearly horizontal layers; now referred to as the principle of original horizontality. Steno also suggested that sedimentary layers and lava flows are laterally extensive in all directions within a depositional basin (area receiving sediment) until they pinch out or terminate at the edges of their depositional basin; now called the principle of lateral continuity (Krumbein and Sloss, 1963; Levin, 1999; Repcheck, 2003; Wicander and Monroe, 2010; Monroe and Wicander, 2011).

William Smith (1769-1839) (Figure 3-4), an English engineer with very little formal education, was the first person to correlate rock units from one area to another. Smith was unaware of the bitter controversy between the Neptunists and Plutonists, and probably didn't care anyway. Smith was more concerned with the practical aspects of correlating rock units from one area to another, both as continuous

Figure 3-4. Portrait of William Smith, the Father of English Geology, by Hugues Fourau about 1835. Smith constructed the world's first geologic map (of England and Wales with part of Scotland) in 1815. (Wikimedia commons. This image is in the public domain.)

formations and also as age equivalent strata (Gillispie, 1951; Krumbein and Sloss, 1963; Reznick, 2010). He published the first geologic map in 1815 entitled "*A Delineation of the Strata of England and Wales with Part of Scotland*." Smith's map (Figure 3-5) showed the aerial distribution of 31 rock units (Krumbein and Sloss, 1963; Winchester, 2001). Smith learned to distinguish different groups of strata based on their fossil content, mineral composition, grain size, color, and position of particular strata within a sequence. By using these criteria, particularly fossil content of the strata, he was able to correlate strata (and formations) from localities that were miles apart. He was the first geologist to use fossils to correlate from one area to another and the first to determine that fossils of each age group of strata were characteristic for that particular age of rocks (Gillispie, 1951). Thus, he was the first to use the principle of fossil succession (Smith referred to this as the principle of faunal succession), which states that fossils succeed one another in a definite and determinable order and the relative age of sedimentary strata can be determined by their fossil content (Krumbein and Sloss, 1963; Levin, 1999; Boggs, 2006; Winchester, 2001; Reznick, 2010; Wicander and Monroe, 2010; Tarbuck, Lutgens, and Tasa, 2013; Wikipedia, William Smith, last accessed 01/10/2018).

Because of his contributions to the science of stratigraphy (the study of and correlation of stratified rocks), Smith became known as William "Strata" Smith, the Father of Stratigraphy. Smith is also known as the inventor of geologic maps and the father of English geology (Krumbein and Sloss, 1963; Winchester, 2001).

By the 1830's enough stratigraphic and paleontologic data had accumulated to kill off Neptunism. Smith, Lyell, and others in Britain, as well as Georges Cuvier, Alexandre Brongniart, and Jean Baptiste de Lamarck (three French naturalists), and others on the continent, had demonstrated a succession of fossil types such that most of the rocks in Europe could be correlated by their fossil content (Krumbein and Sloss, 1963; Reznick, 2010).

Charles Lyell (1797-1875) (Figure 3-6) published the first textbook of geology in 1833 (actually three volumes, the first in 1830 and the last in 1833) entitled *Principles of Geology - Being An Attempt to Explain the Former Changes of the Earth's Surface, By Referencing to Causes Now in Operation*, which emphasized the Huttonian concept of uniformitarianism. Neither Hutton nor Lyell coined the term uniformitarianism; the Cambridge polymath William Whewell did this in about 1833 to contrast Lyell's geological philosophy of uniform, gradual Earth change with Whewell's own catastrophic philosophy of Earth change (Secord, 1997).

Figure 3-5. William Smith's geologic map (in grayscale) of 1815 entitled *"A Delineation of the Strata of England and Wales with Part of Scotland."* This is the first geologic map ever published. (Image from Wikimedia commons. This image is in the public domain. The original map is in color.)

Lyell graduated from Oxford where he studied geology, however, his main studies at Oxford were the classics and mathematics. Upon graduation from Oxford,

he went to law school in London. Lyell practiced law for a short time, but soon his interest in geology prevailed and in 1831 he was appointed Professor of Geology at Kings College, London. However, he resigned the second year there. In the early 1840s, Lyell traveled to the United States to do field geologic studies and to present lectures (Secord, 1997). Lyell was a close friend of Charles Darwin, and without doubt the ideas of each influenced the thought of the other. In fact, Darwin took the first volume of Lyell's *Principles of Geology* along with him as he set out on the voyage of the HMS

Figure 3-6. Portrait of Charles Lyell by G. J. Stodart (date unknown). (Wikimedia commons. This image is in the public domain.)

Beagle (a gift from Captain FitzRoy), and received the second and third volumes while on the voyage (Browne, 1995; Kolbert, 2014). Lyell's uniformitarian approach to geology greatly influenced Darwin in his geologic studies on the voyage, and later on his ideas about slow, gradual evolutionary change of organisms (Browne, 1995, Secord, 1997). Lyell was knighted by Queen Victoria in 1848 and upon his death in 1875 he was buried in Westminster Abbey (Wikipedia, Charles Lyell, last accessed 01/10/2010).

In his studies of European stratigraphy, Lyell had presented a stratigraphic column (geologic formations arranged vertically from oldest below to youngest above) with the units arranged in order of superposition and assigned to groups. Groups were in turn, subdivisions of higher categories termed "periods". Lyell's "periods", Primary, Secondary, and Tertiary, kept some of the characteristics of Werner's classification (in general form only). Lyell's "periods" are generally what we would refer to today as eras (Krumbein and Sloss, 1963) (see Figure 2-1).

Lyell's classification scheme set the pattern for the next 100 years. Other geologists referred to Lyell's groups as systems and the systems were named according to dominant lithological (rock) characteristics or by locality of the typical exposure. For example, the Oolitic System (now referred to as the Jurassic System) was named for the exposures of oolitic limestone in the Jura Mountains of southern France (Krumbein and Sloss, 1963).

Lyell's systems were identified over broad areas and took on added meaning. In the late 1800's and early 1900's, North American stratigraphers could correlate similar sequences of strata with intermittent unconformities to the sequences in Europe. The eastern portion of North America also contained similar fossils to the sections described by Lyell and others in Great Britain and on the European continent. Thus, systems began to have time significance; there had developed a clear concept of the relative ages of the major rock groups, or systems, by superposition and fossil succession (Krumbein and Sloss, 1963).

With Lyell's systems taking on time significance, due to 1) a clear concept of the relative ages of the major rock groups by superposition, 2) fossils of each system had become known, and 3) each system appeared to have a distinct assemblage of fossils; many of the early naturalists of the time (early to late 1800s) reasoned that each system represented a chapter of Earth history ending with some sort of catastrophe, killing most, if not all, forms of life. Those who believed in this concept of catastrophic Earth history became known as catastrophists. Catastrophist views were supported by lithologic discontinuities (unconformities – surfaces representing hiatuses in the rock and/or fossil record) at system boundaries. These unconformities were considered recognizable and as universal relative dates on the geologic time scale (Now we know this is not always true) (Krumbein and Sloss, 1963).

The ideas of Hutton, as adhered to by Lyell and others, implied a vast amount of time to account for the geologic and stratigraphic record as they observed it. Other naturalists, of the late 1700s and into the middle to late 1800s, tried to fit the geologic and stratigraphic record into the Old Testament account of the Deluge (Great Flood) and thus they had to fit the geologic evolution of Earth into the 6000 years as given by Old Testament scholars and clergy. Such a short time implied that geologic processes must take place at a rapid and catastrophic rate. Thus, these diluvianists (believers in catastrophic flooding and other rapid catastrophic events) were, by definition, hard-core catastrophists. Of course, Werner was also a diluvianist, and thus a catastrophist.

Baron Georges Cuvier (1769-1832) (Figure 3-7), a highly respected French naturalist of his time, proposed a solution to the vexing problem of the time question which has become known as Cuvier's Compromise. Cuvier, commonly thought of as the father of comparative vertebrate anatomy and vertebrate paleontology, had been studying the fossil vertebrate remains in Tertiary deposits of the Paris Basin. Cuvier noted a definite succession of fossil types in the Tertiary strata. He also noted that the strata consisted of alternating marine and freshwater fauna, thus possibly implying large-scale movements of land and sea (Levin, 1999). However, Cuvier, being a highly intelligent person, reasoned that perhaps the geologic record did require more time than many naturalists were allowing. Thus, Cuvier came up with his compromise by suggesting that the history of Earth could be divided into three periods: 1) the Diluvian Period (time of Noah's Flood), 2) the Post-Diluvian Period (time after the great deluge), and 3) the Ante-Diluvian Period (time before the great deluge). Cuvier proposed that geology dealt with the Ante-Diluvian Period

Figure 3-7. Portrait of Georges Cuvier, painted by W.H. Pickersgill, 1831. Cuvier is known as the father of comparative vertebrate anatomy and vertebrate paleontology. (Image from Wikimedia commons. This image is in the public domain.)

and that the happenings during this dark, supernatural time were beyond the scope of rational scientific investigation; i.e., the scientific method could not be used to interpret the happenings or time relationships of the Ante-Diluvian Period (Friedman and Sanders, 1978).

Cuvier also believed that there were many lesser floods both before and after the great deluge and that most forms of life were killed during these times of flooding and subsequent uplifting of the continents (Levin, 1999). He believed that species did not change once created, thus new species were created to repopulate Earth. Thus, Cuvier was also the first naturalist to show that some species of organisms have gone extinct at various times in the past (Montgomery, 2012; Wikipedia, Georges Cuvier, last accessed 01/10/2018).

Jean-Baptiste de Lamarck (1744-1829) (see Figure 1-3), a pioneer French biologist, was a contemporary and fellow countryman of Cuvier and also adhered to the idea of the history of Earth as represented by a series of floods, or inundations by a global sea. However, Lamarck thought that geologic time was vast, as exemplified from his writings: "Time is insignificant and never a difficulty for nature. It is always at her disposal and represents an unlimited power with which she accomplishes her greatest and smallest tasks" (Lamarck [as translated by Carozzi, Albert V.], 1964). Lamarck believed that new forms of life were created from species that survived the catastrophic mass extinctions. He believed that evolution of living organisms occurred by inheritance of acquired characteristics (referred to as Lamarckism). Although Lamarck chose the wrong mechanism, he was one of the pioneers of the idea of evolution of organisms, rather than supernatural creation. We have to remember that this was before the ideas of Darwin were published and before the science of genetics was established.

Lyell's publications (including eleven editions of *Principles of Geology*, the last in 1872) (Secord, 1997) eventually convinced the scientific community that Hutton's concept of uniformitarianism was the logical scientific explanation for the interpretation of Earth history and geologic processes. Thus catastrophism (and diluvianism) slowly died out during the last half of the nineteenth century. Following the publication of Lyell's stratigraphic classification in 1833, later stratigraphers added to and modified his original classification until it evolved into the present relative time scale of today (Krumbien and Sloss, 1963; Secord, 1997) (see Figure 2-1).

Before passing on, it is interesting to note the stratigraphic work of two early British stratigraphers, Sir Roderick Murchison (1792-1871) and Reverend Adam Sedgwick (see Figure 1-8), who together are responsible for the naming of four of the present systems (and thus periods; a system is a time-rock unit, whereas a period is the time unit corresponding to the time-rock unit called a system) of the Paleozoic (Cambrian, Silurian, Devonian, and Permian) (Boggs, 2006; Krumbein and Sloss, 1963). Prior to 1835 the geologic column had been described down to the Old Red Sandstone (now known as continental deposits of the Devonian

System, the marine Devonian was not yet known). The rocks below the Old Red Sandstone were known as the "Primitive Series" or lumped into the lower part of Lyell's 1833 Carboniferous Group and his Primary Period. Murchison studied the fossiliferous rocks below the Old Red Sandstone in northern England and southern Wales and found that this system had not been known before. He named this group of strata the Silurian System in 1835. At the same time, Murchison urged his friend and colleague, Sedgewick, to give the name Cambrian System to thick sequences of graywackes (sandstones with considerable clay matrix) (then being mapped and studied by Sedgwick) that occurred below the fossiliferous Silurian strata. However, with more extensive mapping, particularly in northern Wales, the two friends soon realized that their two systems overlapped and they soon became involved in a heated dispute over the boundary between the two systems. No physical or paleontologic hiatus acceptable to both could be established (Krumbein and Sloss, 1963). Murchison believed the Silurian contained the oldest fossils known, thus as the fossiliferous strata of England and Wales were extended lower, he pushed his lower contact downward into Sedgewick's Cambrian System. Sedgewick believed that the Cambrian represented the thicker of the two systems and thus would not accept the changes proposed by Murchison. A heated (and sometimes bitter) controversy between the two raged in scientific meetings and the literature for 30 years (Krumbein and Sloss, 1963; Thrackray, 1976; Levin; 1999).

Murchison was the more convincing of the two, and thus the Silurian reached widespread application before the Cambrian was recognized as a valid system. The dispute was eventually settled in 1879 by Charles Lapworth, who proposed the arbitrary term Ordovician for strata between the Cambrian and Silurian, but Ordovician was not widely accepted until the end of the nineteenth century, particularly by American stratigraphers (Krumbein and Sloss, 1963; Thrackray, 1976; Levin, 1999).

Before Murchison and Sedgwick split ways, they studied rocks above the Silurian rocks and below Carboniferous rocks from western England, where the Old Red Sandstone was absent. They determined that the fossiliferous marine strata in Devonshire were equivalent to Lyell's Old Red Sandstone and they proposed the

Devonian System for these outcrops of rock near Devonshire. This was proof that a system may change in lithologic character from one area to another, but still maintain constant relationships to unconformities. To some this strengthened catastrophism (Krumbein and Sloss, 1963). Now we realize that this was an excellent example of the facies concept (Levin, 1999) (see Figure 3-8).

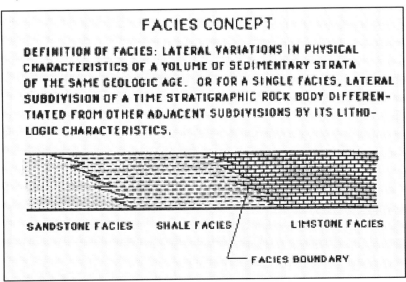

Figure 3-8. Diagram illustrating the facies concept. The diagram shows a typical lateral sequence that would be expected in going from the nearshore marine environment on the left to a deeper marine environment on the right. (Sketch by E. L. Crisp.)

LATERAL AND VERTICAL STRATIGRAPHIC RELATIONSHIPS

Sedimentary strata may change laterally by pinching out at the edge of the depositional basin, or where eroded, or truncated by a fault. Strata may also intertongue (split laterally into thinner units) and pinch out into adjacent strata. Strata may change in composition laterally until they grade into the same composition as adjacent strata (this is called lateral gradation). Distinctive bodies of sediment or sedimentary rock that are characterized by their physical, chemical, and biological parameters (i.e., lithological parameters) are called sedimentary facies (see Figure 3-8 above).

Sedimentary facies are defined as lateral variations in lithological characteristics of a volume of sediment or sedimentary strata of the same geologic age, or for single facies, lateral subdivisions of a time-stratigraphic rock or sediment body differentiated from other adjacent subdivisions by its lithological (physical, chemical, and biological) characteristics. A simple example can illustrate the facies concept. The modern ocean beach and shallow nearshore environment is typically a high, consistent energy environment (due to wave action), thus only relatively coarse material (like sand and/or gravel) can be deposited in the beach or nearshore environment. However, in deeper water, offshore on the continental shelf, the wave and current energy is less and smaller particles (silt and clay sized) can settle out of the water. Typically, the farther from the shoreline in a marine basin, the deeper is the water column and the lower is the current energy; thus, the smaller the particle sizes that will settle out of the water column. In some cases, if the detrital sediment supply (from the continent) is low, or in deep water far from the shoreline, chemical and biochemical carbonates may dominate the sediment accumulating (resulting ultimately in limestone formation). A typical sequence of sediment types as one goes from the beach into a deep offshore environment might be sand, silt mud, detrital clay mud (clay sized sediment), and calcium carbonate mud (as a biochemical and chemical precipitate). Each of these sediment types is considered a sedimentary facies, and if one could walk along the sea floor towards deeper water these sediment types would be encountered. Of course, when these sediments become lithified, sandstone will grade into (or intertongue with) mudrocks (siltstone and/or shale), and mudrocks will grade into limestone. Seas have advanced over much of the area of the continents numerous times in the past and deposited marine sedimentary rocks deep into the interiors of continents. As many times as the marine waters have advanced over the continents to form eperic (epicontinental) seas, they have also retreated from the continents. Marine water advancing onto a continent is referred to as a marine transgression (Figure 3-9), whereas when seas retreat from a continent it is called a marine regression (Figure 3-10).

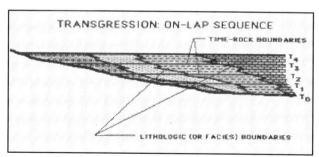

Figure 3-9. A typical marine transgressive sequence, called an onlap sequence. Notice that the facies shift towards the continent as time advances, thus the lithostratigraphic units (for example, the sandstone and shale units) become progressively younger in a landward direction. Notice that at a particular location along this cross-section there would be a fining upward sequence. (Sketch by E. L. Crisp.)

Obviously, when a transgression or regression occurs, adjacent facies become superimposed over each other in a vertical sequence. For example, as a sea transgresses onto the continent, the shallow water (nearshore) facies will be covered with deeper water (offshore) facies. This results in individual rock units (lithostratigraphic units) becoming progressively younger in a landward direction. Also, at a particular location a fining upward sequence will result; typically, with the nearshore sandstone facies at the bottom followed upward by the offshore marine shale facies and that followed by the offshore marine limestone facies (see Figure 3

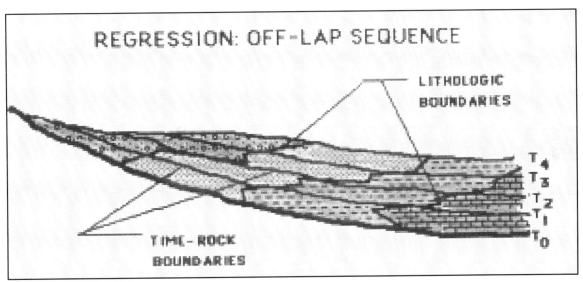

Figure 3-10. A typical marine regressive sequence, called an offlap sequence. In this case the facies shift in a marineward direction and each lithostratigraphic unit (for example, the sandstone) becomes progressively younger in a marineward direction (to the right). Notice that at a particular location along this cross-section there would be a coarsening upward sequence. (Sketch by E. L. Crisp.)

11). Figure 3-12 shows a typical marine transgressive sequence that occurs in southwestern Montana and resulted when an early Cambrian sea transgressed across this region from west to east. The advancing sea deposited continuous layers of sandstone, shale, and limestone as it migrated eastward, however, the sandstone, shale, and limestone become increasingly younger to the east. For example, the equivalent lithostratigraphic unit to the Flathead Sandstone (Early

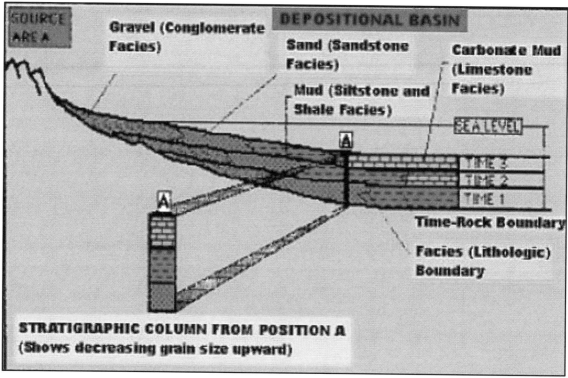

Figure 3-11. A typical transgressive (onlap) sequence illustrating that at a particular location a fining upward sequence will result. Of course, this assumes that there is continuous sedimentation with no significant unconformities present in the sequence. (Sketch by E. L. Crisp.)

Cambrian in age in Southwestern Montana) in the vicinity of the Black Hills of South Dakota is the Deadwood Sandstone, which is late Cambrian in age. A similar onlap sequence comparable to the onlap sequence in SW Montana is also present in the Cambrian rocks of the Grand Canyon in northern Arizona, indicating a major transgression of the sea onto the western portion of the North American continent (or what would become the North American continent) during Cambrian time (Levin, 1999; Wicander and Monroe, 2010). Geologists can often determine whether a relative sea level rise or fall has occurred in a certain area during a particular time interval by determining the vertical sequence of lithologies. For example, in Figure 3-12, the field geologist would observe a marine glauconitic sandstone (glauconite is a green phosphate mineral that forms as small rounded grains, perhaps from fecal pellets of organisms, thought to occur only in marine environments), overlain by marine shale and the shale overlain by a marine limestone. This sequence shows a general decrease in grain size upward and suggests a decrease in environmental energy conditions as time passed. Thus, the sequence may be interpreted as a marine transgressive onlap sequence. In many field geologic studies, geologists have noted transgressive-regressive cycles (Figure 3-13) (and vice versa). In such cases, the maximum shift of the facies (landward or basinward) can often be correlated within the area as a synchronous surface (same age surface) (see Figure 3-13).

The stratigraphic sequence at the position of A in Figure 3-11 above is also important in the respect that it suggests that the vertical distribution of sedimentary facies (lithofacies) is the same as the lateral sequence of lithofacies for a particular time interval. This concept is called Walther's Law and says that lithofacies vary in a similar manner both laterally and vertically about a given point in a sedimentary sequence (see Figure 3-14). Johannes Walther (1860-1937) was the first (in 1894) to clearly recognize the significance of lateral and vertical relationships of facies. Walther's Law is important because it allows a geologist to work out (predict) the lateral changes in facies based on the vertical changes, even if he or she can't always (or it is not practical to) laterally trace out the facies. However, Walther's Law does not apply in situations involving unconformable relationships of the facies being compared (i.e. one cannot connect facies on opposite sides of unconformities using Walther's Law) (Levin, 1999; Wicander and Monroe, 2010).

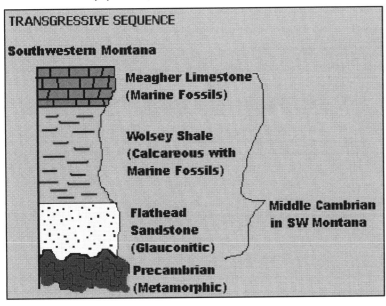

Figure 3-12. This is a vertical stratigraphic sequence that occurs in the Tobacco Root Mountains of southwestern Montana. This is a typical transgressive fining upward sequence representing the transgression of a middle Cambrian sea over exposed Precambrian rocks of what is now southwestern Montana. (Sketch by E. L. Crisp, with thanks to the field instructors of the 1972 Indiana University Geology Field Camp near Cardwell, Montana.)

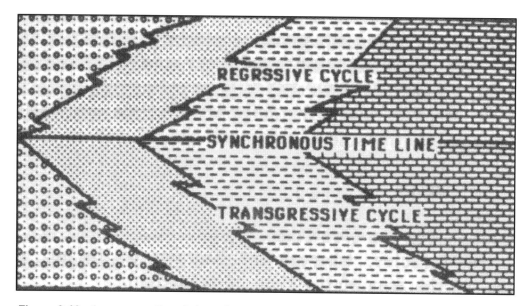

Figure 3-13. A cross-sectional view of a transgressive-regressive cycle. The dips of the rock units are greatly exaggerated. Transgressive-regressive (onlap:offlap) cycles can be symmetrical or asymmetrical. (Sketch by E. L. Crisp.)

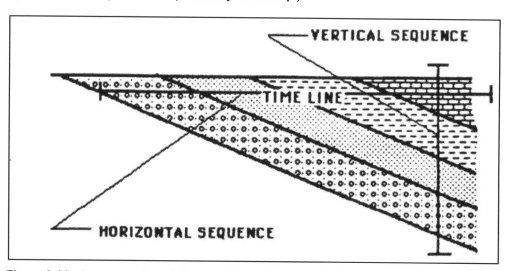

Figure 3-14. A cross-sectional diagram illustrating Walther's Law. Walther's Law states that in a single sedimentary series, the variety of facies laterally that are observed about a given point, reproduce, in general, the variations of facies of strata that follow vertically from the point. More briefly, facies vary in an analogous manner both horizontally and vertically. (Sketch by E. L. Crisp.)

PRINCIPLES OF RELATIVE DATING AND STRATIGRAPHIC CONCEPTS

Let's briefly review the principles of relative dating already discussed. First, we will review these principles: Principle of Superposition, Principle of Original Horizontality, and Principle of Lateral Continuity, (see Figure 3-15 for examples of these principles).   As we discussed earlier, these principles are based on the work of Nicolas Steno (Latinized name for Niels Stensen) (1638-1686), a Danish anatomist. Steno worked in Italy and was curious about how sediment was deposited and how rocks form. He observed sediment transport and deposition during stream flooding near Florence, Italy. Steno's three main principles (1669) are listed and summarized below (see Levin, 1999 and Wicander and Monroe, 2010):

1. **Principle of Superposition**:

In a sequence of sedimentary strata, the oldest layer is at the bottom of the sequence and the strata are progressively younger toward the top of the sequence (see Figure 3-15 and Figure 3-16).

2. **Principle of Original Horizontality**:

Sedimentary strata are originally deposited in a near horizontal manner (see Figure 3-15 and Figure 3-16). Therefore, if sedimentary strata are found to be in a steeply inclined position, some force has altered them from their original position (see Figure 3-16, Figure 3-17, and Figure 3-18).

3. **Principle of Lateral Continuity**:

Sedimentary strata are originally deposited over a laterally extensive area and are continuous until they pinch out at the edge of the ancient depositional basin (or unless removed by subsequent uplift and erosion) (see Figure 3-15).

Figure 3-15. Horizontally bedded sedimentary strata as seen from the North Rim of the Grand Canyon illustrating the immensity of geologic time. It took hundreds of millions of years for these strata to be deposited as layers of sediment that were eventually converted into rock. The Principles of Superposition, Original Horizontality, and Lateral Continuity are illustrated here. The geologic history of the Grand Canyon region can be read from these sedimentary layers (photo by E.L. Crisp, 2002).

Figure 3-16. Horizontal strata of the Jurassic aged Morrison Formation near Cleveland, Utah. These strata were originally deposited as near horizontal layers of river and lake sediment accumulated across this region. Later the sediment was compacted by overlying sediment and cemented by percolating ground waters that lithified the sediments into sandstones, mudstones, and shales (Photo by E. L. Crisp, 2002).

Figure 3-17. The Morrison Formation at Dinosaur National Monument, near Vernal, Utah. Note that the beds are strongly dipping here and have been deformed from an original near horizontal attitude. These strata have been deformed due to internal Earth forces that formed the Rocky Mountains (Photo by E. L. Crisp, 2002).

Figure 3-18. Deformation of once horizontal Paleozoic (Devonian and Mississippian aged) sedimentary strata at Sidling Hill Syncline on I-68 near Cumberland, Maryland (Photo by E. L. Crisp, 2005). Sidling Hill Syncline (down-folded strata) is part of the Valley and Ridge Province of the Appalachian Mountains and was formed by compressional forces as the mountains formed.

The other principle that we have briefly discussed is the Principle of Fossil Succession (or Principle of Faunal Succession) as discovered by William Smith in the early 1800s, and supported by the rediscovery (also in the early 1800s) of this principle by Cuvier and Brongniart. As we stated earlier, this principle states that fossil assemblages succeed themselves in a definite and determinable order and thus we may determine the relative age of strata based on the fossils present in the rocks. Before giving a more detailed discussion of this principle and its use to determine stratigraphic age relationships, let's present three more principles of relative dating. Geologists have used three additional principles of relative dating since the early days of geology. Certainly, Steno and Hutton were aware of these principles. The Principle of Cross-cutting Relationships says that any rock body or discontinuity that cuts across strata must be younger than the strata. This would include igneous (molten rock) intrusions, faults (fractures of Earth's crust along which displacement has occurred), and unconformities (discontinuities due to erosion or non-deposition) that cut across previous rocks. In other words, there had to be a medium for the cross-cutting event and the medium is the older rock material (Figures 3-19 and 3-20).

Figure 3-19. A basalt dike cutting through granite. The basalt dike is youger than the granite. Susan Sowards serves as scale. (Photo taken on Cadillac Mountain, Acadia National Park, near Bar Harbor, Maine by E. L. Crisp, 2005.)

Figure 3- 20. This is a small reverse fault in Allegheny Group rocks along Schultz Road in Pleasants County, West Virginia. Light colored mudstones are adjacent to the asphalt road. The Lower Freeport coal is the dark band in the photo, with the Upper Freeport sandstone at the top. Notice that the coal seam is offset by several feet, thus the fault must be younger than the sedimentary rocks that it cuts across. Dwayne Stone serves as scale. (Photo by E. L Crisp, September 2010.)

The Principle of Inclusions states that inclusions of one kind of rock in another are always representative of the older rock material. For example, if a granitic magma has intruded into a sandstone and chunks of sandstone have been incorporated into the rising magma as xenoliths (chunks of unmelted country rock within a crystallized igneous body [pluton]), as cooling occurs there will be inclusions of sandstone in the granite and the inclusions will represent the older rock.

The third additional principle of relative dating is the Principle of Unconformities. An unconformity is a surface representing a sedimentologic or temporal (time) discontinuity between the strata above and below the surface. The surface represents missing rock material and thus is also representative of a time gap (hiatus) in the sedimentary sequence, i.e., rocks representative of a particular episode of geologic time are missing. The missing rock section was either eroded or was not deposited. There are four basic types of unconformities:

1. **Disconformity**:   The strata are parallel above and below the unconformable surface. The unconformable surface is an erosional surface and can usually be identified by topographic relief or pebbles (inclusions) of the older layer incorporated into the younger rocks.

2. **Paraconformity**:   The strata are also parallel above and below the unconformable surface, but the surface is not an erosional surface. The unconformable surface shows no topographic relief or inclusions, but strata of significantly different age are separated by the unconformable surface. The difference in age (and thus detection of the unconformable surface) of the strata above and below the surface must be accomplished using paleontologic (fossils) and/or radiometric methods.

3.   **Angular Unconformity**: The strata below the unconformable surface are at some angle to the strata above. The younger strata are essentially parallel to the erosional surface, while the older strata are inclined to the erosional surface. This type of unconformity implies tectonic deformation which folded and/or uplifted the older strata, a period of erosion then occurred which planed off the surface, followed by deposition of sedimentary rocks roughly parallel to the erosional surface (Figure 3-21).

4.   **Nonconformity**: The rocks below the unconformable surface are intrusive igneous or high rank metamorphic rocks (usually referred to as basement rocks). The unconformable surface is an erosional surface and the younger sedimentary rocks above the surface typically have inclusions of the igneous or metamorphic rocks. There may be considerable topographic relief on the surface, but the igneous rocks do not intrude the younger sedimentary layers.

Figure 3-21. An angular unconformity on the Olympic Peninsula, Washington, between horizontal Quaternary beds above and older Tertiary beds below. Susan Sowards serves as scale. (Photo by E. L. Crisp, November 2003.)

Now, let's continue our discussion of the Principle of Fossil Succession and its importance in determining the relative time framework of Earth's rocks. Let's look a little deeper at what fossils are and how they are used to place Earth's rocks into a time framework.

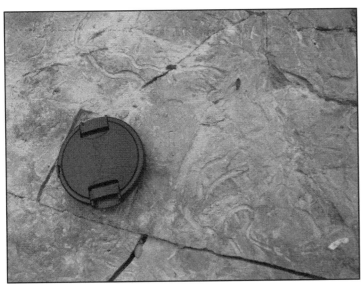

Figure 3-22. Horizontal trails (traces) of worm fossils in a calcareous siltstone of the Upper Ordovician Kope Formation near Maysville, Kentucky (Photo by E. L. Crisp, October 2012.)

Fossils are any remains or evidence of activity of a once living organism (usually restricted to prehistoric time). The study of ancient life via fossils and other rock materials is referred to as paleontology. Paleontologists recognize two major types of fossils: body fossils and trace fossils. Trace fossils represent evidence of biologic activity, but do not include any of the actual remains of an organism. Examples of trace fossils are tracks, trails (Figure 3-22), burrows, coprolites (fossilized feces), gastroliths (stomach stones), etc. Trace fossils often may be used to determine characteristics of the organisms that made the trace fossils and the environment of deposition of the rocks they are found within, but are typically not useful for age determination. Body fossils include either some of the actual remains of an organism (however, usually altered) or molds and casts of the actual remains.

The usual prerequisites for fossilization to form body fossils are the possession of hard parts (bones, teeth, mineralized exoskeleton, etc.) and the rapid burial of the hard parts by sediment (this reduces the amount of oxygen present to very low levels and slows decomposition of the hard parts). Usually soft parts of an organism rot rapidly. Only rarely are soft parts preserved (such as skin impressions for dinosaurs or soft-bodied invertebrates), but under some conditions they are preserved and give paleontologists valuable information that is usually not present in the rock record. After burial, some sort of mineralization typically occurs. Unaltered remains are very rare.

The following are the most common types of alteration that preserve body fossils (see Moore, Lalicker, and Fischer, 1952; Clarkson, 1998; Levin, 1999; Prothero, 2004; and Wicander and Monroe; 2010):

- **Permineralization**: Mineral matter from percolating ground waters is added to pores and cavities in bones, shell, teeth, etc. (Figures 3-23 and 3-24). In this type of preservation, the original material is still present, with new mineral matter added to the voids. Many dinosaur bones and fossil trees are preserved by this method, but other fossils often are also. The mineral matter added is often calcite or silica.

- **Replacement**: Sometimes original hard parts are replaced (sometimes referred to as petrified, which means turned to stone) with new mineral matter of a different composition than the original mineral matter (often at a molecular level, so the microstructure of the original mineral matter is preserved). Silica (as microcrystalline quartz or chert, SiO_2), iron oxide (hematite, Fe_2O_3), and calcium carbonate (calcite or aragonite, $CaCO_3$) are common replacement minerals (they are also common permineralizing agents). Some body fossils are both permineralized and partially replaced. Other replacement minerals may be present in some cases also, for example in a highly reducing postmortem environment an organism's mineralized skeleton may be replaced with iron sulfide (Pyrite). Pyritized fossils (particularly invertebrates, such as brachiopod shells) are fairly common in the rock record.

- **Recrystallization**: The recrystallization of fossils is another common type of preservation in which the original mineral present simply recrystallizes (the original crystals grow larger and fill most of the void space). This is more common in invertebrate fossils (such as bivalves [clams], brachiopods, gastropods, etc.) than in vertebrate fossils. This form of preservation usually destroys or partially obscures the original microstructure of the skeletal material. An example would be the recrystallization of a clam shell originally composed of the mineral aragonite (a metastable form of calcium carbonate) to calcite (the more stable form of calcium carbonate at low temperatures).

- **Carbonization**: Sometimes soft parts and/or hard parts of the body of an organism are compressed by burial before decomposition is complete, such that the volatile substances (such as oxygen, nitrogen, carbon dioxide, water, etc.) are squeezed out leaving behind a film of fairly pure carbon. This is particularly common in the preservation of plant fossils, such as ferns (Figure 3-25) and leaves, and some invertebrates, but also occurs sometimes for vertebrates (for example, fifty million-year-old fossil fish of the Eocene Green River Formation of Wyoming, Colorado, and Utah).

- **Molds and Casts**: Sometimes the hard parts (bone or other material) (and sometimes even soft tissue) of organisms are buried by sediment and even may remain until the sediment is lithified (by compaction and cementation), but are later dissolved by acidic ground waters percolating through the pores of the rock (or decomposed by other processes). This will leave an impression or cavity of the external morphology of the original material that was buried. This is called an external mold (when the external mold is very shallow, it is sometimes just referred to as an impression). If later the mold is filled in with mineral matter or sediment, a cast is formed which mimics the external morphology of the original material (Figure 3-26). Sometimes internal cavities of skeletons (from both invertebrates and vertebrates) may be filled with sediment or mineral matter resulting in a mold of the internal morphology of the cavity that was filled; this is called an internal mold (or steinkern). Internal molds are quite common for some invertebrates (such as for clams and gastropods).

Figure 3-23. A photograph of the trilobite species *Flexicalymene meeki* of Late Ordovician age. Trilobites had calcium carbonate (primarily) and calcium phosphate impregnated chitinous exoskeletons. Minute pores ran through the mineralized portion of the exoskeleton (Moore, Lalicker, and Fischer, 1952). The specimen above may be partially permineralized with additional mineral matter. Much of the chitin may have been carbonized and lost. (Photo by E. L. Crisp.)

Figure 3-24. A photograph from a different angle of the specimen in Figure 3-23. (Photo by E. L. Crisp.)

Now that we know what fossils are and how they form, let's discuss how they can be used to determine the time relationships of stratified rocks. Stratigraphy is the study of stratified rocks. All sedimentary rocks are stratified, but stratification is not as prominent in igneous and metamorphic rocks. Stratigraphers try to correlate rock

Figure 3-25. Photograph of Pennsylvanian aged fossil ferns in a black shale. The ferns are preserved by carbonization. The coin is a quarter. (Photo by E. L. Crisp.)

units that were deposited during the same time interval. They try to correlate all the sedimentary rocks of the world into the correct time sequence, and therefore interpret Earth history in the correct chronologic order. However, there is a dichotomy when conceptualizing geological history from rocks. Time is continuous. Rocks and their contained fossils are a record of what happened during particular time intervals. But rocks at a particular locality do not always represent continuous deposition, and therefore rocks of a certain time interval may not be present at that location (i.e. the rocks may not be conformable).

Stratigraphy is divided into three main categories: Chronostratigraphy, Lithostratigraphy, and Biostratigraphy. Chronostratigraphy involves 1) correlation of rock units deposited at the same time and the placement of rocks into the correct chronologic order and 2) determining the absolute age of rock strata. Therefore, chronostratigraphy involves both relative dating and absolute dating of rocks. Geologists and paleontologists often speak of relative time versus absolute time. Thus, we may speak of the Jurassic Period of time (a relative time period that is younger than the Triassic Period, but older than the Cretaceous Period). Or we may speak of the time from 200 million years ago to 145 million years ago (the time span in absolute years of the Jurassic Period).

Figure 3-26. Late Ordovician brachiopods, *Vinlandostrophia ponderosa* (formerly called *Platystrophia ponderosa*, see Zuykov and Harper [2007]), preserved as casts in the Grant Lake Limestone near Maysville, Kentucky. (Photo by E. L. Crisp.)

We also have to remember that in chronostratigraphy we are correlating rocks rather than time, however, we are correlating rocks of the same age. Chronostratigraphic (or time-stratigraphic) units represent rock sequences that were deposited during a particular interval of time. Thus, the Jurassic System of rocks was deposited during the Jurassic Period of time or the time interval from 200 to 145 million years ago. Chronostratigraphic Units are thus different in concept than Time (Chrono) Units. The basic units of time are (in decreasing magnitude) Eon, Era, Period, Epoch, and Age. The corresponding chronostratigraphic units are (in decreasing magnitude) Eonothem (rarely used in practice), Erathem (rarely used in practice), System, Series, and Stage.

Lithostratigraphy is the study of and correlation of rock units. The basic lithostratigraphic unit (rock unit) is the formation. A formation is a rock unit that can be easily distinguished from rocks above and below, and is thick enough and has adequate lateral distribution as to be a practical map unit to display on geologic maps. A formation may be of a single lithology (rock type, like sandstone) or it may consist of alternating layers of more than one rock type (like sandstone with interbedded shale and mudstone). An example is the Morrison Formation of Jurassic age that is present over a broad area in Colorado, Utah, Wyoming, New Mexico, and Montana (and small portions of some other states). The Morrison Formation consists of alternating red, brown, and yellow sandstones and multicolored mudstones.

Formations may be divided into members or have distinctive lithologies that can be designated as a member of a particular formation. Members have all the characteristics that formations have, but may not be as extensive (cover as large an area). Sometimes similar formations (in terms of lithology or relationships to unconformities) are lumped to form bigger rock units called groups. Groups are often used rather than formations on geologic maps of large regions (such as states).

Biostratigraphy attempts to establish units of strata that have distinctive fossils or fossil assemblages and to correctly order the fossil assemblages (based on the Principle of Fossil Succession – fossil assemblages succeed each other in a definite and determinable order). Another goal of biostratigraphy is to correlate units of rock strata that have the same distinctive fossils or fossil assemblages. This implies that the rock strata are of the same approximate relative age and thus were deposited at about the same time. In a sequence of sedimentary strata, the range (first appearance in the strata to last occurrence in the strata) of distinctive fossils or members of a fossil assemblage is referred to as a biostratigraphic zone or biozone. The most common biozones are range zones (for an individual fossil type), concurrent range zones (the zone of overlap of two or more fossil types), and assemblage zones (fossil types that commonly occur together). Fossil types (also types of modern organism) are referred to as taxa (or taxon, singular). The most accurate biozones use the species level taxon (lowest level of the Linnaean classification system). Distinctive fossils that are used for biostratigraphic correlation are called index fossils or guide fossils. Index fossils have the following characteristics: 1) short geologic time range, 2) wide geographic distribution (implies ecologic tolerance), 3) abundant (relatively easy to find in rocks), and 4) easily recognizable by a trained geologist or paleontologist (not so hard to identify that only a few experts can identify the fossils).

Although biozones are observable units and are not the same as chronostratigraphic units, the tops, bottoms, or both of certain biozones are used to define the boundaries of chronostratigraphic units. Of course, this is because biozones are defined by index fossils that lived during particular time intervals that are represented by specific stratigraphic rock intervals.

PRINCIPLES OF ABSOLUTE DATING OF ROCKS AND EARTH

The relative time scale (and thus relative dating of rocks) is based on the irreversible evolutionary succession of life forms as expressed by fossils in the rock record (i.e., the Principle of Fossil Succession), combined with the Principle of Superposition and other principles of relative dating that we have discussed. With the publication of Darwin's theory of evolution by natural selection (in *On the Origin of Species*), it became clear why successive strata could be identified by their fossils. The Principle of Fossil Succession implies that as new species and higher taxonomic categories evolved into existence, many older species and higher taxonomic categories became extinct (even though Smith, Cuvier, and Brongniart did not imply this evolutionary succession in their work). Darwin's theory also suggested that it would take hundreds of millions of years for the evolutionary succession of organisms as determined from the fossil record. Uniformitarians (such as Hutton, Playfair, and Lyell) also believed it would take hundreds of millions of years for the rock record to have formed. This type of thinking spurred attempts to calibrate the geologic time scale in absolute units.

While writing *On the Origin of Species*, Darwin, in his Chapter IX, On the Imperfection of the Geologic Record, contemplates rates of weathering and rates of sedimentation in estimating the age of certain portions of Earth's crust in Great Britain. His calculations suggested hundreds of millions of years to accumulate (and in some places to weather and erode) sections of sedimentary rocks (see Quammen, 2008, p. 284; Reznick, 2010, p. 277).

Darwin's work spurred other geologists to attempt to determine the absolute age of Earth (Reznick, 2010). A popular method used by many of Darwin's contemporaries was based on rates of sedimentation. This method for determining the age of the sedimentary rock column was based on the idea that if the average rate of accumulation of one foot of sediment was determined and if the total thickness of the sedimentary rock sequence in the geologic record was measured, then one could calculate the age of Earth. Estimates by many geologists varied, from a low of 3 million years to a high of 1.5 billion years, with most estimates in the range of 80 to 100 million years (Wicander and Monroe, 2010). These are great lengths of time to us humans, but the uniformitarians (including Lyell and Darwin) still thought the age of Earth was much older. Estimates of the age of Earth based upon sedimentation rates are fraught with difficulties, such as the following:

1) Different sediment types accumulate at different rates.

2) One type of sediment accumulates at different rates under different conditions of deposition.

3) Different sediment types have different compaction histories to form sedimentary rocks.

4) The geologic column in most areas is full of gaps, as evidenced by unconformities.

5) Some ancient rocks that were originally deposited as sedimentary rocks are now no longer recognized as sedimentary rocks and have been altered to metamorphic rocks or igneous rocks due to processes of metamorphism or melting and recrystallization of rock material.

Another method for estimating the age of Earth was first proposed in 1715 by the Oxford astronomer, Sir Edmund Halley (1656-1742). Halley suggested that the world's oceans formed soon after the origin of Earth and thus would be only slightly younger than the age of the solid Earth. Halley assumed that the oceans initially consisted of fresh water. Therefore, if one estimated the amount of salt in the oceans (based on the salinity of the ocean) and then estimated the rate that salt was being added to the oceans by run-off from the continents and knowing the approximate volume of the oceans, the time required for the ocean to obtain its present salinity could easily be calculated (Levin, 1999). The Irish geologist John Jolly, in 1899, calculated an estimate of 90 million years for the age of Earth using this method (Levin, 1999). However, as we now realize, the salinity of the oceans has remained relatively constant during much of geologic time because steady-state chemical conditions have been maintained. Other serious problems with this method are evaporation from the oceans (which increases salinity), saline deposits (precipitates of salt), and recycling of saline deposits (uplift and dissolution of saline deposits).

As with the average age of Earth due to sedimentary rock accumulation studies, Joly's calculated age was large and lent credence to a vast amount of geologic time. These ages were certainly much larger than Archbishop Ussher's age of 6000 years for Earth, but the age estimates were much less than Lyell, Darwin, and others believed to account for the geologic and evolutionary record as encoded in the rocks.

But the chief proponent of the concept that Earth was too young to account for evolution according to Darwin's theory was William Thomson, later referred to as Lord Kelvin (Reznick, 2010). Thomson, a prominent and preeminent physicist of the middle to late 1800s, calculated the age of Earth based on the concept of a once molten Earth that is in the process of cooling. Based on his model that Earth began as molten rock material that cooled to form a solid crust with a molten interior, he calculated a maximum age of Earth at 400 million years, but later revised this estimate to a maximum age of 100 million years. He also believed that the Sun's heat was gradually diminishing and reasoned that the Sun was too hot during much of the past history of Earth for life to have been present. Based on the cooling of the Sun he estimated that organisms could have inhabited Earth only for about 20 million years to perhaps a maximum as high as 40 million years. (Levin, 1999; Reznik, 2010; Wicander and Monroe, 2010). Thomson argued that this young age for life on Earth was incompatible with Darwin's theory. So, Thomson, after the publication of the first edition of *On the Origin of Species* and prior to the sixth edition, had developed a rather elegant mathematically reinforced cooling model for Earth that was a serious challenge to Darwin's theory of natural selection (Reznik, 2010). Other uniformitarians were also in a quandary, because they also believed Earth to be much older. Reznik (2010, p. 278) states the following: "Darwin modified his fifth edition of the *Origin* to try to reconcile Thomson's calculations with his theory, but he also maintained that even though he could not find fault with Thomson's argument, the details of the geological record were inconsistent with such a young earth or such a short duration of life on earth." So, Thomson's work resulted in a real crisis in geology. Should geologists accept Thomson's age (it was much older than Ussher's age), even though they thought Earth was much older than the age given by Thomson?

Of course, Darwin and other uniformitarians were eventually vindicated and shown to be correct in their reasoning that Earth was hundreds of millions of years old. This process started with the discovery of radioactivity near the end of the nineteenth century and soon yielded evidence that falsified Thomson's age estimate for Earth and ultimately gave geologists a method of experimentally determining the age of Earth at 4.6 billion years (4.6 Ga).

Radioactivity was discovered by the French physicist Henri Becquerel (1852-1908) in 1896. Becquerel's graduate student, Marie Skłodowska-Curie (1867-1934) and her physicist husband Pierre Curie (1859-1906), also isolated and discovered polonium and radium from pitchblende (an ore of uranium) in 1903. Marie was the first to use the term radioactivity for the spontaneous release of electromagnetic radiation from atoms of certain elements. Becquerel, Marie Curie, and Pierre Curie received the 1903 Nobel Prize in Physics for their work on radioactivity. Marie was the first woman to receive the Noble Prize (Levin, 1999; Wicander and Monroe, 2010; Wikipedia, Henri Becquerel, last accessed 01/11/2018; Wikipedia, Marie Skłodowska-Curie, last accessed 01/11/2018; Wikipedia, Pierre Curie, last accessed 01/11/2018).

So, the work of Marie and Pierre Currie (1903) determined that the element radium (and other radioactive elements) gives of energy. This energy liberated inside Earth generates heat, thus Earth has an internal heat source, radioactivity.

This concept basically killed William Thomson's idea of a passively cooling Earth. Our Earth is slowly cooling off, but because of radioactive decay of elements within Earth, it is cooling at a much slower rate than Thomson allowed. In addition to killing Thomson's basic premise, radioactive decay also gave geologists a method to date rocks, and thus to determine the age of various rock units and Earth.

In the early 1900s, absolute dating techniques were developed based on the concept of radioactive decay. Many elements are radioactive and decay with time. Thus, if the rate of decay is established for certain radioactive elements, it is possible to date minerals and rocks that contain those radioactive elements, assuming that the minerals and rocks analyzed have not been severely altered (by metamorphism or weathering processes).

The following statements reveal the time-table for the development of radiometric dating of Earth's rocks during the 20th century and into the 21st century:

1. Henri Becquerel discovered radioactivity in 1896 (Pitchblende, ore of Uranium, gives off some kind of energy that darkens photographic plates) (Levin, 1999).

2. Marie and Pierre Currie (1903) determined that the element radium gives off heat energy. The fact that certain elements, like radium, are radioactive and liberate heat killed Kelvin's basic premise (Wikipedia, Marie Skłodowska-Curie, last accessed 01/11/2018; Wikipedia, Pierre Curie, last accessed 01/11/2018).

3. Ernest Rutherford and coworkers, about 1905, explained how radioactive decay works based on new concepts in atomic theory (U.S. Geological Survey, 2001).

4. Bertram Boltwood of Yale (1907) used Uranium-Lead radiometric methods to measure the age of several old rocks and got ages ranging from about 400 million years to about 2 billion years (Kovarik, 1929; U.S. Geological Survey, 2000; Wikipedia, Bertram Boltwood, last accessed 01/11/2018).

5. Arthur Holmes (from about 1910 to 1927), using Uranium-Lead and Uranium-Helium methods, produced the first calibrated (with absolute ages) geologic time scale (Wikipedia, Arthur Holmes, last accessed 01/11/2018).

6. Many thousands of samples have been radiometrically dated since the early 1900's and the geologic time scale has been revised many times. The absolute dates on the geologic time scale are continually being revised as more radiometric dates are obtained.

So, what is radioactive decay and how is it used to date rocks? Radioactive decay is the emission of particles and energy from the nucleus of an atom, whereby one element is converted to another element. The three main types of radioactive decay are alpha decay, beta decay, and electron capture (Figure 3-27). In all types of radioactive decay, we infer simply that the parent radioactive isotope (parent radioisotope) decays to (sometimes through several intermediate radioactive daughter isotopes) the stable daughter isotope(s), or P (parent) decays to D (daughter).

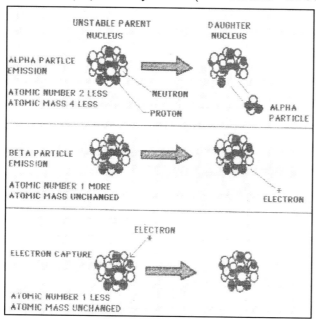

Figure 3-27. Diagram illustrating the three major types of radioactive decay (Sketch by E. L Crisp, after several sources.)

In alpha decay, an alpha particle (just like a helium nucleus), consisting of two protons and two neutrons, is ejected from the nucleus of a radioactive parent isotope, thus the atomic number decreases by two and the atomic mass number decreases by four for the daughter isotope. Beta decay involves the emission of a high-speed electron (beta particle) from the nucleus resulting in the conversion of a neutron to a proton (i. e., a neutron decays to a proton by emitting a beta particle), thereby increasing the atomic number by one without changing the atomic mass number for the daughter isotope. Electron capture decay takes place when the nucleus captures an electron from the inner electron orbits of the atom; thereby converting a proton to a neutron; the atomic number of the daughter isotope decreases by one, but the atomic mass number is unchanged. In addition to the nuclear particles involved in the preceding, gamma radiation is also a usual product of radioactive decay.

How do we relate radioactive decay to geologic time? When speaking of a single radioactive atom, the decay process is spontaneous and unpredictable. However, it is statistically possible to predict how many atoms of a given element will decay in a given amount of time. Thus, we cannot say which atoms will decay, but knowing how many atoms we start with, we can predict how many will decay in a certain amount of time. We can use an analogy of radioactive decay to a Life Insurance Company. The Life Insurance Company can predict (based on past statistics) how many people of a certain age group will die in a certain time interval, but not which individuals will die. However, the analogy breaks down when referring to different age groups of people. Younger age groups will have a smaller percentage of individuals dying for a particular time period (Suttner, 1967). However, regardless of the age of a group of radioactive atoms, each atom within the group still has the same probability as any other atom of that particular isotope of spontaneously decaying.

The number of atoms of a radioactive isotope (radioisotope) that decay during a given time interval is directly proportional only to the number of atoms of the radioisotope present in the sample (i.e., radioactive decay is proportional to the remaining concentration of radioisotope). Thus, the rate of decay, proportional to the remaining amount of radioisotope, is independent of time. This is the basis for the concept of half-life (T1/2). The half-life of a radioisotope is defined as the time taken for any initial number of atoms of a particular radioisotope to be reduced by one half (50%) through decay.

Two implications of the half-life concept are obvious. If we know the ratio of the stable daughter isotope to parent radioisotope in the mineral or rock sample analyzed and if we know the half-life of this particular radioisotope, then we can determine the age of the sample. The relationship can be expressed as follows:

Age = value of half-life for that particular radioisotope times the number of half-lives elapsed.

Or, in equation form: **T = (T1/2)n**, where T is the age, T1/2 is the half-life in years, and n is the number of half-lives elapsed. This concept is shown graphically in Figure 3-28.

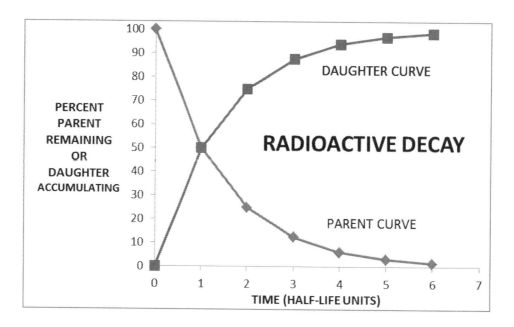

Figure 3-28. Diagram illustrating the exponential decrease in the parent radioisotope with time (as number of half-life units elapsed) and how the daughter isotope accumulates in a mineral crystal as the parent radioisotope decays.

Most methods of radiometric dating are based on measurements of accumulation of stable daughter atoms produced by a parent radioisotope. When a mineral initially crystallizes, there are no radiogenic daughter atoms of the radioisotope contained within the mineral's crystalline structure, i.e., the daughter to parent ratio is equal to zero (D/P=0). If we know the decay constant (k) (which can be determined by measurement in the laboratory) or the half-life (T1/2) of the parent radioisotope, then we can measure the ratio of daughter and parent isotopes (D/P) in a mineral sample (this is done with a mass spectrometer) and compute the age of the sample by the following relationship:

$$T = \left(\frac{1}{k}\right) ln \left[\left(\frac{D}{P}\right) + 1\right]$$

where k is the decay constant (expressed as number of nuclear decays per year) for a particular radioisotope, D is the concentration of the stable daughter isotope, and P is the concentration of the parent radioisotope. If we assume that one half-life has elapsed, then the above equation reduces to:

$$T1/2 = 0.693/k, thus\ k = 0.693/(T1/2).$$

Thus, as you can see above, if we can measure the decay constant in a laboratory setting over a few days to a few months, then we can easily calculate the value of the half-life of the radioisotope. Now if we want to express the general equation in terms of half-life (rather than decay constant), the general equation reduces to the following:

$$T = 1.443 \ (T1/2) \ ln\left[\left(\frac{D}{P}\right) + 1\right],$$

or if we prefer to use log base 10 (rather than the natural log), we can derive the following equation:

$$T = 3.323 \ (T1/2) log \left[\left(\frac{D}{P}\right) + 1\right].$$

For example, assume we want to determine the age of a volcanic rock using the Potassium-Argon dating method. Potassium-40 has a half-life of about 1.3 billion years. If we determine a ratio of daughter to parent (D/P) from mass spectrometer analysis of the volcanic rock to be 3 to 1 (this would be equivalent to 75% daughter and 25% parent), then what is the age of the rock sample? Substituting into the equation gives:

T = 3.323 (1.3 x 10^9 years) log [3 + 1]; therefore, the age of the rock sample is 2.6 x 10^9 years (or 2.6 billion years, or 2.6 Ga).

The following is a summary of the basics of radiometric dating:

1. Usually only igneous rocks are radiometrically dated.

2. When a mineral crystallizes it has trace quantities of radioactive atoms and ideally no stable daughter product(s) of the particular radioisotope/stable daughter isotope pair that we are using.

3. As the parent radioisotope decays, the stable daughter isotope increases in the mineral and cannot get out of the mineral crystal. This is where detecting metamorphism or weathering of the rock or mineral sample is important.

4. The decay rate (which is constant for a particular radioisotope) must be known. There are now many parent/daughter pairs that are routinely used in radiometric dating and we know their decay rates and half-lives very accurately.

5. The ratio of daughter to parent (from mass spectrometer analysis) can then be used to calculate the age of the rock containing the mineral.

6. There is typically a plus or minus 1% (or less) analytical error for radiometric dates (Wicander and Monroe, 2010).

Of course, there are some possible sources of error in radiometric dating, as with all laboratory and field situations involving collecting, analyzing, and measuring. The measurement of parent radioisotopes and stable daughter isotopes in minerals and rocks involves laboratory procedures and mass spectrometer analyses, thus there is always some error in measurement. However, with careful laboratory procedures, this analytical error is minimal (in the less than 1% range). Some of the problems that must be considered in the determination of radiometric dating are listed below:

1. One of the basic premises in radiometric dating is that the minerals and rocks used as samples have remained in a closed system since crystallization. If parent or daughter isotopes have been added or removed from the system, the radiometrically calculated age would not be the true age of the mineral or rock. Samples used for radiometric dating must be chosen that represent unaltered igneous rock (only rarely can sedimentary rocks be dated directly and metamorphic rocks usually give an age younger than their last metamorphic event, sometimes useful in itself). For example, Argon leakage is a common problem with the Potassium-Argon method. This is because Argon (an inert gas) is easily driven out of the crystalline structure due to weathering of the rock or any degree of metamorphism.

2. Any initial non-radiogenic daughter product that was present at crystallization must be corrected for. There are ways to determine this and this is routinely done in certain situations.

3. Because, typically, only igneous minerals and rocks are being dated, inferences about the age of sedimentary rocks are based on field geologic interpretations. Normally, bracketed ages of sedimentary strata are inferred using principles of cross-cutting relationships, superposition, etc.

COMMONLY USED RADIOISOTOPES FOR ABSOLUTE DATING

The radioisotopes most commonly used to date geologic samples, along with the stable daughter products, decay constants, and half-lives are given in Table 3-1.

Potassium-Argon dating has been used for decades to date very old igneous rocks (hundreds of millions to billions of years old). Marine sedimentary rocks containing the authigenic (or early diagenetic) mineral glauconite (an iron potassium mica group mineral) may also, sometimes, be dated using the potassium-argon dating method. Fairly recently the method has been used on much younger rocks, in the range of 100,000 years to 5 million years before present (Johanson and Edey, 1981). This age range is one that other dating methods cannot date accurately. To date such young rocks, a very accurate mass spectrometer had to be developed. James Aronson, then at Case Western Reserve University, Cleveland, Ohio developed a very accurate mass spectrometer to measure small quantities of Argon-40. Aronson developed the very accurate mass spectrometer in order to date lavas and volcanic ash deposits interbedded with sedimentary rocks containing fossil hominins from eastern Africa, in particular, the fossils of "Lucy", an australopithecine fossil hominin from the Afar region of Ethiopia found by Donald Johanson in 1974 (Johanson and Edey, 1981).

Parent Radioisotope	Stable Daughter Isotope	Half-life of Parent (years)	Decay Constant (atoms/year)	Minerals and Rocks Dated
Uranium-238	Lead-206	4.46 Billion	1.55×10^{-10}	Zircon
				Uraninite
				Sphene
				Apatite
				Whole Igneous Rock
Uranium-235	Lead-207	704 Million	9.8×10^{-10}	Same as for U-238
Thorium-232	Lead-208	14 Billion	4.95×10^{-11}	Same a for uranium
Rubidium-87	Strontium-87	48.8 Billion	1.42×10^{-11}	Muscovite Mica
				Biotitie Mica
				Potassium Feldspar
				Whole Igneous Rock
Potassium-40	Argon-40	1.25 Billion	5.81×10^{-11}	Glauconite
				Muscovite Mica
	Calcium-40	1.25 Billion	4.96×10^{-10}	Biotitie Mica
				Hornblende
				Potassium Feldspar
				Whole Volcanic Rock
Carbon-14	Nitrogen-14	5,730	1.21×10^{-4}	Organic Matter

Table 3-1. Table listing the major radioisotopes used in dating geologic samples, along with their half-lives, decay constants, stable daughter products, and source material (minerals and rocks dated). The information in this table came from several sources, but any introductory geology textbook would give similar information.

When Potassium-40 decays, two stable daughter isotopes are formed, Calcium-40 and Argon-40. During the decay process, 89% of the Potassium-40 decays to Calcium-40 by beta decay, while 11% decays to Argon-40 by electron capture. Because Calcium-40 is very abundant in most rocks, it is of no value for radiometric dating, thus only the concentration of Argon-40 and Potassium-40 is used in dating samples. Upon cooling and crystallization of an igneous rock, the individual crystals would have no radiogenic Argon-40. However, as time passes Argon-40 would accumulate in the crystals at a constant rate, thus the ratio of Argon-40/Potassium-40 in the crystals of the rock would allow the determination of the age of the rock. Of course, for volcanic ashes that may contain wind-blown or water transported contaminating grains, the Potassium-Argon method may give erroneous dates when whole rock analysis is used. Fairly recently, a variant of the Potassium-Argon method called Argon-Argon ($^{40}Ar/^{39}Ar$) dating, allows for more accurate dating of hominin fossils in eastern Africa that are related to reworked volcanic ash beds that may contain contaminating detrital grains (Walter, 1997). As Walter (1997, p. 97) proposes in the following statement: "Single-crystal laser-fusion $^{40}Ar/^{39}Ar$ dating has been a major factor in this success. This grain-discrete method now permits precise and accurate ages to be measured on single grains and, thus, contaminating grains can be eliminated."

For the U-238:Pb-206, U-235:Pb-207, and Th-232:Pb-208 parent-daughter systems, it is often possible to use all three methods as a cross-check on each other (Levin, 1999). For example, these three parent-daughter systems are often present in granite and whole rock analyses may be used for dating. Using all three isotope pairs serves as a cross-check on the accuracy and precision of the dates obtained. Certainly, other isotope pairs (such as K-40:Ar-40 or Rb-87:Sr-87) could also be used as cross-checks. In addition, when U-238:Pb-206 and U-235:Pb-207 isotope pairs are present within a rock, the ratio of Pb-207/Pb-206 may be used to determine an age. This is because the half-life of U-235 is much less than the half-life of U-238, so the ratio of Pb-207 to Pb-206 will change systematically with age, and thus may be used to obtain a radiometric age. This is referred to as Lead-Lead dating, rather than Uranium-Lead dating (Levin, 1999).

The Rubidium-87:Strontium-87 radiometric dating method is commonly used on whole igneous rock samples, such as granite (although individual minerals may be analyzed also). It is also sometimes used on specific minerals in metamorphic rocks, such as gneiss, to determine the age of last metamorphism (for most parent-daughter isotopes, high grade metamorphism usually resets the radiometric clock in individual minerals). These igneous and metamorphic rocks commonly contain concentrations of Sr-87 (^{87}Sr) other than that generated by the radioactive decay of Rb-87 (^{87}Rb). Sr-86 (^{86}Sr) is another isotope of strontium that will be present in a mineral or rock if ^{87}Sr is present. The isotope ^{86}Sr is not produced as a byproduct of radioactive decay (it is non-radiogenic), but the isotope ^{87}Sr is produced by the decay of the radioisotope ^{87}Rb. Both isotopes of Sr behave the same way chemically. The initial ^{87}Sr to ^{86}Sr ratio may be obtained by analyzing a mineral within the rock being dated that does not contain any Rubidium (Rb). Therefore, the ^{87}Sr to ^{86}Sr ratio as measured in a mineral or whole rock sample being dated is due to the increase in ^{87}Sr as a result of the decay of ^{87}Rb. However, the following procedure, as outlined below, is typically used in the ^{87}Rb: ^{87}Sr dating method (summarized from Levin, 1999; Zimmer, 2010; Nave, R., Clocks in the Rocks, accessed 03/06/2014). When the rock material was molten magma, prior to crystallization to form the igneous rock, the ^{87}Sr to ^{86}Sr ratio was the same throughout the magma body. Thus, the initial minerals that crystallized would have the same ^{87}Sr to ^{86}Sr ratio as the molten rock (magma) contained. However, as time passed, the more ^{87}Rb that a mineral contained, the more ^{87}Sr is produced as ^{87}Rb decays. Thus, as time passes (for a particular mineral) the ^{87}Rb to ^{87}Sr ratio decreases, whereas the ^{87}Sr to ^{86}Sr ratio increases. If we analyze several samples (either different minerals or whole rock samples) from a rock and plot the ^{87}Rb to ^{87}Sr ratio (on the x-axis) versus the ^{87}Sr to ^{86}Sr ratio (on the y-axis) we should get a straight-line relationship. The straight line is referred to as an isochron. The more time that has elapsed since crystallization of the magma (molten rock material), the steeper will be the slope of the isochron. The value for the slope of the isochron allows a calculation of the age of the rock (note that this age is independent of the initial concentration of ^{87}Rb and ^{87}Sr). If the points on the graph do not form a statistically significant straight line, then the date given by the "best-fit" line is invalid. This may indicate that all of the rock body may not have crystallized at the same time or during magma formation some mineral crystals did not completely melt.

For dating of very recent geologic material, particularly of Late Pleistocene and Holocene age, and for archaeological material, the radiocarbon (Carbon-14) dating technique has been quite successful (Levin, 1999). W. F. Libby and co-workers at the University of Chicago discovered that minute amounts of radioactive carbon (Carbon-14 or ^{14}C) exists in air, natural water, and living organisms and they suggested that this could be used to radiometrically date dead organic material (Grosse and Libby, 1947). Arnold and Libby (1951) published several radiometric dates of archaeological organic material of supposedly known age; their radiocarbon determined ages agreed quite well with the known ages of the archaeological materials. ^{14}C dating may be used to date organic material as old as 100,000 years (Levin, 1999), however, the accuracy and precision decreases significantly beyond about 40,000 years, particularly for accurately dating fossil hominins (for example, ancient *Homo* species older than 40,000 years) (Stringer, 2012). This is because of the short half-life of ^{14}C (5,730 years); such that after 8 or more half-lives have elapsed, the amount of the radioisotope is so low that mass spectrometer readings are not as accurate. However, recently, many technical improvements have been made in radiocarbon dating since the time of Libby and associates and other dating techniques have been developed (such as Argon-Argon dating mentioned earlier in this chapter) to cross-check radiocarbon dates and to extend the time range that may be accurately dated (see Stringer, 2012).

Carbon-14 is being continuously produced in the upper atmosphere by reactions involving cosmic radiation bombardment of Nitrogen-14 (^{14}N) that converts it to ^{14}C. When the nucleus of ^{14}N is hit by a neutron from cosmic radiation, the neutron knocks a proton out of the nucleus, but the neutron stays in the nucleus. Thus, now there are 6 protons rather than 7, and 8 neutrons rather than 7, thus the isotope is now ^{14}C rather than ^{14}N. ^{14}C is unstable (radioactive) and decays by beta emission back to ^{14}N, with a half-life of 5,730 years. Carbon dioxide (CO_2) containing ^{14}C becomes uniformly distributed by wind, rivers, and ocean currents. Therefore, the ratio of $^{14}C/^{12}C$ becomes theoretically constant, thus a natural state of equilibrium (steady state) is established between the amount of ^{14}C generated in the upper atmosphere and the amount of ^{14}C lost to the decay process. However, recently the burning of fossil fuels (which are so old that they contain no ^{14}C) and the explosions of nuclear bombs has upset the balance. The burning of fossil fuels enriches the atmosphere in ^{12}C, whereas, the explosions of nuclear bombs enriches the atmosphere in ^{14}C.

Plants and animals are continually absorbing CO_2 to replenish their tissues. Thus, the $^{14}C/^{12}C$ ratio in their tissues remains constant as long as the organism is alive. However, upon death of the organism there is no further replacement of fresh carbon. ^{14}C begins to diminish and the ratio of $^{14}C/^{12}C$ becomes smaller and smaller with time. Because the half-life of ^{14}C is constant, the decrease in the $^{14}C/^{12}C$ ratio with time may be used to date organic materials. To correct for the recent imbalance in the ^{14}C to ^{12}C equilibrium in the biosphere, wood samples dating from before the industrial era (say 200 to 300 years old) are normally used as the standard of comparison, rather than the present $^{14}C/^{12}C$ ratio. The exact age of the wood must be determined by dendrology (counting annual growth rings) and thus the age obtained by radiocarbon dating is compared to the true age of the wood sample in order to make corrections to older samples (Levin, 1999).

ABSOLUTE AGE DATING OF SEDIMENTARY ROCKS

With a few exceptions (such as the K-Ar dating of sedimentary formed glauconite or volcanic ash deposits mixed with other sediment), radiometric dates are typically determined for igneous rocks and minerals. So, how do we date sedimentary rocks in an absolute way? The relative ages of the geologic events in a particular area must be worked out before radiometric ages can be applied. Once the relative sequence of events in a particular field situation is determined, radiometric dates can be obtained for igneous intrusions and/or extrusions (volcanic rocks) that bracket certain sedimentary strata, thus giving an age range for the sedimentary strata. Also, any index fossils that are associated with sedimentary strata are also dated in an absolute sense when we infer absolute ages for sedimentary strata. Once index fossils have been dated by this method, when we find them in sedimentary strata that are not associated with radiometric dates we know the ages of the strata based on the previous determination of the age of the index fossils.

Figure 3-29 illustrates the procedure for using cross-cutting relations and superposition to determine the absolute age of sedimentary strata. In Figure 3-29, the sequence of geologic events is the following (from youngest to oldest downward; geologist typically think of geologic events from oldest to youngest, with the older layers of strata at the bottom of a sequence):

Erosion of the current land surface (present land surface).   **Youngest Event**
Intrusion of andesite dike.
Deposition of Cenozoic sedimentary sequence.
Extrusion of rhyolite lava flow.
Period of erosion ultimately forming a disconformity.
Intrusion of peridotite dike.
Deposition of Mesozoic sedimentary sequence.
Period of erosion ultimately forming an angular unconformity.
Intrusion of basalt dike.
Folding (deformation) of Paleozoic sedimentary sequence.
Deposition of Paleozoic sedimentary sequence.   **Oldest Event**

Also in Figure 3-29, the basalt dike has been radiometrically dated at 250 million years before the present (Ma), the peridotite dike at 70 Ma, the rhyolite flow at 60 Ma, and the andesite dike at 2 Ma. Now, how do we use the sequence of relative ages and the radiometric dates to give an absolute age for particular events in a sedimentary sequence of strata?

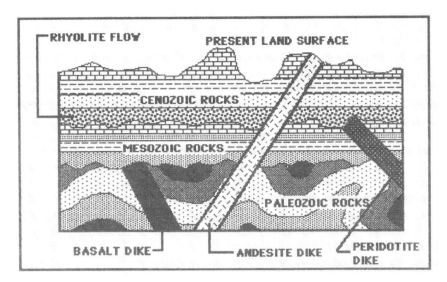

Figure 3-29. Cross-sectional diagram of an area illustrating a sedimentary sequence of strata cut by three igneous intrusions (dikes) and containing an extrusive volcanic flow (rhyolite flow). Note that the boundary between the Paleozoic Rocks and the Mesozoic Rocks is an angular unconformity and the boundary between the Mesozoic Rocks and the Cenozoic Rocks is a disconformity.

The following four examples will illustrate how we can determine bracketed absolute ages for the sequence of events as depicted in Figure 3-29:

    1. What is the bracketed absolute age for the Mesozoic rocks?

ANSWER: By superposition, the Mesozoic rocks are younger than the underlying Paleozoic rocks and also younger than the basalt dike. Since the basalt dike has been radiometrically dated at 250 Ma, the Mesozoic rocks shown here have to be younger than 250 Ma. The peridotite dike has been dated at 70 Ma and cuts through the Mesozoic sequence, thus the Mesozoic rocks have to be older than the peridotite dike. Therefore, the bracketed age for the Mesozoic rocks in this example is 250 Ma to 70 Ma.

    2. What can we determine about the age of the Paleozoic rocks?

ANSWER: The basalt dike cuts across (was intruded into) the Paleozoic rocks shown in this example, thus it is younger than the Paleozoic rocks. So, the Paleozoic rocks have to be older than 250 Ma, which is the radiometric date of the basalt dike.

    3. What is the age of the Cenozoic rocks above the rhyolite lava flow?

ANSWER: By superposition, the rhyolite lava flow is older than the Cenozoic sedimentary sequence. By cross-cutting relationships, the andesite dike is younger than the Cenozoic sedimentary sequence. Based on the radiometric dates given for the igneous rocks, the Cenozoic sedimentary sequence of rocks are between 60 Ma and 2 Ma in absolute age.

4. What is the age of the unconformable surface (disconformity) between rocks of Mesozoic age and rocks of Cenozoic age?

ANSWER: By cross-cutting relationships and truncation, the peridotite dike is younger than the Mesozoic rocks, but has been truncated by erosion, so it is older than the unconformable surface. By superposition, the rhyolite lava flow is younger than the unconformable surface. Therefore, the absolute age of the unconformable surface separating rocks of Mesozoic and Cenozoic age has to be between 70 Ma and 60 Ma.

CHAPTER 4

SOURCES OF VARIATION IN POPULATIONS

INTRODUCTION

In Chapter 1 we briefly discussed the work of Gregor Mendel and the nature of heredity. Now, let's look a little deeper at how traits are passed on within individuals and populations and how variations arise in individuals within populations.

GENES AND CHROMOSOMES

Typically, in eukaryotic organisms only sex cells pass on genetic information, thus variations and mutations of genes can only be transmitted by the sex cells. Genes are located on chromosomes (composed primarily of DNA), thread-like structures in the cell nucleus of eukaryotic cells. In eukaryotic cells, the number of chromosomes is specific for a particular species, thus it varies among different species. Humans have 46 chromosomes. Horses have 64. Chromosomes always occur in pairs, so humans have 23 pairs whereas horses have 32 pairs. Each chromosome is made up of hundreds to thousands of genes.

Each somatic (body) cell in eukaryotic organisms has chromosomes that occur in pairs, as stated above humans have 23 pairs or a total of 46 chromosomes. In each pair, one chromosome comes from the male parent and one from the female parent. Somatic cells are said to have the diploid (2n) number of chromosomes. Sex (germ) cells only have one chromosome and are said to be haploid (n). The genes are aligned on the chromosomes somewhat like beads on a string. The genes themselves consist of certain specific sequences of DNA nucleotides. The gene positions (gene loci, singular locus) on pairs of chromosomes match up and alleles at these gene loci code for the construction of particular proteins (of course, as we discussed in Chapter 1, some alleles may be dominant over others in the same pair). A pair of chromosomes that match-up (have the same genes and gene loci) are said to be homologous chromosomes (Figure 4-1). The number and appearance of chromosomes in the nucleus of a eukaryotic cell is referred to as a

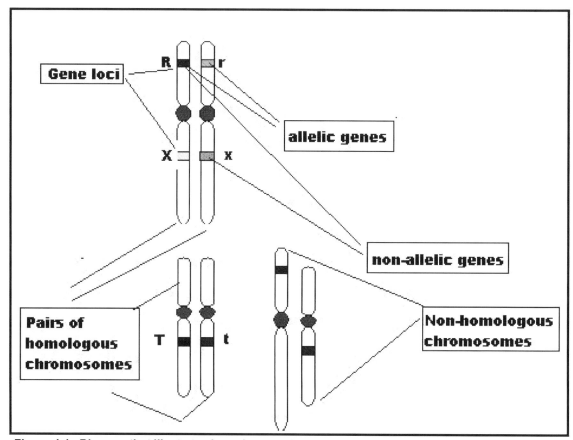

Figure 4-1. Diagram that illustrates homologous chromosomes, gene loci, and allelic genes. The letters R,r; X,x; and T, t represent examples of dominant and recessive alleles for particular genes at specific gene loci. (Sketch by E. L. Crisp.)

karyotype. There are 22 pairs of autosomal chromosomes in the human karyotype and the sex chromosomes are an X chromosome and a Y chromosome. Human females have two X chromosomes and males have an X chromosome and a Y chromosome. Mendelian inheritance occurs for many traits in humans. Pedigree analysis often reveals the patterns of Mendelian genetics in family groups. Many human diseases and disorders follow Mendelian patterns of inheritance.

A common example of an autosomal dominant disorder in humans is Huntington's disease, a degenerative disease of the nervous system (Urry et al, 2014). This disease causes a progressive loss of brain and muscle function. Symptoms usually begin in middle age. Obviously autosomal dominant lethal disorders are not very common, unless they occur later in life, during or after reproductive age. A common example of an autosomal recessive disorder in humans is cystic fibrosis, which is the most common lethal genetic disease in the U. S., with one out of 25 people of European descent (4%) being carriers (heterozygotes) for the allele that causes cystic fibrosis (for recessive homozygotes) (Campbell, 1996; Urry et al, 2014).

MEIOSIS AND MITOSIS

Sex cells in Eukarya have only one-half of the number of chromosomes as somatic (body) cells. Sex cells undergo a type of division called meiosis (see any high school or college introductory biology textbook). This reduces the number of chromosomes to one-half (called the haploid number, n). Sex cells end up with one each of the paired homologous chromosomes. Somatic cells undergo a different type of cell division called mitosis (see any high school or college introductory biology textbook). In mitosis, each daughter cell has the same number of chromosomes as the parent cell (called the diploid number, 2n).

MENDEL AMENDED

Genes on the same chromosomes stay together during meiotic division (i.e., they are linked together, referred to as gene linkage), thus they remain linked together as they go into the next generation. Therefore, these genes are assorted interdependently, not Mendelian independent assortment. However, the exception to this is genes far apart on the chromosome that may become unlinked during meiotic cross-over (see any high school or college introductory biology textbook). There may be multiple alleles for a particular gene in a population of organisms. The site of residence of a particular gene on a chromosome is called the gene locus. Of course, each individual has only two alleles for a particular gene locus (on homologous chromosomes from each parent). Thus, although in a diploid (2n) organism the two sites on homologous chromosomes may be occupied by only two alleles, with multiple alleles for a particular gene the two specific combinations of alleles carried by an individual may be many. For most traits in organisms (including humans), multiple genes control each trait. This is referred to as polygenic inheritance. Human height, weight, and skin color are examples of polygenic inheritance.

POPULATION GENETICS

A population is a localized group of interbreeding individuals belonging to the same species. A species is a group of populations of potentially interbreeding organisms (Biologic Species Concept) or a group of populations that have diagnostic characteristics that are distinct from other groups of populations and which have a unique evolutionary history (Phylogenetic Species Concept) (Note: we will discuss more on the species concept later, in another chapter).

Population genetics integrates Mendelian genetics with Darwin's Theory of Natural Selection. In fact, during the 1940s the modern synthesis of evolutionary theory established the importance of natural selection and Mendelian genetics. The modern synthesis brought together discoveries and ideas from many different fields of study, such as taxonomy, embryology, comparative anatomy, biogeography, paleontology, etc., and of course, population genetics. Populations are the units of evolution. Natural selection works on individuals within the population (except, perhaps in the case of social insects, such as ants, where altruistic kin selection may be important and result in group selection), but individuals do not evolve, populations evolve.

In a population, a particular individual has two alleles (one on each pair of homologous chromosomes) at specific gene loci, one from the male parent and one from the female parent. All the alleles at all the gene loci for the entirety of individuals in a population are called the gene pool. A particular allele may sometimes become fixed at a particular gene locus if all the individuals in the population become homozygous for that specific allele and the trait that it codes for.

However, in most cases there are two or more alleles at a particular gene locus, each with a certain relative frequency within the gene pool.

Let's take a look at an example. The following example and discussion of population genetics was modeled after Freeman and Herron (2007) with some modifications (primarily in the method used to randomly pick the gametes). Assume that we have a population of mice that have two possible alleles at a particular gene locus, with one dominant for the resulting trait and the other recessive. In our population of mice, we will use *A* for the dominant allele and *a* for the recessive allele. We will assume that the adult mice mate at random (of course, this is not always the case). Let's also assume that each adult mouse produces gametes such that 60% of the sperm and eggs received a copy of the *A* allele, and 40% received a copy of the *a* allele. Thus, the frequency of the *A* allele in the population is 0.6 and the frequency of the *a* allele is 0.4. What will be the genotype frequencies of the zygotes (and eventually the juveniles and adults) produced as the sperm fertilize the eggs? We will assume that 100% of the zygotes reach adulthood. Once the zygotes become adults, and the adults produce gametes (we will assume that all the adults donate the same number of gametes to the gene pool), what will be the frequency of alleles *A* and *a* in the gene pool of the next generation?

Now we devise a method to choose the gametes that will unite to form zygotes. We place several thousand marbles in a cylindrical container. Sixty percent (or 0.60) of the marbles placed in the container have an *A* printed on them, and forty percent (or 0.40) of the marbles placed in the container have an *a* printed on them. We can rotate the container with a crank to randomly mix the marbles. We assume that both sperm and eggs for our mice are represented here, each with the above frequencies. Now, assuming that we have cranked the container to randomly mix the marbles, we open the container, close our eyes and choose a marble. We record which allele (*A* or *a*) we have picked and assume that this will represent a sperm. Then we mix the marbles again and then pick another marble that will represent an egg. We now assume that the sperm and egg unite to form a zygote. By repeating this process for 50 to 100 times, we could record the genotypes that we get each time for the zygotes.

Using a similar method as presented above, I chose both a sperm and an egg and repeated the process 100 times to represent 100 zygotes. I determined the following genotypes for the 100 resulting zygotes: 34 had genotype *AA*, 56 had genotype *Aa*, and 10 had genotype *aa*. Remember, we are assuming that all zygotes develop into juveniles, and that all juveniles survive and become adults. We are also assuming that when the adults reproduce they will donate the same number of gametes to the gene pool. We assume that each adult will supply 10 gametes to the gene pool (to make the math easy). We won't worry about which adults supply sperm and which eggs. So, counting gametes we have:

- 34 *AA* adults make 340 gametes
    - 340 carry allele *A*, none carry allele *a*
- 56 *Aa* adults make 560 gametes
    - 280 carry allele *A*, 280 carry allele *a*
- 10 *aa* adults make 100 gametes
    - none carry allele *A*, 100 carry allele *a*

Summing up, the gene pool of our next generation now has 620 gametes of allele *A* and 380 gametes of allele *a*. This gives a frequency of 0.62 for allele *A* and a frequency of 0.38 for allele *a*. Remember we started with an allele frequency of 0.60 *A* and 0.40 *a*. Thus, our population has evolved. The allele frequencies from one generation to the next have changed. If the allele frequencies change, the population has evolved. If we repeat this process we would probably get numbers close to what we have above, but probably different. This is the luck of the draw, i. e. blind luck. The evolution of a population by blind luck is referred to as genetic drift. If we had chosen 10,000 sperm and 10,000 eggs to form zygotes, we would probably have gotten results very close to the initial allele frequencies of 0.60 *A* and 0.40 *a*.

What if blind luck (genetic drift) had played no role in our simulations. By using a Punnett square, we can predict exactly the combining of eggs and sperm to form zygotes to get the frequencies of alleles in the next generation. The Punnett square, invented by Reginald Crundall Punnett (Henig, 2000) is usually used to predict the genotypes among the offspring of a specific male-female cross. In that case, with 2 gametes possible from each, the probability of a particular allele from the sperm or egg is 0.5 for each. For a population, we can write the alleles for the eggs and sperm along the side and top of the Punnett square in the proportions that reflect their frequencies in the gene pool of the population. For our example of **A** and **a**, with frequencies of 0.60 and 0.40, respectively, we can show the four possible combinations of eggs and sperm and then calculate the probability of each genotype (Figure 4-2).

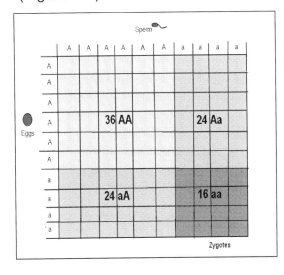

Figure 4-2. A Punnett square predicting the allele frequencies in the next generation as 0.60 A and 0.40 a and the genotypes as 0.36 AA, 0.48 Aa, and 0.16 aa (after Freeman and Herron, 2007). Calculations of genotypes as follows:

Egg		Sperm	Zygote	Probability
A	&	A	AA	0.6 x 0.6 = 0.36
A	&	a	Aa	0.6 x 0.4 = 0.24
a	&	A	aA	0.4 x 0.6 = 0.24
a	&	a	aa	0.4 x 0.4 = 0.16

Thus, by using the previous Punnett square we have predicted that the allele frequencies in the next generation will be 0.60 **A** and 0.40 **a**, and the genotype frequencies will be 0.36 **AA**, 0.48 **Aa**, and 0.16 **aa**. So, we have removed blind luck (genetic drift). This would be the case in an infinitely large population. (Note: The **Aa** heterozygotes were formed by combining either an **A** egg and an **a** sperm or an **a** egg and an **A** sperm.). Notice in the above example that 0.36 + 0.48 + 0.16 = 1. So, we have accounted for all of the zygotes. When these zygotes grow to adulthood, we will have the genotype frequencies shown above and the allele frequencies shown above.

The previous example has satisfied what is called the Hardy-Weinberg Equilibrium Principle. In the simple situation where only two alleles (**A₁** and **A₂**) are present for a particular trait, the following equation defines Hardy-Weinberg Equilibrium for genotype frequencies: $p^2 + 2pq + q^2 = 1$, where p represents the frequency of allele **A₁** and q represents the frequency of allele **A₂**. The Hardy-

Weinberg Equilibrium Principle assumes random combination of alleles (i.e. random mating), no selection, and no additions or changes to gene pool frequencies of alleles by migration or mutation. The principle also assumes that no genetic drift is occurring (thus we have a large population). In other words, the population is not evolving. If the Hardy-Weinberg Equilibrium is violated in a population, then the population is evolving.

THE GENOME

The totality of characteristic genetic material within each cell in an organism is popularly termed an organism's genome (DeSalle and Yudell, 2005; Kardong, 2008). We can also define the genome of an organism as all the DNA in a cell of a particular organism. Also, the DNA is typically the same in every somatic cell (but, sex cells have one-half of the full component of DNA) of each individual organism (except, in eukaryotic organisms, for DNA within the mitochondria and in the chloroplasts of plants). Eukarya have both a nuclear genome and a mitochondrial genome. In 2001, the first draft of the human genome sequence (nuclear genome) was completed (Carey, 2014) (actually completed in 2000, but published in 2001) by the biotech company Celera (published in *Nature*, February 2001) and the U. S. government via the Human Genome Project (published in *Science*, February 2001), with final completion in April 2003 (DeSalle and Yudell, 2005) (see also the following website of the National Human Genome Research Institute at http://www.genome.gov/).

The human genome has more than 3 billion nucleotide base pairs; however, because each individual has two of each chromosome (basically two genomes), each human has ovier 6 billion nucleotide base pairs), but only about 2% of these base pairs are genes that code for a particular protein or RNA. Further the human genome has only about 24,000 genes (Carey, 2015a). So, only 2% of the genomic DNA is coding, with the remaining 98% sometimes referred to as "Junk DNA" (noncoding) (Carey, 2015a). But we are learning more about this noncoding DNA all the time and some of it (maybe much of the noncoding DNA) may be involved with gene regulation (Carey, 2015a).

We have mapped the complete genomes of several other organisms also. Recently the chimpanzee genome and the Neanderthal genome have been sequenced (Pääbo, 2014). The sequencing of the human genome and the genomes of other organisms has initiated the genomic revolution. There are tremendous potentials for this type of data in the medical, biotechnology, pharmaceutical, conservation, biological, and paleontological fields. Also, we can legitimately infer that organisms that share more of their DNA have a more recent common ancestor (i. e., they are more closely related in an evolutionary sense). So, we can use comparisons of genomes or portions of genomes amongst organisms to work out phylogenetic (evolutionary) relationships. This has important applications in biology and paleontology.

The following is an example of the DNA sequence of *Pan paniscus* (bonobo chimp) for Mitochondrial Cytochrome C (Cox1) (involved in respiration) from the National Center for Biotechnology Information (NCBI) at the following website:

http://www.ncbi.nlm.nih.gov/genbank/

Pan paniscus mitochondrion, complete genome

NCBI Reference Sequence: NC_001644.1

GenBank Graphics

>gi|5835135:5322-6863 Pan paniscus mitochondrion, complete genome

```
ATGTTCACCGACCGCTGACTATTCTCTACAAACCACAAAGATATTGGAACACTATACCTACTATTCGGCACAT
GAGCTGGAGTTCTGGGCACAGCCCTAAGTCTCCTTATTCGAGCTGAATTAGGCCAACCAGGCAACCTTCTA
GGTAACGACCACATCTATAATGTCATTGTCACAGCCCATGCGTTCGTAATAATCTTTTTCATAGTAATACCTAT
CATAATCGGAGGCTTCGGCAACTGGCTAGTTCCCTTGATAATTGGTGCCCCCGACATGGCATTCCCCCGTA
TAAACAACATAAGCTTCTGACTCCTACCCCCTTCTCTCCTACTTCTACTTGCATCTGCCATAGTAGAAGCCGG
CGCCGGAACAGGTTGGACAGTCTACCCTCCCTTAGCAGGAAACTATTCGCATCCTGGAGCCTCCGTAGACC
TAACCATCTTCTCCTTGCACCTGGCAGGCGTCTCCTCTATCCTAGGAGCCATTAACTTCATCACAACAATCAT
TAATATAAAACCTCCTGCCATAACCCAATACCAAACACCCCTTTTCGTCTGATCCGTCCTAATTACAGCAGTC
TTACTTCTCCTATCCCTCCCAGTCCTAGCTGCTGGCATCACCATACTATTAACAGATCGTAACCTCAACACTA
CCTTCTTCGACCCAGCTGGGGGAGGAGACCCTATTCTATATCAACACTTATTCTGATTTTTTGGCCACCCCG
AAGTTTACATTCTTATCCTACCAGGCTTCGGAATAATTTCCCACATTGTAACTTATTACTCAGGAAAAAAAGAA
CCATTCGGATACATAGGCATGGTTTGAGCTATAATATCAATTGGTTTCCTAGGGTTTATTGTGTGAGCACACC
ATATATTTACAGTAGGAATAGACGTAGACACACGAGCCTATTTCACTTCCGCTACCATAATCATTGCTATTCC
TACCGGCGTCAAAGTATTCAGCTGGCTCGCTACACTTCACGGAAGCAATATGAAATGATCTGCCGCAGTACT
CTGAGCCCTAGGGTTCATCTTTCTCTTCACCGTGGGTGGCCTAACCGGTATTGTACTAGCAAACTCATCATT
AGACATCGTACTACACGACACATATTACGTCGTAGCCCACTTCCACTACGTCCTATCAATAGGAGCTGTATT
CGCCATCATAGGAGGCTTCATTCACTGATTCCCCCTATTTTCAGGCTATACCCTAGACCAAACCTATGCCAA
AATCCAATTTGCCATCATATTCATTGGCGTAAACCTAACCTTCTTCCCACAACACTTCCTTGGCCTGTCTGGA
ATGCCCCGACGTTACTCGGACTACCCTGATGCATACACCACATGAAATGTCCTATCATCCGTAGGCTCATTC
ATCTCCCTAACGGCAGTAATATTAATAATTTTCATAATTTGAGAAGCCTTTGCTTCAAAACGAAAAGTCCTAAT
AGTAGAAGAGCCCTCCGCAAACCTGGAGTGACTGTATGGATGCCCCCCACCCTACCACACGTTCGAAGAAC
CCGTGTACATAAAATCTAGA
```

Data similar to this may be obtained at the above site for many different organisms using the mitochondrial genome. This data can be compared between several organisms to construct phylogenetic trees (or cladograms) that show evolutionary relationships. Of course, the nuclear genome gives us much more information for showing the evolutionary relationships among organisms (Pääbo, 2014). We are now sequencing more and more nuclear genomes of organisms as time passes. We completed the sequencing of the human genome in 2003 and have since sequenced the genome of the common chimpanzee (*Pan troglodytes*) and even the Neanderthal man (*Homo neanderthalensis*) in 2010 (Pääbo, 2014). Let's discuss the human nuclear genome and compare it to the chimp and some other organisms.

GENOMIC NUCLEAR DNA AND VARIETY

Even though a large part of the human genome is similar to the genome of the common chimpanzee (*Pan troglodytes*), the recent sequencing of the chimpanzee genome has also shown some significant differences. As stated by DeSalle and Yuddel (2005), the DNA nucleotide sequence comparison between humans and the chimp differs by only about 1.2% (i.e. the nucleotide sequences are 98.8% identical). These numbers are based on the initial sequencing of the West African captive-born chimpanzee (subspecies *Pan troglodytes verus*) named Clint, with the data published in September 2005 in the journal *Nature* (The Chimpanzee Sequencing and Analysis Consortium, 2005). Further, the sequence comparisons show that the chimpanzee and human genomes differ by another 2.7% due to gene duplications or deletions. Thus, there is a total difference in the DNA nucleotide sequences of about 4%. Chromosomal differences are also present between the two genomes, with humans having 23 pairs of chromosomes, and chimpanzees 24 pairs of chromosomes. Results of the chimpanzee genome project suggest that chimpanzee chromosomes 2A and 2B fused to form the human chromosome 2, with no genes lost from the fused ends of 2A and 2B. However, at the site of fusion in human chromosome 2, there are about 150,000 base pairs of sequence not found in chimpanzee 2A and 2B. But even with these differences, about 29% of homologous proteins in chimpanzees and humans are extremely similar and with the typical homologous protein differing by only two amino acids (The Chimpanzee Sequencing and Analysis Consortium, 2005). Thus, the chimpanzee genome and the human genome are 96% identical in terms of the nucleotide base sequences and the proteins that are produced are very similar. So, why are humans so different from chimpanzees and yet at the same time more closely related to chimpanzees than any other organism on Earth?

The DNA sequence in the human genome has about 3 billion base pairs (Carroll, 2005), so the approximately 4% differences due to single nucleotide polymorphisms (SNPs) (differences at the sites of single nucleotides) and nucleotide insertions and deletions correspond to about 120 million base pair differences. Assuming that about one half of these are chimp specific and one half are human specific changes (just for ball park numbers) (note: I got this idea from Carroll, 2005; although he only used the 1.2% differences in nucleotide sequences between the chimpanzee and human genomes), that means that 60 million changes in the nucleotide bases have happened since our last common ancestor with the chimpanzee. Since (as far as we currently know) only about 5% of the DNA in our genome is involved in coding or gene regulation (Carroll, 2005), that leaves roughly 3 million coding nucleotide bases that separate us from chimpanzees. But which of the 3 million or so coding bases and their associated genes makes us so different from the chimps?

It is interesting that we share 89% of our DNA nucleotide sequence with mice (rodents) (DeSalle and Yuddel, 2005), and the rodent line and the primate line had a most recent common ancestor between about 90 million to 75 million years ago (DeSalle and Yudell, 2005; Carroll, 2005). In fact, Carroll (2005) points out that when we compare mouse and human genomes that greater than 99% of all genes in the human have a mouse counterpart, and vice versa. Further, according to Carroll (2005) 96% of all genes in the human genome are found in the exact same relative order in human chromosomes as in mouse chromosomes. This is pretty remarkable and points out that in the millions of years of evolution that have taken place since the last common ancestor of the mouse and the human that the DNA sequence, gene number, and gene organization have not played much of a role in the origin of humans or primates (Carroll, 2005). So again, what causes the big differences between rodents and primates and between primates and humans, and of course even between chimpanzees and humans?

We have already seen above that with the chimpanzee and human genomes there are also no big differences in the proteins that are made by the genes in the two genomes. Most proteins do not affect form, but instead carry out physiologic roles (Carroll, 2005). Certainly, this is important, and there are differences here between the roles of these proteins in humans as compared to other mammals, including the chimpanzee, such as smell, immunity, and reproduction. However, this does not affect the form and appearance of mammals, such as mice, chimpanzees, and humans (Carroll, 2005). So, again what accounts for the big differences in form?

However, a relatively small number of the genes in organisms are involved in regulatory functions and many scientists think that human and chimpanzee biology differ (including anatomical and behavioral differences) in the way that genes express themselves, thus the way that regulatory genes direct and regulate the type, amount, and timing of the production of proteins (DeSalle and Yudell, 2005; Carroll, 2005). These regulatory genes turn switches on and off that direct the development of size, shape, and fine-scale anatomy of structures (Carroll, 2005). Thus, mutations in regulatory genes will result in differences in the timing and amount of protein production. If these mutations are adaptive, then positive natural selection will fix them in a population and evolutionary changes will take place. The following statement from Carroll (2005, p. 270) emphasizes this point: "Because human evolution is largely a matter of the evolution of size, shape, and fine-scale anatomy of structures, and of timing in development, it is only logical that switch evolution would be important in the evolution of humans as well. Everything in our bodies is a variation on a mammalian or primate template. Thus, I believe that the weight of the genetic evidence is telling us that the evolution of primates, great apes, and humans is due to changes more in the control of genes than in the proteins the genes encode."

Three big evolutionary changes in body form and function that have occurred in the human lineage over the last 6 million years, or so, that make us different from the chimpanzee lineage are first, the evolution of bipedalism, second, the evolution of large brains (relative to body mass), and third, the evolution of speech. The first two, bipedalism and increase in brain size, may have evolved due to selection pressure related to climate change and associated changes in habitat conditions (this is another story in and of itself, so I will not discuss that here). Once humans had a large and complex brain, that may have supplied the selection pressure for the evolution of speech. The evolution of speech and the accompanying thought processes have really made humans what we are and enhanced the differences between humans and chimpanzees. But how did these changes take place? Comparisons of the human and chimp genomes may give us information that we need to answer this. As we continue to study the differences in the genomes, more information will come to light that will help us answer this question. However, we are already seeing some of the evidence in the comparison of the chimp genome and the human genome that may give us some partial answers to this question.

Let me just give one example of some of the studies that are currently taking place relative to the chimpanzee and human genomes which highlights regions of the human genome that are different than the chimp genome and which show recent (over the last 6 million years) positive selection. Katie Pollard, an associate professor of biostatistics at the Gladstone Institute at the University of California, San Francisco, in an interview with NOVA relative to "What makes us human?" and how we differ from chimpanzees and other great apes, explains that there are major switches in our genome that when turned-on by regulatory genes control structural genes that are responsible for producing proteins involved in brain development. Pollard found 18 base pairs in a stretch of 118 base pairs that were different in the human genome than in the chimpanzee genome. She further determined that there are only 2 nucleotide base differences between a chicken and chimp for this same stretch of 118 nucleotide bases. The last common ancestor of the chicken (or any other bird, such as a crow) and chimp lived greater than two hundred fifty million years ago and was some sort of reptile (Gee, 2013), thus within the last six million years (the time since the last common ancestor between chimps and humans), these nucleotide base differences have evolved in the human. Pollard determined that this region of the genome contains a major switch for turning on genes that are responsible for brain development. Pollard, in her words, stated it this way: "This fast-evolving sequence was exciting in and of itself. But when I looked around on the internet and databases that talk about things that people know about little pieces of our genome, lo and behold, it turned out this sequence is active in the human brain." And she further stated: "It turns out that the vast majority of these fast-evolving sequences are not genes, the parts of our genome that encode proteins. The pieces that have changed the most in our DNA look like they are switches, switches that turn nearby genes on and off. So, what makes a human different from a chimp isn't that we're made up of different building blocks, different genes, but instead that we're using those pieces in different ways." (see Nova, 2009, at http://www.pbs.org/wgbh/nova/evolution/dna-human-evolution.html)

So, what makes us so different in form from even other humans? Most humans are about as closely related to any other human as any human is to one of their second cousins (DeSalle and Yudell, 2005). The DNA nucleotide sequence between any two humans is 99.9% identical (DeSalle and Yudell, 2005). Thus, the concept of race is really not meaningful. In fact, there is more difference within so-called races than between races (DeSalle and Yudell, 2005). Most of the DNA sequence differences between any two human genomes are the result of single nucleotide polymorphisms (SNPs), and most SNPs appear to have no effect on the proteins that are produced by our genes. However, a few do result in changes in proteins or result in no protein produced. These SNPs may result in diseases and genetic disorders and are important to identify and study in the human genome (DeSalle and Yudell, 2005). The human genome has 3.2 billion nucleotide base pairs of DNA. So, even though we only differ by 0.1 % in DNA nucleotide bases, that is still a large number of base differences, about 3.2 million. Also, because of the environment that a particular person has experienced, the gene interactions and expressions may be different.

GENES CODE FOR BUILDING OF ORGANISMS

DNA (deoxyribonucleic acid) in the cell contains the code that builds an organism. The DNA in the cell is the code for assembly of the phenotype. Although environmental influences may control when certain portions of a DNA strain become activated, DNA ultimately controls the phenotypic traits of an organism. DNA is composed of nitrogenous bases (purines and pyrimidines), a 5-carbon sugar (deoxyribose), and a phosphate backbone.

The similarity of DNA, blood proteins, and other organic molecules among organisms is strong evidence that organisms share a common ancestor. DNA carries the code that is the blueprint to build each organism. The DNA molecule consists of chains of nucleotides. Nucleotides consist of a sugar (deoxyribose in DNA and ribose in RNA [ribonucleic acid]), joined to a phosphate group on one end of the sugar and a nitrogenous nucleotide base (nucleobase) on the other end of the sugar. In DNA, the nucleotide bases are adenine, guanine, cytosine, and thymine (thymine is replaced by uracil in RNA). Guanine and adenine are referred to as purines, whereas cytosine and thymine (and also uracil in RNA) are pyrimidines. Adenine always links with thymine between the two strands of DNA via two hydrogen bonds, while guanine always links with cytosine between the two strands via three hydrogen bonds (Watson, 1968) (Figure 4-3). Each double-stranded helical molecule (Watson and Crick, 1953a; Watson and Crick, 1953b; Watson, 1968) (see Figure 4-4) is wrapped in protein to form a chromosome.

DNA is composed of two specialized strands of nucleic acid wound in a spiral helix (see Figure 4-4) (Watson and Crick, 1953a; Watson and Crick, 1953b; Watson, 1968). Between the double strands of DNA, the nitrogenous bases pair specifically and preferentially: adenine (A) with thymine (T), and guanine (G) with cytosine (C). In large part, this is the basis for coding of information within DNA (and allows for the reliable duplication of DNA) (Watson and Crick, 1953a; Watson and Crick, 1953b; Watson, 1968), and later transcription to RNA. Each end of a DNA strand is polarized—one end is the 3' end, the other the 5' end. During DNA duplication, the new strand of DNA lengthens in the 5' to 3' direction. During replication of DNA, the two strands unravel (with the aid of an enzyme, DNA polymerase) and new complementary strands forms with each of the original strands, with complementary pairing always between adenine and thymine and between guanine and cytosine. The replication process always moves from the 5 ' (5 prime) end of the single new strand towards the 3' (3 prime) end of the single new strand. By this replication process and the complimentary base pairing of the nucleotide bases, an exact copy of the DNA code can be made (Campbell, 1996; Griffiths, Gelbart, Lewontin, and Miller, 2002).

DNA is the template that carries the genetic code. By complimentary bases that can attach to a DNA strand (with uracil substituting for thymine in RNA), the mRNA forms a sequence of nucleotide bases that are complimentary to the DNA strand that is being read: DNA to RNA (note that RNA is a single strand) such that uracil links to adenine, adenine to thymine, cytosine to guanine, and guanine to cytosine. With the assistance of transfer RNA (tRNA) and ribosomal RNA (rRNA), mRNA translates the coded information from the DNA to form proteins (linked amino acids). Each set of three mRNA nucleotides forms a codon that is specific for the synthesis of a particular amino acid. For example, GGC (guanine-guanine-cytosine) would code for the amino acid glycine.

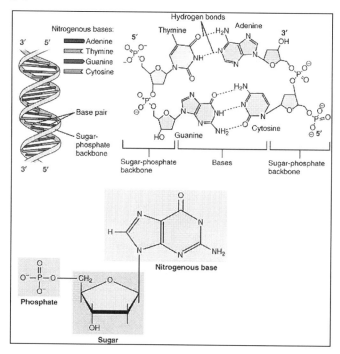

Figure 4-3. This illustration shows the nucleotide base pairing in a portion of a DNA molecule (upper right), the double helical structure of the DNA molecule (upper left), and the general structure of a complete DNA nucleotide (lower) (the nitrogenous base, the deoxyribose sugar, and the phosphate

Thus, the organized sequence of bases in DNA is read in transcription to produce a complementary, chemically matched, pre-mRNA molecule. In eukaryotic cells, the transcribed message in the pre-mRNA from the DNA template in a gene sequence contains regions that are coding, exons, and other regions of the sequence that are noncoding, introns. The pre-mRNA must be processed (by enzymes and small nuclear RNA [snRNA]) into mRNA by removing the introns in the transcribed sequence and splicing the remaining exons together. The processed mRNA (much shorter in terms of its nucleotide sequence) now consists of spliced exons that carry the coded message from the gene into the cytoplasm for translation into a protein. This translation is aided by ribosomal RNA (rRNA) in ribosomes within the cell and transfer RNA (tRNA) which carries amino acids to the ribosome to attach to the mRNA. In turn, three bases each form a codon in the mRNA which matches an anticodon on individual tRNA molecules that carry specific amino acids. Sequentially, codon by codon, the mRNA is a cipher program used in translation to produce a connected chain of particular amino acids, the protein molecule (Campbell, 1996; Griffiths, Gelbart, Lewontin, and Miller, 2002).

So, as we have described above, translation is the process in which the nucleotide sequence of mRNA (inside a ribosome) directs the sequential assembly of the amino acids in the growing protein molecule (polypeptide chain). There are four different nucleotides in mRNA that can be read three at a time as codons, thus there are 64 (4x4x4) different possible codons. This assembly of the 64 possible codons is called the genetic code (Figure 4-4) and this is the genetic code used by almost all organisms on Earth. There are only 20 common amino acids that make up most of the proteins of organisms; therefore, some amino acids are encoded by more than one of the 64 possible codons (see Figure 4-4) (Griffiths, Gelbart, Lewontin, and Miller, 2002).

Standard genetic code								
1st base	**2nd base**							**3rd base**
	U		**C**		**A**		**G**	
U	UUU	(Phe) Phenylalanine	UCU	(Ser) Serine	UAU	(Tyr)	UGU	(Cys) Cysteine	U
	UUC		UCC		UAC	Tyrosine	UGC		C
	UUA	(Leu) Leucine	UCA		UAA	Stop	UGA	Stop	A
	UUG		UCG		UAG	Stop	UGG	(Trp) Tryptophan	G
C	CUU		CCU	(Pro) Proline	CAU	(His)	CGU	(Arg) Arginine	U
	CUC		CCC		CAC	Histidine	CGC		C
	CUA		CCA		CAA	(Gln)	CGA		A
	CUG		CCG		CAG	Glutamine	CGG		G
A	AUU	(Ile) Isoleucine	ACU	(Thr) Threonine	AAU	(Asn)	AGU	(Ser) Serine	U
	AUC		ACC		AAC	Asparagine	AGC		C
	AUA		ACA		AAA	(Lys) Lysine	AGA	(Arg) Arginine	A
	AUG	(Met) Methionine	ACG		AAG		AGG		G
G	GUU	(Val) Valine	GCU	(Ala) Alanine	GAU	(Asp)	GGU	(Gly) Glycine	U
	GUC		GCC		GAC	Aspartic acid	GGC		C
	GUA		GCA		GAA	(Glu)	GGA		A
	GUG		GCG		GAG	Glutamic acid	GGG		G

Figure 4-4. The standard genetic code. The codon AUG both codes for methionine and serves as an initiation site: the first AUG in a <u>mRNA</u>'s coding region is where translation into protein begins.

CHAPTER 5

SYSTEMATICS, CLADISTICS, AND TREE THINKING

INTRODUCTION

One of the goals of science is to recognize patterns and order in the natural world. Are there patterns in the biota (sum total of all living organisms that have ever lived) (Fastovsky and Weishampel, 1996)? Can we recognize any patterns that may be present and use the patterns to order the biota? The answer to both of these questions is yes. Can we use the patterns to help us understand Earth and biotic processes that account for the diversity of life on Earth, both in the past and at present? Again, the answer is yes (Fastovsky and Weishampel, 1996; Fastovsky and Weishampel, 2005).

The key to understanding the evolution and diversity of Earth's organisms is to determine their patterns of relationships (how they are related to each other). In order to do this, we need to understand some of the principles of evolution, taxonomy, and phylogeny (Greek: phylon + geneia = tribe or race + birth or origin). The following terms and concepts will be useful for our discussion of biotic patterns and their meaning:

- **PHYLOGENY**: The history of descent of organisms (evolutionary relationships).

- **TAXONOMY**: The classification of organisms (the science of classifying organisms). A grouping of organisms is called a **taxon** (**taxa**, plural). Taxonomy is not just naming groups of organisms; species and higher taxa reflect evolution.

- **BIOLOGIC (ORGANIC) EVOLUTION**: The change of organisms over time. As Darwin (1859) put it: "Descent with modification".

- **DIVERSITY**: The different types of organisms. Diversity of the biota (all organisms alive and extinct) is measured both in time and in space. Diversity in time is reflected by evolutionary change. Diversity in space is reflected by the geographic distribution of organisms (biogeography).

- **BIOGEOGRAPHY**: The geographic distribution of organisms. For the distribution of ancient organisms we use the term **PALEOBIOGEOGRAPHY**.

Systematics refers to the combination of the above terms. Systematics goes beyond just traditional taxonomy (naming and classification of organisms). Systematists (scientists who study systematics) look at past and present geographic distributions of organisms (biogeography and paleobiogeography), diversity of both modern organisms and past organisms, evolutionary history (phylogeny), and the total pattern of natural diversity to provide the basic framework for all of biology and paleontology. The following statement from Prothero (2004, p. 47) states the importance of the systematist quite well: "The systematist uses the *comparative approach to the diversity of life to understand all patterns and relationships that explain how life came to be the way it is.* Put this way, systematics is one of the most exciting and stimulating fields in all of biology and paleobiology." (italics by Prothero)

We can organize the biota into a hierarchy (rank or order of the features of the biota). For example: living organisms - things that are alive; vertebrates - living organisms that have a backbone; mammals - living organisms that have a backbone and have fur and mammary glands. Thus, mammals are a subset of all animals that have backbones. Taxa of the biota (both past and present organisms) are related by the sharing of features in a hierarchy. Most organisms could be described as having a primitive body plan with variations (but the original, unmodified body plan is always present in the biota). Life is not really infinitely diverse, but is connected by the sharing of certain features in a hierarchy (Fastovsky and Weishampel, 2005).

To recognize the hierarchy, we must identify features of organisms. These features are referred to as characters. The distribution of characters among a selected group of organisms has meaning, but a single feature of a specific organism does not have much meaning (except to separate it from other organisms) (Fastovsky and Weishampel, 2005; Futuyma, 2009). Thus, shared characters among organisms are important in classifying them as belonging to groups of related organisms. Ancestral characters, also called primitive characters, are characters of larger groups that are not specific to smaller groups within the larger group. For example, birds have a backbone; this is not specific to birds because frogs also have a backbone and both birds and frogs inherited a backbone from a common ancestor that had a backbone (Fastovsky and Weishampel, 2005). Specific characters, also called derived characters (Futuyma, 2009) (or sometimes evolutionary novelties [Lucas, 2005]) are usually restricted to a smaller group within a larger group, thus feathers are specific to birds (and now we know that some dinosaurs also possessed feathers), which belong to the larger group vertebrates; frogs, which are also vertebrates, do not have feathers and thus cannot be grouped with birds using the specific character of possession of feathers (Fastovsky and Weishampel, 2005). Primitive characters (traits) are inherited from ancestors of a particular group (taxon). Derived characters are unique to a particular group (taxon).

PHYLOGENY, HEIRARCHY, AND CLADISTICS

Cladistics, also called Phylogenetic Systematics, is a form of systematics that attempts to determine the phylogenetic relationships of organisms based on unique shared derived characters. Cladists construct cladograms. Cladograms are branching diagrams to show a hierarchical distribution of shared derived characters. We can group anything using shared characters, thus it is not restricted to living organisms. However, when we group living organisms or fossils (once living organisms) into a hierarchy based on shared derived characters, we are implying that the organisms have an evolutionary relationship (i.e. they share a common ancestor). As we stated earlier, the history of descent of a group of organisms is referred to as a phylogeny. The phylogeny of a group of organisms shows the evolutionary relationships within the group. Phylogenies are determined by constructing cladograms.

Each branch (or bundle of branches) of a phylogeny is called a clade (from clados meaning branch). A divergence is a split on a cladogram. Convergence is the evolution of similar features in two unrelated (or more distantly related) clades. Groups of organisms shown on a phylogeny are called taxa (singular taxon). For very detailed work, the species level taxon is used. Many (if not most) biologists define species as a population of naturally interbreeding organisms (in other words, members of the same species share a common gene pool). Of course, paleontologist cannot use this definition directly. Paleontologists base their work on the morphologic species concept. A morphologic species (morphospecies) is defined by similarity of anatomical or morphological characters within a fossil group (we will discuss more on the definition of a species later). So, the manner in which organisms are related is defined as their phylogenetic relationships.

In order to understand the history of life, we have to understand the patterns of evolution. Darwinian evolution (as modified by the modern synthesis into neo-Darwinism, as we discussed earlier) is the most accepted theory of evolution today and was first proposed by Charles Darwin (1859) in *On the Origin of Species* (except for the brief papers on evolution by natural selection presented by Darwin and Wallace in 1858, as we discussed in Chapter 1). This concept is sometimes expressed as "Survival of the Fittest", however, this expression is often misunderstood and perhaps its use should be avoided. Natural selection is the primary mechanism of evolution and really means that the organisms that survive to reproduce, or reproduce more offspring than other members of their species, will selectively pass on more of their traits to their offspring, thus a change will occur in the gene frequencies of the gene pool and evolution will take place. For neo-Darwinian evolution, we now use cladistics (phylogenetic systematics) to show evolutionary relationships of ancestors to descendants. The German entomologist, Willi Hennig, first developed phylogenetic systematics or cladistics. He published a book in 1950 (written in German) that dealt with his ideas on phylogenetic systematics. His book was translated into English and published by the Illinois

University Press (Urbana, Illinois) in 1966. Phylogenetic systematics then quickly became the preferred method to show evolutionary relationships amongst organisms (Prothero, 2004; Fastovsky and Weishampel, 2005).

Similarity in anatomical (morphological) features is one line of evidence for the evolutionary relationship of organisms (but so are genetic, molecular, and DNA nucleotide similarities). When two anatomical (or genetic, molecular, or DNA nucleotide) structures can be traced back to a single structure in a common ancestor, we say that the two structures are homologous. Thus, homologous structures are called homologues (or homologies). So, homology refers to two or more features that share a common ancestry. For example, our hands (as with all mammals) are homologous to the wrist bones and digits (carpals, metacarpals, and phalanges) on dinosaur forelimbs and the common ancestor to both mammals and dinosaurs had wrist bones and digits on the forelimb.

Analogues (analogous structures), on the other hand, perform a similar function in two different organisms, but may or may not trace back to a common ancestor (i. e., analogous structures may or may not be homologues). For example, the wings of an insect and the wings of a bird are not homologues, but are analogues; they also cannot be traced back to a single structure in a common ancestor (thus, they have a different embryological origin).

Other characters in organisms often look similar, but may or may not be analogous or homologous. These are referred to as homoplastic characters or structures. Sometimes organisms evolve structures that look similar to structures in other organisms, but these structures cannot be traced back to a similar structure in the ancestors of two different organisms; the structures may also not perform the same function in two different organisms, although they may look superficially similar. One example of homoplasy is when organisms evolve structures that mimic the structures on other organisms, like large spots on the wings of a moth or butterfly that resemble eyes (Figure 5-1), perhaps to fool (or startle) a potential predator.

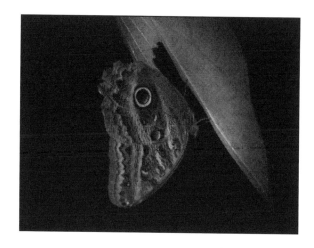

Figure 5-1.   A butterfly of the genus *Caligo* with large eyespots. (Photographed in Costa Rica by E. L. Crisp, 2008.)

Understanding evolution requires the recognition of homologous structures or characters (including homologous molecular structures).   Obvious (but often ignored) evidence of evolution is the hierarchical distribution of homologous characters in nature. Some homologous characters are present in all organisms (such as DNA and/or RNA and cell membranes). Some homologous characters are present in smaller groups. And some homologous characters are restricted to very small groups.

If evolution has occurred (and it has), there must be a single phylogeny.   We want to reconstruct evolutionary patterns.   Cladograms are hierarchical branching diagrams that allow us to show shared derived homologous characters (synapomorphies) that presumably relate organisms. A cladogram is a testable hypothesis of evolutionary relationships amongst the organisms being compared (Carlson, 1999; Fastovsky and Weishampel, 2005).

If derived characters are shared between two taxa, then cladistics argues that the two taxa are related. Shared primitive characters do not reveal phylogenetic similarities.   Shared   derived   characters   result   in   cladograms   that   are monophyletic. A monophyletic group includes the common ancestor and all the descendants of the common ancestor. Polyphyletic groups do not share a most recent common ancestor (Fastovsky and Weishampel, 2005).

How do we identify derived characters? It is not always easy. But, when a new taxon originates it inherits features from its ancestor. These inherited characters are ancestral or primitive characters (plesiomorphs) (Carlson, 1999). Features that arise for the first time in a new taxon are advanced characters or derived characters (apomorphies) (Carlson, 1999). These derived characters unite organisms (or fossils) into closely related groups, but only if the derived characters arose only once in related groups. If the derived characters arose more than once (in unrelated groups) then the features are not representative of closely related groups.

In fact, evolutionary convergence is where derived characters have arisen more than once in different distantly related organisms. For example, wings in birds, insects, and bats. These groups are not closely related, but share derived characters (wings). Of course, if we recognize that these are analogous structures, rather than homologous structures, we know the derived character of possessing wings does not necessarily relate these organisms. So, we only want to compare homologous shared derived characters (synapomorphies) to show phylogenetic relationships (Carlson, 1999; Fastovsky and Weishampel, 2005; Futuyma, 2009). Convergent evolution of characters presents the greatest threat to cladistic analysis. We must recognize that convergence has occurred (Fastovsky and Weishampel, 2005).

So, cladistics is a form of systematics that attempts to determine the phylogenetic relationships of organisms based on unique homologous shared derived characters (synapomorphies) (Carlson, 1999; Fastovsky and Weishampel, 2005); Futuyma, 2009). Systematists construct cladograms. Cladograms are branching diagrams to show a hierarchical distribution of homologous shared derived characters and that illustrate patterns of phylogenetic (evolutionary) relationships (Carlson, 1999). Of course, systematists must be careful to recognize derived characters that are truly homologous within the group (taxon) in question and have not evolved more than once as analogous characters in that particular clade. However, homoplastic characters may be present and useful as separate derived characters within sister groups (two monophyletic groups or clades that are each other's closest relative are called sister-groups [Cracraft, 2005]). In other words, due to convergent evolution, analogous and homoplastic characters may complicate the construction of cladograms. In addition, derived characters may be lost from a lineage as time passes (Fastovsky and Weishampel, 2005; Futuyma, 2009). Because of these problems, alternate cladograms may be constructed for the same characters and taxa. According to Cracraft (2005), Fastovsky and Weishampel (2005), and Futuyma (2009), the cladogram to show the best estimate of the true phylogenetic (evolutionary) relationships is the cladogram that is the most parsimonious ("...the one that requires us to postulate the fewest evolutionary changes" [Futuyma, 2009]). Futuyma (2009, p.28) further states the following: "Thus parsimony holds that the best phylogenetic hypothesis is the one that requires us to postulate the fewest homoplasious changes." Thus, a cladogram is a hypothesis of evolutionary relationships amongst the organisms being compared in a cladogram. We can make predictions of evolutionary relationships amongst taxa using a cladogram and then test the cladogram by adding more taxa and characters to the cladistic analysis. The characters that we use on cladograms may be morphologic, behavioral, or molecular. Multiple independent sources of data that support a hypothesis (in this case a cladogram or phylogenetic tree) make a hypothesis much stronger and more robust. However, sometimes the different sources of data do not give the same picture. Then of course, we have to modify the hypothesis, or test the hypothesis further with additional information. For example, until recently the cladogram for cetacean (whale) phylogeny based on DNA analysis of modern mammals did not agree with that based on the paleontologic and morphologic data (Prothero, 2007; Zimmer and Emlen, 2013). However, recent fossil finds have found intermediate fossils with morphologic traits that are more consistent with the DNA data. So, further testing of the phylogeny of whales (whale cladogram) based on DNA was complimented by using additional fossil data that gave more support to the DNA-based phylogeny (Prothero, 2007; Zimmer and Emlen, 2013). There are numerous other examples of situations like this and as more data (both molecular and morphologic, and from new fossil finds and restudy of old fossils) are gathered, phylogenetic hypotheses (cladograms) will become stronger.

So, as you see from the above discussion, we can construct cladograms (phylogenetic trees) to infer evolutionary relationships amongst organisms (both past and present). Once we learn the ages of fossils, we can sometimes even put a timescale on the vertical axis of a cladogram to show the timing of the evolutionary history. Sometimes we can even make estimates of the timing of divergence of two sister clades (two sister monophyletic groups) from analysis of mutational rates in the DNA of modern organisms using so called "molecular clocks" (Futuyma, 2009). Mutations in the DNA (or RNA for viruses) in the non-coding portions of the genome (which is the bulk of the DNA in most organisms) are not acted on by natural selection, so they should occur at a fairly constant random rate (Griffiths, Gelbart, Lewontin, and Miller, 2002; Futuyma, 2009). These are sometimes referred to as neutral mutations and may only become fixed in species by genetic drift (a random process) (Zimmer, 2010). So, evolutionary changes in neutral DNA should occur according to a molecular clock that is ticking at a certain rate for particular DNA sequences. Thus, if we know the rate of mutation, we would be able to determine the absolute age (rather than relative age) of divergence of two species from a certain common ancestor by determining the number of nucleotide differences between two species. Griffiths, et al (2002) state this concept as follows: "The constant rate of neutral substitution predicts that, if the number of nucleotide differences between two species is plotted against the time since the divergence from a common ancestor, the result should be a straight line with a slope of μ." They show a plot on p. 639 for the β-globin gene that illustrates this. This concept could also be used with amino acid substitutions in proteins. Griffiths, et al (2002, p. 639) also show a plot of time (x-axis) versus number of amino acid substitutions per 100 residues in vertebrates for three different proteins. But as you can see from this concept, the molecular clock has to be calibrated by knowing the age of divergence. This can often be estimated by using fossils that have already been dated for absolute age and that we are reasonably sure are the common ancestors for a group of organisms that we are interested in studying and determining the younger (than the fossil used for calibration) points of divergence in the lineage (Futuyma, 2009). Futuyma (2009) gives an example for the migrations of Hawaiian honeycreepers in the past as new islands arose from the sea as the Pacific Plate drifted northward over the Hawaiian hot spot (mantle plume). As new islands emerged from the sea, honeycreepers would fly from older islands to the new islands (it is assumed that this colonization was fairly rapid). Because these new islands are volcanic (basaltic igneous rock), the rocks that make-up the islands can be dated radiometrically. Thus, when we look at the phylogenies and sequence differences of several of the birds on two of the islands (a younger island and an older one) to determine that there is correlation between sequence distance (for DNA; in this case mitochondrial DNA for the birds) and the time of separation of the birds (time of migration of some birds to the younger island). There are several other examples similar to this in the literature that illustrate that the rate of mutation of neutral DNA nucleotide sequences is fairly constant for similar DNA sequences in different organisms.

So, from several studies it appears that the rate of neutral mutations and thus the rate of our molecular clock for particular genetic sequences are relatively constant (Futuyma, 2009). Futuyma (2009) gives an example of how to calculate the average rate of base pair substitution in a lineage if we know the absolute time of divergence for one of the organisms in the lineage. The example is based on a study by Donoghue and Benton (2007). They determined (based on fossil data) the time of divergence between the rhesus monkey (genus *Macaca*) and hominoids (apes, including humans) at 25 million years ago (based on fossils of the oldest cercopithecoid monkeys [Old World monkeys] represented by the rhesus monkey). By looking at the DNA sequence differences for the ψη –globin pseudogene (a neutral gene) and determining the number of base-pair substitutions from the rhesus monkey to the modern human common ancestor, the rate of the molecular clock (the mutation rate) calculates to be 1.24×10^{-3} per million years (Futuyma, 2009, p. 34). Others have done similar studies. Thus, it does look like the rate of mutation in neutral gene sequences is relatively constant for a particular section of the genome and thus when compared between different species within a lineage, sometimes a molecular clock can be developed to determine the divergence points (points of common ancestry) on a phylogenetic tree (cladogram).

The Tree of Life (TOL) (Cracraft, 2005; Tree of Life Web Project, accessed 04/15/2014) that shows the history and phylogenetic relationships for the three domains of life on Earth (Bacteria, Archaea, and Eukarya) is a phylogenetic tree that uses the methods of cladistics to show the evolutionary relationships and evolutionary history of life on Earth. This is an ongoing project that is being refined more as time goes by and as new discoveries of fossils and extant organisms are made. Of course, this is a hypothesis of relationships based on our current understandings and will be refined more in the future. As we can see from the TOL, life has had a long and eventful history on Earth, with, in general, increasing diversity and complexity from one-celled prokaryotic cells that evolved somewhere in the interval from 3.5 to 4.0 Ga, to more complex eukaryotic single celled organisms by about 2.0 Ga, to multicelled algae by about 1.0 Ga, to multicelled animals by 600 to 700 Ma, and to a real evolutionary explosion of life forms in the Cambrian Period of the Paleozoic Era beginning about 540 Ma. This increase in fossils at the beginning of Cambrian time is because marine invertebrates started secreting hard shells at that time, so the fossil record is much better beginning with that age of rocks; and also, because oxygen levels in the atmosphere and oceans had increased dramatically by then such that there was a rapid increase in the rate of evolutionary emergence of new groups of invertebrate animals (Cracraft, 2005; Ward, 2006; Wicander and Monroe, 2010; Erwin and Valentine, 2013).

The organisms currently living on Earth only represent a small fraction of all the organisms that have ever lived. As Cracraft (2005) indicates, paleontologists estimate that 90 to 99% of all organisms that have ever lived are extinct. Cracraft (2005) also states that only about 1.7 million living organisms have been named and described, but that may represent a fraction of the several million more that may exist on Earth.

Why is knowing phylogeny important? Knowing the phylogeny of a group of organisms is important for many reasons, other than to give the history of life on Earth and as evidence of evolution. Organisms that have evolved from common ancestors and that are more closely related may react more similar to medications (such as antibiotics), pesticides, and other medical and agricultural treatments or uses. Knowing the phylogeny of humans may help target particular diseases that affect a certain subset of the human population. The more closely related organisms are to each other (which we can determine by phylogenetic comparisons), the better the biological comparisons between the organisms. This concept may have many applications in medicine, agriculture, conservation, criminal investigations, etc. (Donoghue, 2005).

Darwinian and neo-Darwinian evolution is described by descent with modification. In order to understand the history of life, we have to understand the patterns of evolution. We use phylogeny to show relationships of ancestors to descendants; therefore, phylogeny explains the history of descent of organisms. In modern phylogenetic methods, we use cladograms to show monophyletic groups (natural groups that descended from a common ancestor). Polyphyletic groups are groups that do not share a closest common ancestor, and thus are not of value in determining phylogeny. Paraphyletic groups are groups that do not include all the descendants of a common ancestor.

CONSTRUCTING A CLADOGRAM

The first thing that we want to do to show the evolutionary relationships of a group of organisms is to choose characters and construct a character matrix for the organisms that we have chosen. We will construct a fairly easy example here, but the method will be valid for more complex examples. Because of the large number of characters and taxa that are often used to construct cladograms, computer analysis is usually required. However, that will not be necessary for our example here. For our example, let us choose the following organisms for which we want to show the evolutionary relationships by constructing a cladogram: a clam (bivalve), a shark (cartilaginous fish), a bluegill (boney fish), a salamander (amphibian), an iguana (reptile, lizard), an alligator (reptile, archosaur), a crow (bird and also reptile, archosaur), a raccoon (mammal), and a human (mammal).

Now let us choose the characters that we are going to use to show the evolutionary relationships. We will choose the following characters: backbone (vertebral column or possession of vertebrae), bony skeleton, four limbs (2 pairs of appendages with digits at the end - the tetrapod condition), amniotic egg (egg with membrane and/or mineralized shell around an amniotic fluid that baths the embryo), hair or fur, two openings in the skull behind the eye socket (diapsid reptilian skull), and an opening in the skull in front of the eye socket (antorbital fenestra). (Note: we would need some introductory knowledge of comparative anatomy of animals to choose these characters.)

Taxa ⟶ Characters ↓	Clam	Shark	Bluegill	Salamander	Iguana	Alligator	Crow	Raccoon	Human
Backbone	0	1	1	1	1	1	1	1	1
Bony Skeleton	0	0	1	1	1	1	1	1	1
Four Limbs	0	0	0	1	1	1	1	1	1
Amniotic Egg	0	0	0	0	1	1	1	1	1
Hair	0	0	0	0	0	0	0	1	1
Two openings behind eye	0	0	0	0	1	1	1	0	0
Opening in front of eye	0	0	0	0	0	1	1	0	0

Table 5-1. Matrix of taxa versus characters for the construction of a cladogram.

Now we will construct a table (matrix) (Table 5-1) of the taxa versus the characters. If the organism has the character (derived condition) we will place a 1 in the appropriate box, but if it doesn't (primitive condition) we will place a 0 in the box. We will place our taxa as the columns on our table and the characters as the rows (this may be reversed, and often is). One of the organisms that we have chosen above (the clam) will have none of the derived characters that we have chosen; we will refer to this as the outgroup (a closely related group, but not part of the group or groups of interest). An outgroup in cladistics is used to polarize the characters that we choose; we choose one of the organisms (the clam, an invertebrate) such that it will have all ancestral (primitive) characters. The remaining organisms would be referred to as the ingroup and this is the group of interest (in this case all vertebrates) and we score the characters of the ingroup in comparison to the ancestral (primitive) characters of the outgroup.

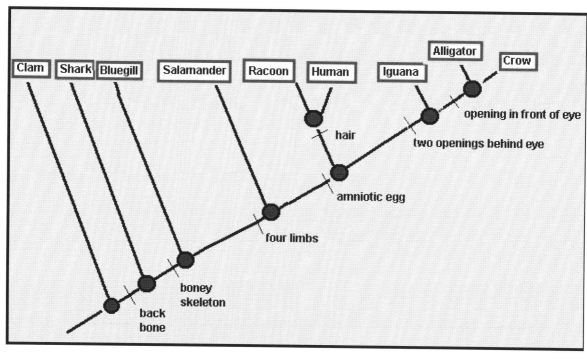

Figure 5-2. A cladogram that clusters the taxa that have shared derived characters. Note that the marks before each node (large dot at position of each divergent branch) represent derived characters that each ancestor (at the position of the nodes) possesses and that is shared with all taxa upward and to the right of that node.

Now we have a cladogram that is a hypothesis of the evolutionary relationships of the organisms that we have chosen. Notice that the alligator and crow have a more recent common ancestor than the alligator and iguana, therefore, the alligator and the crow are more related in an evolutionary sense than is the alligator and iguana. The clam does not share any of the derived characters (that we have chosen) with the other taxa and is considered the outgroup (for fixing [polarizing] the derived characters). We could choose other characters to separate the human and raccoon (like bipedalism and large brain for humans and not for the raccoon); however, this is not necessary just to group them together (as we see from the above cladogram). We could also choose other characters to separate the alligator from the crow (like possession of feathers for the crow and not the alligator), but again, this is not necessary at this stage.

If fewer steps in a cladogram provide an explanation of the derived characters, then we assume it is the correct cladogram until we have evidence to the contrary. This (as stated earlier) is referred to as parsimony. So, we start with the simplest hypothesis and consider it in the context of new or independent evidence (such as adding new characters or new taxa to our cladogram). We consider our cladogram of evolutionary relationships as our hypothesis. We test our hypothesis with new or independent evidence (i.e. we consider more derived characters and more taxa and whether they fit the predictions made by the cladogram) (Fastovsky and Weishampel, 2005). Thus, cladograms are hypotheses. They are more robust if they survive falsification attempts. The addition of characters may result in the rejection of a certain cladogram (if the addition results in a character distribution which is not the most parsimonious) (Fastovsky and Weishampel, 2005).

Now let us test the cladogram that we have presented above. Let us predict that the Late Jurassic meat-eating (theropod) dinosaur *Allosaurus* is more closely related to the crow (thus birds) than to the alligator. We will look at the characters that we have already looked at, but we will need to add some more characters to test this hypothesis. So, let us add *Allosaurus* to our table (Table 5-2) with the new characters added also. We will add the following characters: hole in the hip socket (perforated acetabulum), 4th and 5th fingers on the hand lost, and three-toed foot (with digits 2, 3, and 4).

Based on the additional characters that we have added, our original hypothesis is supported by the new data. Thus, based on the data we have looked at, we would conclude that *Allosaurus* has a more recent common ancestor with the crow than with the alligator. Our cladogram (see Figure 5.3) may now be modified to include *Allosaurus*.

Characters	Clam	Shark	Bluegill	Salamander	Iguana	Alligator	Crow	Raccoon	Human	*Allosaurus*
Backbone	0	1	1	1	1	1	1	1	1	1
Bony Skeleton	0	0	1	1	1	1	1	1	1	1
Four Limbs	0	0	0	1	1	1	1	1	1	1
Amniotic Egg	0	0	0	0	1	1	1	1	1	1
Hair	0	0	0	0	0	0	0	1	1	?
Two openings behind eye in skull	0	0	0	0	1	1	1	0	0	1
Opening in front of eye in skull	0	0	0	0	0	1	1	0	0	1
Hole in Hip Socket	0	0	0	0	0	0	1	0	0	1
4th and 5th fingers on hand lost	0	0	0	0	0	0	1	0	0	1
Three-toed foot	0	0	0	0	0	0	1	0	0	1

Table 5-2. Based on the additional characters that we have added, our original hypothesis is supported by the new data. Thus, based on the data we have looked at, we would conclude that *Allosaurus* has a more recent common ancestor with the crow than with the alligator. Our cladogram now may be modified to include *Allosaurus*.

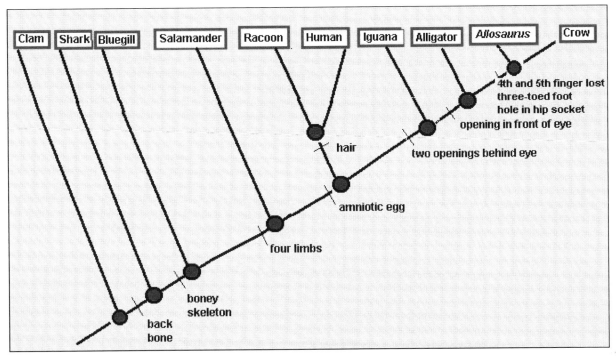

Figure 5-3. New cladogram including *Allosaurus*. Any other position on the cladogram for *Allosaurus* would require more evolutionary steps and thus would be less parsimonious.

TAXONOMY

Taxonomy, a part of systematics, is the process of classifying organisms into groups based on their similarities and of naming organisms. Carolus Linnaeus (1707-1798) devised the present system of classification of organisms into major groups. The Linnaeus system of classification is a hierarchical scheme, as one proceeds up the classification ladder the categories become more inclusive. Cladistics may be used to get the evolutionary relationships – then the organisms can be placed into the Linnaeus classification scheme. However, strict cladists prefer to name clades rather than to place the clades into the Linnaeus system. The following is an example of the Linnaeus classification system for *Tyrannosaurus rex*, the ferocious late Cretaceous theropod dinosaur:

Major Subdivisions	Classification
Kingdom	Animalia
Phylum	Chordata
Subphylum	Vertebrata
Class	Reptilia
Order	Theropoda
Family	Tyrannosauridae
Genus	*Tyrannosaurus*
Species	*Tyrannosaurus rex*

Linnaeus also said that each organism should have two names (a binomen) to define it, the generic (genus) name and the specific (species) name, for example, *Tyrannosaurus rex* or *Homo sapiens*. Linnaeus, although not trying to show evolutionary relationships, lumped organisms that had similar traits into the same groups. Of course, this implies phylogenetic relationships. In fact, Darwin (1859) used Linnaeus' classification system as one line of evidence in support of his evolutionary theory of common descent.

Modern biologists and paleontologists use cladistics to relate both modern and fossil organisms in an evolutionary sense (i.e., determine their phylogenies). Many (if not most) still name organisms based on the Linnaean system and may place their phylogenetic groupings into the Linnaean hierarchy. However, biologists and paleontologists recognize the arbitrary nature of the Linnaean categories (for example, some paleontologists might refer to Saurischia as an order of the Dinosauria, whereas others may consider it to be a superorder), and thus may prefer that groupings on cladograms not be placed in formalized Linnaean categories. On the other hand, some biologists and paleontologist do prefer to use the Linnaean system once the evolutionary relationships have been worked out using cladistics.

However, biologists and paleontologists always name organisms at the genus and species level according to the Linnaean system and must follow international codes of zoological and botanical nomenclature (for example, the International Code of Zoological Nomenclature). The International Code of Zoological Nomenclature provides the rules that must be used when naming animals (a similar code exists for naming plants). Names at the genus and species level are Latinized and italicized (or underlined). The specific name always includes the genus name preceding it (i.e., the specific name can never stand alone). Particular endings are required for different Linnaean categories (for example: order usually has the suffix "a"; family has the suffix "idae", etc.). However, there is much freedom in the naming of organisms. For example: a big carnivorous dinosaur (that was different than all other carnivorous dinosaurs known then) found by Barnum Brown, of the American Museum of Natural History, in Montana in 1905, and described by his supervisor, Henry Fairfield Osborn, was named *Tyrannosaurus rex*, meaning tyrant lizard + king, or king of the tyrant lizards (McGinnis, 1984; Horner, 2001; Fastovsky and Weishampel, 2005). This is the type specimen (holotype) for *T. rex*, to which all others must be compared, and is now housed at the Carnegie Museum of Natural History in Pittsburgh, PA (McGinnis, 1984).

Priority of the name is another rule of naming organisms. No two different organisms (extant or extinct) can have the same scientific name (binomen). Also, if two organisms belong to the same taxon, they cannot be given different names; the one that was named first is the correct name. For example, the Yale paleontologist O.C. Marsh in 1877 named a partial sauropod dinosaur skeleton (found in Colorado) to the genus *Apatosaurus* (deceptive + lizard). A couple of years later (1879) he found an almost complete skeleton of a sauropod dinosaur in Wyoming and gave it the genus name *Brontosaurus* (thunder + lizard). Many years later, it was determined by paleontologists that the two skeletons were of the same creature, thus *Apatosaurus* was ruled to be the correct genus name by priority.

CHAPTER 6

THE EVIDENCE FOR EVOLUTION

INTRODUCTION

If we, as scientists and students of science, are capable of understanding the world around us and the ways of science, then organisms have changed over time and have descended with modification from common ancestry. Therefore, biologic evolution is a fact (Fastovsky and Weishampel, 2005; Pennock, 2005). I particularly like the following statement by Pennock (2005): "Evolution is not a belief that is taken on faith, it is a fundamental scientific discovery that has been empirically confirmed by the most rigorous of observational tests." Overwhelming evidence supports the Darwin-Wallace Theory (and modern neo-Darwinian Theory) of evolution by natural selection. Thus, natural selection is the primary means by which evolutionary change takes place. The bringing together of evidence from many different fields of study in biology (ecology, comparative anatomy, physiology, genetics, molecular biology, microbiology, zoology, botany, embryology, etc.), paleontology, and geology converges on a conclusion that is almost indisputable; organisms have evolved over time, have descended from common ancestors with modifications, and have done so primarily by natural selection.

The November 2004 issue of National Geographic Magazine had large bold letters on the cover page saying the following: "**WAS DARWIN WRONG?**" Inside the issue in an article by David Quammen, also in large bold letters, was the response: "**NO.** The evidence for biologic evolution is overwhelming." Quammen (2004) goes further in the first paragraph of the article to tell us that evolutionary theory (Darwin's life work) is an explanation "about the origin of adaptation, complexity, and diversity among Earth's living creatures." He further explains that a scientific theory is "not a dreamy and unreliable speculation, but an explanatory statement that fits the evidence." He further compares evolutionary theory with Einstein's theory of relativity, Copernicus' heliocentric theory of the solar system, atomic theory, plate tectonic theory, and electrical theory. Quammen (2004) states the following relative to these theories and evolutionary theory: "Each of these theories is an explanation that has been confirmed to such a degree, by observation and experiment, that knowledgeable experts accept it as fact."

THE PALEONTOLOGICAL RECORD (OR THE FOSSIL EVIDENCE)

The rock record shows that organisms have changed over time. Different layers of rock, and of different ages, have different fossils. Fossil groups from the past show differences as compared to modern organisms. This is sometimes called the Principle (or Law) of Fossil Succession (first proposed by William Smith, 1812) (as we have previously discussed), which basically states that fossil groupings in successively younger strata (layers of rock) succeed themselves in a definite and determinable order and the age of the strata can be determined by the fossils present in the rock layers. Thus, fossil groups from the past show differences as compared to modern organisms. The youngest rocks contain the fossils that are most similar to modern organisms and contain more complex and diverse forms. Also, fossils of species of past organisms occur in a particular order and are not random in their appearance. "Each species or group is preceded by a logical and related ancestor: descent with modification." (Kardong, 2008). However, this does not mean that evolution is progressive in a goal oriented sense, from simple forms to more complex. Natural selection does not guide evolution progressively, but instead allows organisms to evolve to be suited (based on their inherent constraints of body plan, molecular chemistry, etc.) for their environment; in other words, they evolve to adequately adapt to their present environmental circumstances. But, as time passes diversity will have a tendency to increase. Organisms will find new ways to exploit their habitats and occupy new niches.

Archaeopteryx, The First Bird and an Intermediate Between Nonavian Dinosaurs and Modern Birds

There are also numerous examples in the rock record of intermediate fossils (so called "missing links") that show how major lineages of organism have evolved from more primitive (ancestral) to more advanced (derived) organisms. An excellent example of an intermediate between the traditional reptiles (and also the nonavian dinosaurs) and modern birds, and a type of intermediate fossil that Charles Darwin in *On the Origin of Species* had predicted would be found in the rock record, is the crow-sized fossil *Archaeopteryx* (from: archaeo = old or ancient, pteryx = wing). *Archaeopteryx* remains were first found in 1860 (and described by Hermann von Meyer in August 1861), as an impression of a feather, from the 150 million-year-old late Jurassic lithographic (very fine grained, micritic) Solnhofen limestone in a quarry near the town of Solnhofen, Bavaria, Germany (Marshall, 1999; Fastovsky and Weishampel, 2005; Pond, et al, 2008; Tudge, 2008; University of California Museum of Paleontology at Berkley, *Archaeopteryx*: An Early Bird, accessed 2014). In 1861, the first complete skeleton of *Archaeopteryx* was excavated from the Solnhofen limestone quarry. This specimen (specimens of *Archaeopteryx* may be viewed at the following website, http://en.wikipedia.org/wiki/Specimens_of_Archaeopteryx), which shows both feather impressions and also skeletal bones, resulted in much excitement because it had impressions of bird feathers coexisting with reptilian features such as a long bony tail, wings supporting hands with claws, reptilian-like teeth, and other characters intermediate between birds and reptiles (Fastovsky and Weishampel, 2005; Tudge, 2008). It looked like a small coelusaurian theropod dinosaur, such as *Compsognathus*. In fact, one specimen (found in 1951) that lacked well preserved feather impressions was originally described as the small dinosaur *Compsognathus* (Shipman, 1998; Prothero, 2007; University of California Museum of Paleontology at Berkeley, *Archaeopteryx*: An Early Bird, accessed 2014). The 1861 specimen was obtained by a local doctor that sold it to Sir Richard Owen (Figure 6-1) and the Trustees of the British Museum in London (the specimen is now in the world famous Natural History Museum in South Kensington, London due to the campaigning of Richard Owen for a new home [established in 1881] for the natural history specimens in the British Museum [Wikipedia, Richard Owen, accessed 2014]); thus, this specimen became known as the London specimen (Fastovsky and Weishampel, 2005). Sir Richard Owen, the top comparative anatomist in the world at the time and one of the top paleontologists, scientifically described the fossil. He also described the first dinosaur and coined the term Dinosauria in 1842 (Martin, 2004; Fastovsky and Weishampel, 2005).

Figure 6-1. Sir Richard Owen (1804-1892) in the 1870s. A renowned comparative anatomist and paleontologist at the British Museum in London. He described the first dinosaur and coined the term Dinosauria in 1842. He also described the first skeleton of *Archaeopteryx* (the London specimen) in 1861. (Public Domain. Source: Wikimedia commons.)

The next *Archaeopteryx* specimen to be found was retrieved from Solnhofen in 1877 and ended up being sold to the Museum für Naturkunde in Berlin, Germany. Therefore, this specimen is referred to as the Berlin specimen. The Berlin specimen is indeed beautiful and has complete feathers and the wings spread out naturally, with the skull revealing teeth, and with a long bony tail surrounded by feather impressions (Fastovsky and Weishampel, 2005). The wing feathers in all of the fossils of *Archaeopteryx* (including the lone feather fossil) are asymmetrical vane feathers (thus, flight feathers) and according to Fastovsky and Weishampel (2005, p. 312), "Moreover, as in modern birds, the asymmetrical vanes produce an airfoil cross-section in the wing." This would allow for lift and is one line of evidence that *Archaeopteryx* could fly. For modern birds, only birds that fly have asymmetrical vane feathers. Although most paleontologists think that *Archaeopteryx* could flap its wings and thus had powered flight, they don't think that it was a strong flyer and it certainly did not fly like modern birds (Fastovsky and Weishampel, 2005).

Over the years, ten more partial to complete skeletal remains have been found from the Solnhofen limestone and most (but not all) paleontologists studying the remains place all the fossils in the same species, *Archaeopteryx lithographica* (although there is some debate about this) (Wikipedia, *Archaeopteryx*, accessed 2014).

Why are these specimens of the first birds from the Solnhofen limestone so unique and have such exquisite detail of skeletal features and preserved feathers? Paleontologists refer to a locality with a formation that preserves "a gold mine of paleontological treasures" or a horde of beautifully preserved fossils as a *lagerstatte* (*lagerstatten*, plural) (Martin, 2004). The Burgess Shale of Alberta Canada with its trove of lower Cambrian fossils, many of which are of soft bodied marine invertebrate animals; and the Pennsylvanian-aged Mazon Creek fossil deposits in the Mazon River area of Illinois, that are beautifully preserved plants and animals in ironstone (siderite) concretions, are examples of lagerstatten. The Solnhofen Limestone of Bavaria, Germany is also a lagerstatte (Martin, 2004). The limestones of the Solnhofen area are very fine-grained limestones that tend to break-out in sheets and were used as building tiles, but also used in the lithographic printing industry during the 1800s. Thus, the species name for *Archaeopteryx lithographica* is derived from this use of the Solnhofen limestone (Martin, 2004; Fastovsky and Weishampel, 2005). During late Jurassic time, this part of Germany consisted of a group of islands (and coral and sponge reefs) separated by shallow water stagnant lagoons that were depositing fine carbonate sediment and with anoxic bottom conditions. Animals could not live in the anoxic bottom waters. Thus, if a dead or injured animal from land or the open ocean washed into the lagoonal waters or fell from above, they would sink to the bottom and become covered by fine carbonate sediment and would not be quickly destroyed by scavengers or current action. Many exquisitely preserved animals have been found in the Solnhofen Limestone, such as horseshoe crabs, jellyfish, beetles, dragonflies, small alligators, the small dinosaur *Compsoganthus*, pterosaurs, etc., and of course *Archaeopteryx* with finely preserved feather impressions (Shipman, 1998; Martin, 2004; Fastovsky and Weishampel, 2005; Prothero, 2007; University of California Museum of Paleontology at Berkeley, Solnhofen Limestone, accessed 08/02/14; Wikipedia, Solnhofen Plattenkalk, accessed 08/02/14).

Darwin could not have asked for a better transitional fossil in support of his theory of evolution by natural selection than *Archaeopteryx*, a fossil transitional between dinosaurs (and thus reptiles) and birds. Although he mentioned *Archaeopteryx* in his fourth edition (and later editions) of *On the Origin of Species* (see below) in 1866, he did not stress the transitional nature of this fossil. However, an examination of some of Darwin's correspondence after the discovery of *Archaeopteryx* shows his pleasure with the discovery, and there is evidence that indicates that Darwin predicted the discovery of an intermediate fossil like *Archaeopteryx* over two years before its formal description (see Kritsky, 1992).

Darwin (1866) stated the following in the fourth edition of *On the Origin of Species* (p. 367):

> ...Until quite recently these authors might have maintained, and some have maintained, that the whole class of birds came suddenly into existence during the eocene period; but now we know, on the authority of Professor Owen, that a bird certainly lived during the deposition of the upper greensand; and still more recently, that strange bird, the Archeopteryx, with a long lizard-like tail, bearing a pair of feathers on each joint, and with its wings furnished with two free claws, has been discovered in the oolitic slates of Solnhofen. Hardly any recent discovery shows more forcibly than this how little we as yet know of the former inhabitants of the world.

The upper greensand referred to by Darwin is of lower Cretaceous age and occurs in the southwest of England. Notice how Darwin spells *Archaeopteryx* and refers to Solnhofen as slates rather than limestone.

It was already known, even prior to the publication of *On the Origin of Species*, that Darwin had proposed "that birds had scaly legs not because God was tidy-minded and economical with his designs but because somewhere among the ancient reptiles was the ancestor of modern birds" (Tudge, 2008, p. 31). The great Swiss-American professor of zoology and geology at Harvard University, Louis Agassiz (1807-1873), who in 1859 established Harvard's Museum of Comparative Zoology, was known to have said relative to Darwin's ideas about reptiles and birds "...that such a creature that was half bird and half reptile could not possibly exist. Birds were just too peculiar. There was, said Agassiz, no plausible route by which reptiles could evolve into birds" (Tudge, 2008, p. 31). Agassiz did not like Darwin's account of evolution at all (Tudge, 2008; Irmscher, 2013), and Darwin was one of his main antagonists (Irmscher, 2013). In fact, Agassiz was a creationist and did not accept any type of evolution (Tudge, 2008; Irmscher, 2013).

Figure 6-2. Photograph of Thomas Henry Huxley (1825-1895) ("Darwin's Bulldog"). Huxley is 32 years old in this image. Photo by Maull & Polyblanc in 1857. (Public Domain. Source: Wikimedia commons.)

Although Darwin was cautious publically and with his statement (as above) in *On the Origin of Species*, his colleague, friend, and the defender of Darwin's theory (or theories, see Mayr, 1991, p. 36-37), Thomas Henry Huxley (1825-1895) (Figure 6-2), often referred to as "Darwin's Bulldog", certainly was not. Huxley had already been keeping himself busy defending Darwin's theory, by making speeches and writing and publishing works in support of it (Prothero, 2007). Here now (in 1861) was a fossil, *Archaeopteryx* that the great Harvard scientist Agassiz said could not exist, with a mosaic of reptilian and bird characteristics. Richard Owen, at the British Museum, had described the fossil and declared that it was without doubt a bird with some unbird-like characteristics (Tudge, 2008). In his 1863 monograph on *Archaeopteryx* he did not stress the reptilian and dinosaurian anatomical features of *Archaeopteryx* and even gave the fossil a different name, *Archaeopteryx macrura* (Owen) (of course, in order to get more credit for himself) (Tudge, 2008). Owen was also of orthodox religion, did not like Darwin's evolutionary ideas, and simply did not like Darwin (Tudge, 2008). One of Owen's chief rivals was none-other than Thomas Henry Huxley. Huxley had been one of the first to perform anatomical studies on modern birds and had been studying a number of dinosaurs, such as *Compsognathus*; thus, he could not help but notice that *Archaeopteryx* appeared to be a classic "missing link" between dinosaurs and birds (Paul, 1988; Prothero, 2007). In fact, he published his own account of *Archaeopteryx* in 1868, referring to the fossil by its rightful name *Archaeopteryx lithographica* (von Meyer) and pointed out critical errors that Owen had made in his description of the fossil, such as mixing-up the right leg with the left, getting the pelvis backwards, and misaligning the furcula (Tudge, 2008). According to Tudge (2008, p. 33-34) he could not resist adding the following statement: "It is obviously impossible to compare the bones of one animal satisfactorily with those of another, unless it is clearly settled that such is the dorsal and such is the ventral aspect of the vertebra, and that such a bone of the limbs belongs to the left and another to the right side" etc.

According to Prothero (2007, p. 258), "At a famous presentation in front of the Royal Society in 1863, he [Huxley] proposed that birds were descended from dinosaurs and listed thirty-five features shared only by nonavian dinosaurs and birds (seventeen of these are still used by modern paleontologists)." During the early portion of the twentieth century, Huxley's ideas fell into some disfavor because some proposed that the features shared between dinosaurs and birds was due to convergent evolution and that the ancestry of birds rose from deeper into the past from within more primitive archosaurs (from within a group of archosaurs sometimes, incorrectly, referred to as thecodonts) (Paul, 1988; Fastovsky and Weishampel, 2005; Prothero, 2007). Today, because of the similarity of so many characters, most dinosaur paleontologists think that convergence seems unlikely to explain all the character similarities. In addition, the so called "thecodonts" are a polyphyletic clade, so it does not appear that the ancestry of birds lies in archosaurs older than dinosaurs.

In the early 1970s, there was a re-awakening of interest in the relationship between theropod dinosaurs and birds due to the work of renowned paleontologist John Ostrom (1928-2005) of Yale University (Paul, 1988). He documented the relationship between *Archaeopteryx* and coelurosaurian theropod dinosaurs and this convinced him (and eventually others) that *Archaeopteryx* was the first bird and that birds evolved from dromaeosaurid theropod dinosaurs similar to *Velociraptor* and *Deinonychus* (which he discovered in 1964) (Paul, 1988; Fastovsky and Weishampel, 2005; Prothero, 2007).

So, are birds dinosaurs? Yes, *birds are dinosaurs!* Birds have the characteristics that are specific to the monophyletic clade called Dinosauria. If birds really do have the characteristics that are unique to dinosaurs, then by diagnosis, they are dinosaurs. Let's look a little deeper into what constitutes a bird. Much of the following discussion has been paraphrased from Fastovsky and Weishampel, 2005 and Prothero, 2007, but also various other sources with overlapping content.

Birds are obviously vertebrates. They have a dorsal nerve cord and backbone. They are diapsid reptiles and are also archosaurs. Birds have a unique set of highly evolved features superimposed on the basic archosaur body plan. Birds do not resemble crocodiles and alligators very much, the other great living group of reptilian archosaurs. But we can no longer say that birds are unique in the possession of feathers, because some nonavian dinosaurs are now known to also have possessed feathers. Birds, as members of the monophyletic clade Avialae, are defined as *Archaeopteryx* plus living birds and all the descendants of their most recent common ancestor, living and extinct. The clade Aves is a term restricted to more modern birds, with a key derived character being the loss of teeth.

What are the features of Birds? Birds have feathers for sure (but so do nonavian dinosaurs). Feathers are fingernail-type material (keratin). They have a hollow central shaft (rachis). Radiating from the shaft are barbs, barbs linked together with small hooks (barbules). Barbs and barbules form the vane of the feather. The limb bones of birds are hollow and have thinner walls than nonavian theropod dinosaurs. Within the limb bones braces and struts provide support. The long bones are pneumatic; air sacs pass through a small opening (pneumatic foramen) into the bones. Thus, air passes into these air sacs and unidirectionally flows through the lungs when birds respire. The air sacs help lighten the body, supplement the lungs, and help cool the fast avian metabolism.

In the bird skeleton, many bones are fused to form more rigid structures. The carpometacarpus (the semilunate carpal bone [a wrist bone] is fused to metacarpals (hand bones); all the metacarpals are fused) in the forelimb and the tarsometatarsus (distal tarsals [ankle bones] fused to metatarsals [foot bones] and all metatarsals fused together in the hindlimb) are fused limb structures. The astragalus and calcaneum (two proximal ankle bones) are fused to the tibia. The pelvic bones are fused to form the synsacrum (the sacrum is fused to the pelvic bones and the pelvic bones are fused). In the shoulder region, the distal portions of the clavicles are fused to form a furcula (wishbone; this also is the case in some theropod dinosaurs, even as far back as *Allosaurus*). The distal caudal vertebrae are greatly reduced and are all fused to form a pygostyle.

There are modifications of the forelimb to form wings. The sternum is large and keeled. The coracoid is long and forms a brace with the sternum to anchor large muscles for flight. The glenoid fossa (where the humerus makes contact with the scapula and coracoid) is shifted to a lateral and dorsal position (as opposed to a backward and more ventral position in most nonavian dinosaurs). There is an absence of teeth in living birds. The jaws are covered with keratin to form a beak (rhampotheca). Birds have relatively large brains and advanced sight. The larger brain may be the result of the evolution of endothermy (complex controls to maintain endothermy require a larger brain) and because of complex motor control needed for flight. Well-developed vision may also be associated with motor activites of flight.

Archaeopteryx shared many features with small coelurosaurian theropod dinosaurs, like *Compsognathus*. Both are small, with *Archaeopteryx* about the size of a crow and *Compsognathus* about the size of a chicken. Very few bones are fused (dinosaur features rather than avian features). *Archaeopteryx* had hollow limb bones, as do theropod dinosaurs. However, theropod limb bones had thick walls and fewer pneumatic foramens (openings to air sacs). *Archaeopteryx* had a small and flat sternum, so it had an inferior (compared to modern birds) main anchor for powered flight, but the forelimbs indicate it was capable of weak powered flight. The first bird had a furcula, which was also present in some of the related theropod dinosaurs. Like modern birds, *Archaeopteryx* expressed the opisthopubic (or at least semi-opisthopubic) condition (the pubis runs downwards and somewhat backwards below the ischium), as did also some coelurosaurian theropods (such as *Deinonychus*, *Velociraptor*, and other dromaeosaurid theropods). Adult modern birds are completely opisthopubic, with the pubis running downwards and backwards, parallel to the ischium. However, it is interesting and significant that in embryological development of modern birds, "the pubis begins directed forward (the primitive condition) and rotates backward as the embryo develops." (Fastovsky and Weishampel, 2005, p. 323)

The forelimb of *Archaeopteryx* is theropod-like, whereas the hindlimb is bird-like, with a reduced fibula, proximal tarsals (ankle bones) fused to the tibia, distal tarsals (ankle bones) fused to the metatarsals (foot bones), and with long metatarsals partly fused to each other. The pes (foot) of *Archaeopteryx* has three slender forward-facing digits (2,3, and 4) and a fourth (digit 1) that faced backward.

Other features of *Archaeopteryx* are the following:

- Stiff tail – zygopophyses of the caudal vertebra are elongate and interconnected (so in life they were locked-up to form a stiff tail).
- Gastralia (belly ribs) are present - this is a primitive feature.
- Lacks a synsacrum – the pelvic bones are not fused to the sacrum.
- Long arms - greater than 70% the length of the legs.
- The wrist contains a semilunate carpal bone (characteristic of birds)
- The manus (hand) is long, and digits 1,2, and 3 are unfused. There are claws on the end of digits.

So, *Archaeopteryx* could be described as a feathered dinosaur capable of flight.

According to Fastovsky and Weishampel (2005), Jacques Gauthier (1986) applied cladistic analysis to the origin of birds. He outlined 100 characters that indicated that *Archaeopteryx* (and thus other birds) are coelurosaurian theropod dinosaurs (reference for this study: Gauthier, Jacques; 1986; Saurischian Monophyly and the Origin of Birds: in Padian, Kevin, ed.; The Origin of Birds and the Evolution of Flight; *Memoirs California Academy of Sciences*; No. 8.; p. 1–55.). Thus, birds bear the diagnostic features of theropod dinosaurs (Gauthier, 1986). Birds are also maniraptorian theropods (based on hand structure and fused clavicles). Birds are also dromaeosaurid theropods, which have an opisthopubic (backwards projecting) pubis, stiff tail, interconnected zygopophyses (interlocking projections on adjacent vertebra) on caudal vertebrae (Gauthier, 1986), etc. So, *Archaeopteryx* is closely related to *Deinonychus* and *Velociraptor*. Since *Archaeopteryx* is a bird, other birds are closely related to dromaeosaurid theropod dinosaurs, such as *Deinonychus* and *Velociraptor*.

Many dinosaur paleontologists believe that *Archaeopteryx* was one of the first feathered coelurosaurian theropods to modify feathers for true flight. This is primarily because the asymmetric flight feather first appeared in *Archaeoptheryx* (Prothero, 2007). But feathered coelurosaurian theropod dinosaurs were, no doubt, ancestral to *Archaeopteryx*. By the middle of the 1990s, "spectacular feathered theropod dinosaurs from the Lower Cretaceous Yixian Formation, Liaoning Province, China, began to be recovered" (Fastovsky and Weishampel, 2005, p. 319). The fine lake shales and siltstones interbedded with volcanic rocks of this formation preserved amazing features in these fossils, such as feathers, body outlines, and complete articulated skeletons (Fastovsky and Weishampel, 2005; Prothero, 2007). So, these feathered theropods from China, that possessed feathers that were not adapted for flight, show that feathers were widespread in these coelurosaurian theropods and that feathers did not evolve for flight, but served presumably for insulation and were later modified for flight (Fastovsky and Weishampel, 2005; Prothero, 2007). *Archaeopteryx* is still the oldest theropod dinosaur to have possessed flight feathers (Prothero, 2007).

Since we have shown that birds are theropod dinosaurs, and scientists have classified dinosaurs as diapsid (two openings in the skull behind the orbit) reptiles, then birds must also belong to the clade Reptilia (birds must be reptiles). If we do not include birds in the clade Reptilia, then the clade would become paraphyletic (i. e., all the descendants of a common ancestor for the clade would not be included). Remember, to show evolutionary relationships we are striving to construct monophyletic clades. Most modern biologist (particularly most modern systematists) and vertebrate paleontologists have now redefined Reptilia to include the birds (Prothero, 2007).

If birds are classified as dinosaurs, then the dinosaurs are not extinct. Certainly, it is hard to think of those songbirds flitting around in our back yard as dinosaurs, but technically speaking, they are. Of course, the nonavian (non-bird) dinosaurs are extinct. However, the clade Dinosauria currently contains 9600 living species, the birds (Dingus and Rowe, 1998). Darwin's theory of evolution started a revolution and showed that lineages like Dinosauria could evolve (Dingus and Rowe, 1998). Both Huxley and Ostrom (and others), based on the theropod dinosaurs *Compsognathus* and *Deinononychus* (and others) and comparisons to the first bird,

Archaeopteryx, revealed that not only *could* dinosaurs evolve, but that dinosaurs *did* evolve (Dingus and Rowe, 1998). Dinosaurs evolved both birdlike size, with sustained size reduction of theropod dinosaur ancestors for about 50 million years (Benton, 2014; Lee, et al, 2014), and birdlike features (Dingus and Rowe, 1998). But it is still hard for people, including dinosaur paleontologists, even with so much evidence showing the evolutionary relationships between birds and other theropod dinosaurs, to avoid agreeing that dinosaurs are extinct. As Dingus and Rowe (1998, p. 196) point out in the following statement:

> "The belief that dinosaurs are extinct is one of the great ironies of paleontology. Richard Owen is sometimes reviled for fighting throughout his life against evolution. Yet, even though modern science recognizes Darwin as the victor in this battle, the world did not go on to adopt a Darwinian view of dinosaurs. If Owen is looking down from the Hereafter, he must be gratified despite the bad press. Most people still accept his anti-evolutionary view, that dinosaurs are extinct."

Tiktaalik roseae, An Intermediate Fossil in the Rise of Tetrapods from Fish

Many people, in particular many creationists, have the misconception that in science we can only test hypothesis that deal with the present, such as performing a chemistry or physics experiment in the laboratory to test a specific hypothesis. Or perhaps administering a drug to an experimental group of mice to test if the hypothesized effect of the drug is as predicted, of course with a control group that was not administered the drug. However, in science we can also empirically test hypotheses that predict events that occurred in the past. For example, there are numerous examples of paleontologists predicting that certain types of fossils should be found in a certain region and in strata of a specific age. An excellent (and fairly recent) example of this (and an example of the finding of an intermediate fossil between freshwater fish and the first amphibian) is the 2004 discovery of *Tiktaalik roseae* by Neil Shubin, Professor of Anatomy at the University of Chicago, and associates. This 375 million-year-old Devonian fossil of a large freshwater fish was discovered in Arctic northern Canada on Ellesmere Island. However, this was not just any fish. This fish could bend its head without bending the rest of its body and it could bend its front fins at the wrist, even though it did not have any digits (phalanges) on its front fins. In other words, it had a neck and a wrist. Shubin and his team predicted the discovery of this fish fossil. Shubin explains how this discovery took place in his 2008 book, *Your Inner Fish – A Journey into the 3.5 Billion-year History of the Human Body*. It turns out that this fossil is an important anatomical (or morphologic) intermediate (transitional form) between crossopterygian lobe-fined fishes, such as *Eusthenopteron* of the Devonian Period about 380 million years ago and some of the first tetrapods, such as the Devonian-aged *Acanthostega* at about 365 million years old. Shubin and his team predicted that they should find this fossil in the arctic islands of northern Canada. Shubin and some of his associates had been looking for intermediates between lobe-finned fishes and tetrapods for several years and had already looked at rocks of near the right age in Pennsylvania and some other locations with limited success (only fragmentary pieces of fish fin bones and not many of those). But they noticed a geologic map in an introductory geology text that showed rocks of the right age for intermediate fossils between freshwater lobe-finned fish and tetrapods located in arctic islands of northern Canada. They also determined that the rocks located on those islands consisted of freshwater stream deposits. There was already good evidence that the first tetrapods lived in freshwater streams and coastal swamps (rather than on land) and used digits on their limbs to help negotiate around rocks, branches, and other litter in stream beds and swamps as they hunted for fish to eat (Zimmer, 1999; Steyer, 2012). However, they could move onto land for short periods of time to escape predators and most likely had lungs for breathing air (Zimmer, 1999; Steyer, 2012). So, Shubin and his team predicted that if they went to these islands they should be able to find intermediates between advanced lobe-finned fishes and the first tetrapods. It took them six years of expeditions to these arctic islands (with very rough field conditions), and they were about to give up, but in 2004 they struck pay-dirt (so to speak) on Ellesmere Island (after four seasons of searching on that island). Here is a direct quote from Shubin's book that highlights the importance of this fossil find: "It took us six years to find it, but this fossil confirmed a prediction of paleontology: not only was the new fish an intermediate between two different kinds of animal, but we had found it also in the right time period in earth's history and in the right ancient environment. The answer came from 375-million-year-old rocks, formed in ancient streams" (Shubin, 2008, p. 24). The discovery was announced in 2006 and about three articles were published in the British science journal *Nature* in 2006. Here is a reference to one of the articles: "A Devonian tetrapod-like fish and the evolution of the tetrapod body plan," 2006, by E.B. Daeschler, N.H. Shubin, F.A. Jenkins, Jr. *Nature* Vol. 440. p. 757-763. The following link at the University of Chicago tells about *Tiktaalik* and has other resources, including a slide presentation: http://tiktaalik.uchicago.edu/index.html.

For several years, I used this example to illustrate to my college geology, evolution, and earth science students how scientists can make predictions about the past and then go into the field and test those predictions in the rock record. This would not be the case if our understanding of evolution and the history of life were incorrect. This also illustrates that science and the nature of scientific inquiry does not deal with just experiments that we can do in the laboratory today, but that we can also make hypothetical predictions about past events and test those predictions by observing the rock and fossil record.

I have used this concept in the field with my students also. For six years, from the summer of 1998 through the summer 2003, I led students to east central Utah to dig for dinosaur bones on government land (I had the proper exploration and excavation permits from the federal Bureau of Land Management). Within this area of Utah there is a band of late Jurassic rocks that date to about 150 million years ago called the Morrison Formation (sandstones and mudstones of river and floodplain deposits). A famous dinosaur quarry located near Price, Utah in the Morrison Formation gave up about 10,000 bones that were excavated there in the 1960s. We visited that quarry, the Cleveland-Lloyd Dinosaur Quarry. Afterwards I showed my students a geologic map of the region and explained to them how the Morrison Formation trends to the south-southwest from the Cleveland-Lloyd Dinosaur Quarry and that we should find similar dinosaur genera (and species for that matter) along that trend in the Jurassic-aged Morrison Formation. I explained to them that if we get out of that trend, into younger Cretaceous rocks or older Triassic rocks, we would not find the same species of dinosaurs because they did not live during those time periods. We did follow that trend and camped in the desert looking for dinosaur bones. We were fairly successful over the six-year period, finding the partial remains of *Stegosaurus*, *Diplodocus* (both an adult and juvenile at one locality) and *Apatosaurus*; plus, isolated bones of other Jurassic dinosaurs (like *Allosaurus* and *Camarasaurus*) (see Powell, Burton, Crisp, and Stone, 2001; Spencer and Crisp, 2002).

Some of my former students would often ask me, why don't we explore for dinosaur sites in West Virginia? Of course, I told them that there are no sedimentary rocks of the right age in West Virginia and that we would not find any dinosaur bones. In fact, amniotes (vertebrates with an amniotic egg) were at a fairly primitive stage and even the precursors to dinosaurs would not have been present during the time that the Paleozoic sedimentary rocks in West Virginia were deposited. Dinosaurs did not evolve until late Triassic time, several tens of millions of years younger than the youngest rocks in West Virginia (other than Pleistocene fluvial deposits within river valleys).

We would also be flabbergasted if we found real human footprints in rocks alongside of dinosaur footprints in West Virginia or any other place in the world, as some creationists claim for Cretaceous deposits along the Paluxy River in Texas. It has been shown by real scientists looking at these supposed human tracks that they are either underprints (deformation of sedimentary rocks well below the once real dinosaur tracks) or human footprints carved by the locals to draw tourists to the locality. More than sixty million years separate humans (or anything near human) from dinosaurs (except for the avian dinosaurs – the birds).

Summary of the Fossil Evidence

We could continue for some time with examples of intermediate and transitional fossils and more examples of fossils representing morphologic series. An excellent example is the fossil series illustrating the bushy tree of the evolution of the horse from an Eocene-aged (about fifty-five million years old), Boston Terrier-sized, four-toed hoofed front limbed and three-toed hoofed rear limbed animal with low-crowned teeth for browsing soft leaves, and possessing relatively small brains. One branch of this bushy tree ultimately evolved into big-brained, single-hoofed, large grazing animals with high-crowned teeth for eating gritty grasses. This branch of horses finally culminated in the modern genus *Equus* (MacFadden, 2005; Prothero, 2007). Indeed, according to MacFadden (2005), the 55-million-year phylogeny of the family of horses, Equidae, is a prime example of definitive evidence for macroevolution (evolution above the species level). MacFadden (2005) goes on to say that the "speciation, diversification, adaptations, rates of change, trends, and extinction evidenced by fossil horses exemplify macroevolution." Another impressive example of a branching tree with abundant intermediate fossils is the evolution of modern whales from fifty million-year-old Eocene land dwelling animals (see diagrams in Zimmer, 1999, p. 203; Prothero, 2007, p. 319; Coyne, 2009, p. 50; Zimmer and Emlen, 2013, p. 10). Both whale evolution and horse evolution are "poster childs" that dramatically support biologic evolution.

The logical deductive conclusion from the evidence of the fossil record is that species and groups of organisms have succeeded themselves in a logical and predictable manner as we move from older rocks to younger rocks and ultimately to the present. This natural order would not be expected if an intelligent designer created species separately, one at a time. If this were the case, we would expect a haphazard order to the fossil record, with perhaps a fossilized skeleton of a wolf occurring in older rocks than a trilobite, or a eukaryotic cell being found in older rocks than the first prokaryotic cells. This is definitely not the case.

Or if species were created "all at once" by an intelligent designer, we would expect both simple and complex species to be found in the oldest fossil bearing rocks. Such is not the case. Thus, the logical conclusion is that organisms have descended with modifications from ancestral organisms that lived in the past; i. e. organisms have evolved over time from common ancestors.

COMPARATIVE ANATOMY

Evolution (in morphology, anatomy, genetic make-up, molecular chemistry, etc.) by natural selection involves modification such that ancestral (primitive) features (characters) are retained and new (derived) features are evolved. When two anatomical structures (or other features) can be traced back to a single structure (or feature) in a common ancestor, we say that the two structures (or features) are homologous. Thus, homologous structures are called homologues (or homologies). Therefore, homology refers to two or more features that share a common ancestry.

This concept may also be used in comparative molecular biology. The similarity of DNA, blood proteins, and other organic molecules among organisms must be related to organisms that share a common ancestor. In fact, homology of molecular make-up of organisms is a strong line of evidence in support of evolution via common descent and is in general agreement with the evolutionary patterns as determined by homology of anatomical structures.

But let's look at anatomical homology first. As we discussed at some depth in Chapter 5, organisms that share derived homologous features (synapomorphies) share a common ancestry. Therefore, the more derived homologous features that are shared between different organisms, the closer their evolutionary relatedness. Our forelimbs (as with all mammals) are homologous to dinosaur forelimbs and the common ancestor to both mammals and dinosaurs had the same homologous anatomical structures in the forelimb. In fact, this would be the general case for all tetrapods (four-limbed vertebrates). However, the forelimb may have been modified for different functions. Although the vertebrate species differ, the underlying pattern of the forelimb is fundamentally the same because of shared ancestry.

Examples of a morphological series within a group in the fossil record are important lines of evidence for evolution because they show the sequential change of homologous structures, from ancestor to descendant, to perform different functions (examples: evolution of the single horse hoof from multi-toed ancestors, amphibian legs from the fleshy fins of lobe-finned fish, etc.).

COMPARATIVE EMBRYOLOGY

"Early events of embryonic development retain current clues to distant evolutionary events" (Kardong, 2008, p. 108). So, developing embryos may contain clues to more primitive traits possessed by our ancestors (Figure 6-3). During the mid-1800s this was referred to as the biogenetic law by German biologist Ernst Haeckel (1834-1919) (Figure 6-4). Haeckel was not only the primary supporter of Darwinism in Germany, "but he went so far as to argue that we could see all the details of evolutionary history in embryos and reconstruct ancestors from embryonic stages of living animals" (Prothero, 2007, p. 108). In fact Haeckel stated this as "ontogeny recapitulates phylogeny," or, in other words, during embryological development the individual will pass through the evolutionary stages of its ancestors. Today we know that Haeckel overstated the completeness of this proposition (there were lots of exceptions), but there is a preservation of genetic material from ancestor to descendant. Thus, some of these genes may show similar expression in embryological development of different organisms, but more derived organisms may modify the development of the embryo as a result of the action of more derived genes.

Michael K. Richardson and Gerhard Keuck (2002) state the following (from a portion of an abstract of a paper entitled "Haeckel's ABC of Evolution and Development"; in Biol. Rev. Camb. Philos. Soc. 2002 Nov; 77(4): 495-528.) related to Haeckel's embryological studies:

> Haeckel recognized the evolutionary diversity in early embryonic stages, in line with modern thinking. He did not necessarily advocate the strict form of recapitulation and terminal addition commonly attributed to him. Haeckel's much-criticized embryo drawings are important as phylogenetic hypotheses, teaching aids, and evidence for evolution. While some criticisms of the drawings are legitimate, others are more tendentious...Despite his obvious flaws, Haeckel can be seen as the father of a sequence-based phylogenetic embryology.

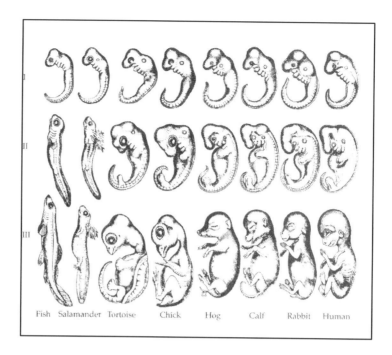

Fish   Salamander   Tortoise   Chick   Hog   Calf   Rabbit   Human

Figure 6-3. G. J. Romanes' 1892 copy of Ernst Haeckel's controversial embryo drawings. (Public domain. Source: Wikimedia commons.)

Figure 6-4. Photograph of Dr. Ernst Haeckel (1844-1919), German biologist and naturalist. Photograph by Nicola Perscheid and published in *Photographische Gesellschaft*, 1906. (Public domain. Source: Wikimedia commons.)

Before Darwin published *On the Origin of Species* in 1859, embryonic studies were being investigated by a contemporary of Haeckel, Karl Ernst von Baer (1792-1876). Darwin certainly used von Baer's embryological studies as evidence of evolution in *On the Origin of Species* and embryological studies have grown in importance for evolutionary biology (Prothero, 2007). Von Baer stressed that all vertebrate embryos show a common pattern and develop from the general to the specific (Kardong, 2008). Regardless of their adult form, all vertebrate embryos attain a stage with a long tail, well developed gill slits, and many other fish-like characteristics (Prothero, 2007). In fish, the tail and gills further develop, but in humans (for example) these features do not continue to develop and are lost before birth (Prothero, 2007). Von Baer pointed out that Haeckel's biogenetic law did not really reveal during embryological development all the steps of evolution from ancestors to descendants. He further stressed that what we observe, at best, during embryological development is a correspondence between the early embryos of descendants and the embryos of their ancestors, not between the descendant embryos and the ancestral adults (Kardong, 2008).

Although Haeckel was perhaps over enthusiastic relative to the meaning of his embryological studies, there is an element of conservatism in embryonic ontogeny even if it does not give a complete picture of evolutionary history of adult forms (Prothero, 2007; Kardong, 2008). Certainly, gill slits in young embryos of mammals, birds, and reptiles that never develop into functional respiratory devices in the adult forms must be telling us that the embryos of vertebrates start off in a similar general fashion, but then develop more specific attributes as they become adults. But surely this conservatism must be a supporting line of evidence for evolution and common ancestry. In this context, Kardong (2008, p. 110-111) states the following:

> It would be odd to attribute this embryonic preservation to a special creative force stamping out separate species one at a time, but with similar redundant features, unless we imagine this feeble force to be running out of new ideas and recycling the old nostrums. Instead, homologous features, regardless of whether in embryo or adult, support our prediction that similarities between species are expected if new species arise from old species – evolution.

Although Haeckel's original embryological diagrams contained errors and oversimplifications, sophisticated modern photographic images of vertebrate embryological development show similar results. Prothero (2007, p. 111) and Kardong (2008, p. 108) show some of these photographic images. Creationists, of course, jump on any slight error or oversimplification in studies that support evolutionary theory because they have no data to support their own ideas, so must try by default to debunk evolutionary theory. Jonathan Wells (2000), in his *Icons of Evolution: Science or Myth?*, attacked the work of Haeckel and attempts to persuade the unaware public that this is another nail in the coffin of evolutionary theory. Wells is one of the leading advocates of Intelligent Design creationism. Nick Matzke on the Talk Origins Archive website (at http://www.talkorigins.org/faqs/wells/iconob.html#haeckel-embryo) reviews and debunks Wells' unfair treatment of this issue and the meaning of Haeckel's overenthusiasm.

HUMAN GENETIC AND GENOMIC EVIDENCE

The completion of the Human Genome Project is a great advancement for global human society. It will revolutionize (and is revolutionizing) the medical field and significantly contributing to other areas of biology (bioinformatics, genetics, genomics, phylogenetics, etc.). Many medical doctors and scientists feel that it is imperative that the information in individual human genomes be shared amongst medical and scientific personal for further study to understand how biological mechanisms operate in organisms, of course in particular for medical purposes, in humans. This sharing of and comparing of information about the human genome is referred to as comparative genomics. Many biological and medical researchers are of the opinion that comparative genomics will, in the future, lead to a much deeper understanding of biological systems and in particular, we will understand the nature of the systems that make us uniquely human.

Although it is very important to study the big picture of biological systems and bioinformatics, there is much information that is contained in the genome currently based on the study of single nucleotide polymorphisms (SNPs). Although about

97% of our DNA is thought to consist of apparent non-coding DNA, the SNPs in the protein coding DNA has very important implications for detecting, preventing, and curing disease (DeSalle and Yudell, 2005). As we know, a single mutation in a coding sequence of DNA can sometimes result in major genetic disorders or diseases because some of these mutations alter the protein that is produced or result in non-functionality of the protein or no protein produced at all (DeSalle and Yudell, 2005). The interaction of different proteins due to variation in SNPs between individuals may result in different reactions to drugs by individuals. Even though a mutation occurs in a single nucleotide position, the protein that the gene encodes for may not change (because the amino acid coded by the single change may not have changed). However, it appears that slight changes such as this may cause people to have different side effects to drugs. This type of research may eventually lead to designer drugs that are optimized for the specific genome of each single person. This would mean of course, that we would have a genomic ID card that we would present to a pharmacist that could be used to give each person an alternate form of the prescribed drug that would result in fewer side effects. Of course, our personal doctors would also have the genomic information that he-she would use when writing the original prescription. Hmmmmm…Do you see how the information in our genome is going to be relatively widespread? Could others get to that information to use it for different purposes, even though those different purposes might not be legal or ethical?

The Human Genome Project and the study of human genomic variation have other applications also. One of those is the study of our ancestry and migration routes of human populations (Stringer, 2012; Pääbo, 2014). The mapping of the human genome and the recent sequencing of the chimpanzee, the Neanderthal genome, and other vertebrate genomes (such as the chicken and the mouse) allow comparisons that are very important in the study of human evolution and what makes us different from our closest living relatives, the chimpanzees, and closest extinct relatives, the Neanderthals.

Several years ago (November 11, 2010) I attended a workshop at the National Science Teachers Association (NSTA) in Baltimore entitled "Why Evolution Matters in the 21st Century." One of the presenters at that workshop was Daniel J. Fairbanks, former Dean of Undergraduate Education and Karl G. Maeser Professor of General Education at Brigham Young University and currently an Associate Dean of the College of Health and Science at Utah Valley University in Orem, Utah. Dr. Fairbanks is a geneticists and professor, and has written several books dealing with the evidence of human evolution in our genome. One of his most recent books is *Relics of Eden – The Powerful Evidence of Evolution in Human DNA* (2007). In his presentation, Dr. Fairbanks informed us that with the mapping of the human genome (combined with the comparison to the chimpanzee genome), that the human species became the single species that gives us the most evidence supporting the fact of evolution and the explanations of evolutionary theory (even though the supporting evidence is voluminous without this). He presented some of that evidence to us at the workshop and much of it is contained in his 2007 book listed above. But as he pointed out, much of the evidence that supports our evolutionary history can be deciphered from variations in our DNA.

One example that Dr. Fairbanks presented at the workshop (and that is discussed in chapter 1 of his 2007 book), was evidence made available from sequencing of a portion of the human genome (part of chromosome 2) in 1991 by two scientists from Yale University (the Human Genome Project began in 1990 and was completed in 2003, U. S. Department of Energy [DOE], 2010) and the recent sequencing of the chimpanzee genome. This was the fusion of the chimpanzee chromosome 2A and 2B to form human chromosome 2. According to Fairbanks (2007), it has been accepted since 1982 (with the publication of a landmark paper by Jorge Yunis and Om Prakash in the journal *Science*, that supported some earlier studies) that chromosomes from humans, chimpanzees, gorillas, and orangutans are very similar and can be aligned with each other to show a strong correlation. However, chimpanzees, gorillas, and orangutans have 48 chromosomes (24 pairs), whereas humans only have 46 chromosomes (23 pairs). DNA sequencing of human chromosome 2 and chimpanzee chromosomes 2A and 2B revealed strong correlations of nucleotide base pairs of human chromosome 2 with chimpanzee chromosomes 2A and 2B. The evidence points to and supports the head-to-head fusion of chimpanzee chromosomes 2A and 2B at their short-armed (p-portion of the chromosome) telomeric ends (Fairbanks, 2007). If you ever read the book by Fairbanks (2007), there is a good illustration of this on page 23. More recently, 2012, Dr. Fairbanks came out with another book entitled *Evolving – The Human Effect and Why It Matters*, which discusses the head-to-head fusion of chimpanzee chromosomes 2A and 2B to form human chromosome 2 on pages 135-139.

It turns out that every telomere in humans and great ape chromosomes have tandem repeats of fifty to 100 times (or more, sometimes several thousand repeats) of the 6 base-pair sequence TTAGGG (Fairbanks, 2007; Carey, 2015b), with the complementary strand having the sequence AATCCC. When the middle of chromosome 2 in humans was sequenced, it was found to contain 158 copies of the tandem repeat sequences found in telomeres (Ijdo, et al, 1991; Fairbanks, 2007). So, it appears that sometime after the split of humans from chimpanzees about 6 to 7 million years ago, "the telomere on one chromosome fused head-to-head with the telomere of a different chromosome" (Fairbanks, 2007). Fairbanks (2007) gives the following nucleotide sequences of tandem repeats (of the six base-pair tandem repeats) for a short section of DNA near the fusion site (I am only showing one of the double strands, but you can figure out the complementary strand):

...............TTAGGGG/TTAGGG/TTAGfusionCTAA/CCCTAA/CCCTAA/...............

Notice that the sequence after the fusion site is an inverted (rotated 180 degrees) complement of the sequence before the fusion site, except at one position that is a mutation (an insertion – do you see the position of the insertion to the left of the fusion site?). Since this region is no longer active with a telomeric function, the region is no longer subject to repair and is not conserved as well as active telomeric sequences, thus mutations slowly creep into the sequence (only a few for the time since the split of chimpanzees and humans from a common ancestor) (Fairbanks, 2007). This is extremely strong evidence for a common ancestor between the human lineage and the chimpanzee lineage.

To further complicate things, because of the fusion of chimpanzee chromosomes 2A and 2B at the short arms of their telomeric ends to form human chromosome 2, the centromeres on chimpanzee chromosome 2A matches up exactly with the centromere on human chromosome 2. However, since chromosomes only have one centromere, the centromere that is on chromosome 2B of chimpanzees is not present in human chromosome 2 (Fairbanks, 2007). But in all humans and great apes, centromeres contain very specific repeating sequences that consist of 171 base pairs and are called alphoid sequences (Fairbanks, 2007). Alphoid sequences are present at the site in human chromosome 2 where the remnants of this centromere should be (Fairbanks, 2007). Again, this is extremely strong evidence for a common ancestor between chimpanzees and humans.

There are many other lines of evidence in our genomes that support our evolution from a common ancestor with chimpanzees. The two books mentioned above by Dr. Fairbanks discuss many of these lines of genomic evidence and are highly recommended to the reader.

VESTIGIAL AND ATAVISTIC STRUCTURES

Why do we have an appendix that can be removed without much ill effect? Why do dogs have tiny toes, dew claws, on their forefeet that do not reach the ground and are obviously nonfunctional (Figure 6-5)? Why do whales have a pelvis, but no hind limbs (usually not, but occasionally rudimentary hind limbs may be present also)? Why do most people either have no wisdom teeth, or wisdom teeth that do not erupt from the gums? Why do flightless birds like the African ostrich or the New Zealand kiwi have vestiges of wings? There are many more features that organisms may have that are remnants of functional structures that their ancestors possessed, but are no longer functional or have reduced functionality. Such remnants are referred to as vestigial structures, features, or organs (Prothero, 2007; Kardong, 2008; Coyne, 2009; Wicander and Monroe, 2013).

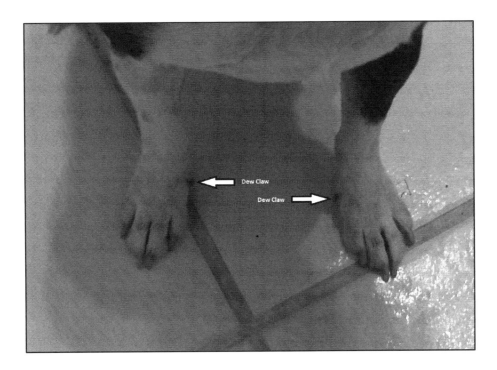

Figure 6-5. Photograph of the forelimbs of my Boston Terrier, Darwin, showing how the first digits (thumbs) have been reduced and serve no useful function. These reduced digits found in the forelimbs (and sometimes hindlimbs) of dogs are vestigial structures referred to as dew claws. The ancestors of dogs had five functional toes on both the forelimbs and hindlimbs, but these have been reduced by evolution to four functional toes on both the forelimbs and hindlimbs. (Photo by E. L. Crisp, August 2014.)

The human appendix is a very small, dead-end vermiform tube of about 10 cm (approximately 4 inches) length and radius of 3-4 mm that extends from the pouch-like cecum of the colon (large intestine) near the junction of the small intestine and the large intestine. In humans, it often becomes infected and must be surgical removed before it ruptures. In ancestral primates, and some modern primitive primates, such as the Woolly monkey (see Kardong, 2008, p. 112), their more expanded cecum harbored (or harbors) bacteria that could (or can) digest cellulose in leafy plant material. As hominin ancestors of modern humans evolved, the larger cecum was no longer as necessary because tough leafy plant material was less important as a major portion of the diet, so mutations that decreased the size of the blind end of the cecum (the appendix) were selected for and the appendix became a vestigial leftover from a more functional organ by our ancestors (Kardong, 2008). As Kardong (2008, p. 111) points out, the appendix may serve as a small part of our immune system involved in neutralizing foreign (bad) bacteria in the food that we eat. It also may serve some function in harboring good digestive bacteria that can replenish our supply after an intestinal infection that may kill much of our good bacteria (Smith, et al, 2009). However, Charles Darwin (1871) first classified the vermiform appendix in humans as a vestigial feature, as he well should have and he *was not wrong* about this (as some creationist argue) "...because it is greatly reduced compared to the homologous organs in non-human relatives, and because it currently exhibits a great range of variation, which is apparently non-functional" (Meyers, 2009). The following is Darwin's full quote relative to the vermiform appendix in *Descent of Man* (1871, Vol. 1, p. 27-28):

> With respect to the alimentary canal, I have met with an account of only a single rudiment, namely the vermiform appendage of the caecum. The caecum is a branch or diverticulum of the intestine, ending in a cul-de-sac, and is extremely long in many of the lower vegetable-feeding mammals. In the marsupial koala it is actually more than thrice as long as the whole body. (34. Owen, 'Anatomy of Vertebrates,' vol. iii. pp. 416, 434, 441.) It is sometimes produced into a long gradually tapering point, and is sometimes constricted in parts. It appears as if, in consequence of changed diet or habits, the caecum had become much shortened in various animals, the vermiform appendage being left as a rudiment of the shortened part. That this appendage is a rudiment, we may infer from its small size, and from the evidence which Prof. Canestrini (35. 'Annuario della Soc. d. Nat.' Modena, 1867, p. 94.) has collected of its variability in man. It is occasionally quite absent, or again is largely developed. The passage is sometimes completely closed for half or two-thirds of its length, with the terminal part consisting of a flattened solid expansion. In the orang this appendage is long and convoluted: in man it arises from the end of the short caecum, and is commonly from four to five inches in length, being only about the third of an inch in diameter. Not only is it useless, but it is sometimes the cause of death, of which fact I have lately heard two instances: this is due to small hard bodies, such as seeds, entering the passage, and causing inflammation. (36. M. C. Martins ("De l'Unite Organique," in 'Revue des Deux Mondes,' June 15, 1862, p. 16) and Haeckel ('Generelle Morphologie,' B. ii. s. 278), have both remarked on the singular fact of this rudiment sometimes causing death.)."

Atavisms (or atavistic structures or features) are the return (or partial return) of ancestral features in an individual organism (MacFadden, 1992; Kardong, 2008; Coyne, 2009). These features are sometimes referred to as evolutionary "throwbacks" (MacFadden, 1992) or as stated by the originator of the term atavism, Stephen Jay Gould (1983, p. 180), atavisms are "reversions to previous evolutionary states." Sometimes structural genes are preserved in the DNA of an organism, but the structural gene has been deactivated by mutations in regulatory genes or mutations that result in new genes that override the old gene; thus, a particular feature that was present in an ancestor no longer develops. However, rarely the old gene is reactivated in embryological development of an individual if a regulatory malfunction occurs. For example, modern horses (members of the order Perissodactyla, odd-toed ungulates) have only one functional toe (digit 3) with digits 2 and 4 vestigial remnants, but ancestors of the modern horse had 3 functional toes (digits 2, 3, and 4). As horses evolved, and as the climate became cooler from the Eocene to the present, more grasslands were formed at the expense of forests. This resulted in horses evolving into grazers of grasses and eventually (via natural selection) resulted in reduced digits 2 and 4 to have a single-toe (digit 3) and longer legs for more speed in the open steppes to be able to run faster than their predators. However, occasionally single-toed modern horses are born with atavisims, additional hoofed toes (usually digit 2 or 4, or both) representing the lost toes of their ancestors (Gould, 1980; Gould, 1983; MacFadden, 1992; Prothero, 2007; Kardong, 2008; Coyne, 2009).

Of the many, a couple of other examples of atavisms are hen's teeth and human post-anal tails. Fairly recently experimental atavistic teeth, similar to alligator teeth have been grown in embryonic chickens, even though the last bird to have reptilian teeth became extinct 70 to 80 million years ago (Harris, et al, 2006). In this experiment, researchers were first working with a type of mutant chicken embryo (a talpid[2] mutation, that is recessive and lethal during embryonic development) that was observed to grow conical alligator-like teeth. The researchers then reactivated the normally silent talpid[2] gene in the mouth of normal chicken embryos and they grew reptilian (archosaurian) alligator-like teeth (Harris et al, 2006). In the words of Harris et al (2006, p. 371) themselves, they state the following:

> Here, we describe the formation of teeth in the talpid[2] chicken mutant, including the developmental processes and early molecular changes associated with the formation of teeth. Additionally, we show recapitulation of the early events seen in talpid[2] after in vivo activation of β-catenin in wild-type embryos. We compare the formation of teeth in the talpid[2] mutant with that in the alligator and show the formation of decidedly archosaurian (crocodilian) first-generation teeth in an avian embryo.

Now, let us briefly discuss the human atavism of a post-anal tail. Of course, humans must have the genetic make-up for post-anal tails because our embryos contain tails during our ontogenetic development. Not only that, over 91% of our genes are identical to Old World monkeys and they certainly have post-anal tails, but of course the apes and humans do not (apes and humans have about 95% to 98% identical DNA, depending on whether we are comparing gibbons or bonobo chimps, respectively – thus our structural genes are mostly very similar) (Prothero, 2007). But, during embryological development, the human tail stops growing and is absorbed and not present at birth. However, rarely a regulatory mistake occurs and humans are born with an atavistic post-anal tail (see Prothero, 2007, p. 344-345; Coyne, 2009, p. 63).

Why do organisms retain genes that form vestigial or sometimes atavistic features? If organisms have been specially created, why would they have nonfunctional or subfunctional features that other organisms possess that are (or were) functional? Or why would apparent ancestral features suddenly reappear in a younger species? Isn't it deductively obvious that these features must have been present as functional features in ancestors of the organisms expressing them – descent with modification (Kardong, 2008). Would an intelligent designer leave the old outdated or nonfunctional parts when making a new organism? Most intelligent people would see these vestigial and atavistic features as strong evidence of evolutionary change of organisms over time from common ancestry.

BIOGEOGRAPHY AND PALEOBIOGEOGRAPHY

Charles Darwin in his opening statement in the introduction of *On the Origin of Species* wrote the following in 1859 (p.1): "When on board H.M.S. 'Beagle,' as naturalist, I was much struck with certain facts in the distribution of the inhabitants of South America, and in the geological relations of the present to the past inhabitants of that continent. These facts seemed to me to throw some light on the origin of species—that mystery of mysteries, as it has been called by one of our greatest philosophers." Darwin (1859) noted that organisms succeed each other geographically. This is referred to as the Principle of Geographic Succession. As stated by Freeman and Herron (2004), "Fossil and living organisms in the same geographic region are related to each other and are distinctly different from organisms found in other areas." Thus, modern organisms living in a particular region are more similar to recent fossils found in the same region than they are to recent fossils of more distantly related organisms found in other regions. This is strong evidence for descent with modification (Freeman and Herron, 2004).

Darwin (1859) felt that the biogeographic evidence for evolution was so strong that he devoted two chapters (Chapters XI and XII) to the topic of Geographical Distribution in *On the Origin of Species*. The following is a quote from Darwin (1859) of several paragraphs, near the end of Chapter XII (p. 408-410), which summarizes his thoughts on the importance of the geographic distribution of modern organisms and fossils as evidence of evolution (thus, biogeographic and paleobiogeographic evidence):

> If the difficulties be not insuperable in admitting that in the long course of time the individuals of the same species, and likewise of allied species, have proceeded from some one source; then I think all the grand leading facts of geographical distribution are explicable on the theory of migration (generally of the more dominant forms of life), together with subsequent modification and the multiplication of new forms. We can thus understand the high importance of barriers, whether of land or water, which separate our several zoological and botanical provinces. We can thus understand the localisation of sub-genera, genera, and families; and how it is that under different latitudes, for instance in South America, the inhabitants of the plains and mountains, of the forests, marshes, and deserts, are in so mysterious a manner linked together by affinity, and are likewise linked to the extinct beings which formerly inhabited the same continent. Bearing in mind that the mutual relations of organism to organism are of the highest importance, we can see why two areas having nearly the same physical conditions should often be inhabited by very different forms of life; for according to the length of time which has elapsed since new inhabitants entered one region; according to the nature of the communication which allowed certain forms and not others to enter, either in greater or lesser numbers; according or not, as those which entered happened to come in more or less direct competition with each other and with the aborigines; and according as the immigrants were capable of varying more or less rapidly, there would ensue in different regions, independently of their physical conditions, infinitely diversified conditions of life,—there would be an almost endless amount of organic action and reaction,—and we should find, as we do find, some groups of beings greatly, and some only slightly modified,—some developed in great force, some existing in scanty numbers—in the different great geographical provinces of the world.
>
> On these same principles, we can understand, as I have endeavoured to show, why oceanic islands should have few inhabitants, but of these a great number should be endemic or peculiar; and why, in relation to the means of migration, one group of

beings, even within the same class, should have all its species endemic, and another group should have all its species common to other quarters of the world. We can see why whole groups of organisms, as batrachians and terrestrial mammals, should be absent from oceanic islands, whilst the most isolated islands possess their own peculiar species of aërial mammals or bats. We can see why there should be some relation between the presence of mammals, in a more or less modified condition, and the depth of the sea between an island and the mainland. We can clearly see why all the inhabitants of an archipelago, though specifically distinct on the several islets, should be closely related to each other, and likewise be related, but less closely, to those of the nearest continent or other source whence immigrants were probably derived. We can see why in two areas, however distant from each other, there should be a correlation, in the presence of identical species, of varieties, of doubtful species, and of distinct but representative species.

As the late Edward Forbes often insisted, there is a striking parallelism in the laws of life throughout time and space: the laws governing the succession of forms in past times being nearly the same with those governing at the present time the differences in different areas. We see this in many facts. The endurance of each species and group of species is continuous in time; for the exceptions to the rule are so few, that they may fairly be attributed to our not having as yet discovered in an intermediate deposit the forms which are therein absent, but which occur above and below: so in space, it certainly is the general rule that the area inhabited by a single species, or by a group of species, is continuous; and the exceptions, which are not rare, may, as I have attempted to show, be accounted for by migration at some former period under different conditions or by occasional means of transport, and by the species having become extinct in the intermediate tracts. Both in time and space, species and groups of species have their points of maximum development. Groups of species, belonging either to a certain period of time, or to a certain area, are often characterised by trifling characters in common, as of sculpture or colour. In looking to the long succession of ages, as in now looking to distant provinces throughout the world, we find that some organisms differ little, whilst others belonging to a different class, or to a different order, or even only to a different family of the same order, differ greatly. In both time and space the lower members of each class generally change less than the higher; but there are in both cases marked exceptions to the rule. On my theory these several relations throughout time and space are intelligible; for whether we look to the forms of life which have changed during successive ages within the same quarter of the world, or to those which have changed after having migrated into distant quarters, in both cases the forms within each class have been connected by the same bond of ordinary generation; and the more nearly any two forms are related in blood, the nearer they will generally stand to each other in time and space; in both cases the laws of variation have been the same, and modifications have been accumulated by the same power of natural selection.

Of course, Darwin recognized that organisms that occupy a particular habitat and niche may superficially resemble other organisms that occupy a similar habitat and niche, but are separated from the organisms in question by a great distance or by some barrier. For example, the modern marsupials of Australia resemble in phenotypic appearance many placental mammals in Eurasia and North America; however, they are not closely related to these animals. Marsupial mammals evolved greater than 80 million years ago in North

America (Coyne, 2009). They expanded southward and reached the tip of South America about 40 million years ago, then crossed Antarctica to reach Australia by about 30 million years ago (Coyne, 2009). How did they cross the ocean between South America and Antarctica? The answer is that the southern portion of Gondwana (the large southern continent that started breaking apart during the Mesozoic Era) had not separated yet via plate tectonics, so there was no ocean at the time separating South America, Antarctica, and Australia (Coyne, 2009). After the ancestors of marsupial mammals made it to Australia, that continent separated from Antarctica via plate tectonics and moved to its present position isolated from other continents. Placental mammals had evolved elsewhere and dominated the mammal species on other continents (Coyne, 2009). However, in similar habitats different animals may evolve similar features to occupy similar niches. This is referred to as convergent evolution. For example, many of the marsupial mammals of Australia have evolved similar features as their placental cousins in America; such as the Sugar Glider in Australia and the Flying Squirrel in America, the Banded Anteater in Australia and the Anteater in America, and the Marsupial Mole in Australia and the Mole in America (see Coyne, 2009, p. 93) (also see Prothero, 2007, p. 112 for additional examples of the convergence of marsupials from Australia with placentals from other countries). There are many other examples of convergent evolution. In convergent evolution, organisms that are not closely related evolve similar features (analogous features) that allow them to function well in a particular habitat and niche. Remember, an organism's habitat is where it lives; but its niche is its role or place in the ecosystem, or what it does for a living).

Although, for the most part, Darwin could see that the distribution of organisms, both past and present, supported the idea of a place of origin for particular types of organisms and modification of those organisms by evolution as they spread outward from their place of origin. That is to say, organisms originate in a particular area and tend to predominant in the areas where they first evolved. If each species was created separately, one by one, we would expect to find those species and their descendants more evenly spread across the world (Kardong, 2008).

Alfred Russel Wallace (Figure 6-6), the co-discoverer of natural selection with Darwin, is recognized perhaps even more so than Darwin in terms of his study and understanding of the significance of biogeographic changes in organisms and is often thought of as the father of biogeography (Quammen, 1996). After spending four years exploring and collecting animals and plants of the Amazon and Rio Negro lowlands tropical rainforests of South America, Wallace spent eight years collecting specimens in the Malay Archipelago, now known as Indonesia (Quammen, 1996; Raby, 2001). Wallace published several papers and books that dealt with biogeography to some extent, but his two major works were his two volume *The Geographical Distribution of Animals* published in 1876 and *Island Life* published in 1880 (Quammen, 1996; Raby, 2001). *Island Life* was the first major volume published on island biogeography (Quammen, 1996). Relative to *The Geographical Distribution of Animals*, Raby (2001, p. 215-216) states the following: "The two large volumes of *The Geographical Distribution of Animals* were widely praised. *Nature* called it 'the first sound treatise on zoological geography'. Darwin expressed 'unbounded admiration' – it would be the basis 'of all future work on Distribution.' " (Internal citation deleted.)

One major discovery that is attributed to Wallace during his eight years (1856-1864) of study within the Malay Archipelago is what became known as Wallace's Line (Quammen, 1996; Raby, 2001). As stated by Quammen (1996, p. 25-26) relative to Wallace's Line:

> As drawn in those days, it split the gap between Borneo and Celebes. Continuing southward, it split also the narrower gap between Bali and Lombok. As far as the pioneer biogeographers were concerned, these two sibling islands, Bali and Lombok, belonged to two different realms. West of the dividing line lived tigers and monkeys, bears and orangutans, barbets and trogons; east of the line were friarbirds and cockatoos, birds of paradise and paradise kingfishers, cuscuses and other marsupials including (farther east in New Guinea and tropical Australia) the ineffable tree kangaroos, doing their best to fill niches left vacant by missing monkeys.

It turns out that the secret that explains Wallace's Line is deep water (Quammen, 1996). Bali shares most of its plants and animals with the Malaysian peninsula and Java, Borneo, and Sumatra, while Lombok, and the islands east of Wallace's line rise from deep water rather than an extension of the southeastern Asian continent and are really at sea (Quammen, 1996).

Continental Drift

Even though biogeographic realms were identified early by scientists such as Darwin and Wallace, there were some puzzling distributions of fossils that were noted in the 1800s and early 1900s. Fossils show evidence that during past time periods some organisms were more widespread and occurred on several continents, now separated by vast oceans. Identical species or very similar related species of certain fossil vertebrates and plants have been found on widely separated continents, particular on the southern continents (Benton, 2003; Coyne 2009; Wicander and Monroe, 2013; de Queiroz, 2014). For example,

Figure 6-6. A photograph of Alfred Russel Wallace taken in Singapore in 1862. The author of the photograph is unknown. (Public domain. Source: Wikimedia commons.)

Mesosaurus was an Early Permian (about 290 Ma) coastal aquatic (lived in fresh or restricted water bodies, rather than being fully marine) reptile that lived in the southern portion of South America and southern Africa (see Figure 6-7). It was about one meter long and was probably incapable of swimming across the present Atlantic Ocean. Two other land vertebrates, *Lystrosaurus* and *Cynognathus*, have also been found on southern continents that are separated by vast oceans (see Figure 6-7). *Lystrosaurus* was a dicynodont therapsid (a synapsid amniote related to mammals) that lived from Late Permian to Middle Triassic time about 255 Ma to 225 Ma (Benton, 2003; Guerrero and Frances, 2009). *Cynognathus* lived in Early to Middle Triassic time and was a cynodont therapsid closely related to mammals. Both *Lystrosaurus* and *Cynognathus* were one to two meters long land vertebrates that certainly could not swim across oceans. Fossils of *Cynognathus* have been found in both central South America and central to southern Africa, and more recently in Antarctica (Guerrero and Frances, 2009), whereas fossils of *Lystrosaurus* have been found in Africa, India, and Antarctica (see Figure 6-7), and more recently in all the other southern continents (Benton, 2003; Guerrero and Frances, 2009).

In addition to these vertebrates, plant fossils of *Glossopteris* and related genera (the *Glossopteris* flora, initially discovered in 1828 in India by French paleobotanist Adolphe Brongniart [Guerrero and Frances, 2009]) have been found on all the southern continents (Prothero, 2007; Coyne, 2009; Guerrero and Frances, 2009; Wicander and Monroe, 2013). *Glossopteris* means "tongue fern" (for the shape of the leaves), but despite this name it produced seeds and it is now thought to be a gymnosperm and related to the conifers (Guerrero and Frances, 2009). During the Permian Period, *Glossopteris* (a shrub to small tree, 4-8 meters in height) was the dominant plant across much of the southern hemisphere (Guerrero and Frances, 2009). Today the continents containing these fossils have widely different climates (Wicander and Monroe, 2013), from tropical rain forests to the glaciers of Antarctica. So, how do we explain these fossil distributions across the southern hemisphere continents? The vertebrates *Mesosaurus*, *Lystrosaurus*, and *Cynognathus* are very unlikely to have been able to swim across the oceanic expanses that we are talking about here; and it is equally unlikely that the plant *Glossopteris* could have had such a widespread distribution across most of the southern hemisphere land environments if the continents were distributed as they are today. A more logical explanation for this fossil distribution is that the southern continents must have been connected during the late Paleozoic Era (Permian Period) and into the Mesozoic Era (Triassic Period) (Wicander and Monroe, 2013). In fact, this was one of the major lines of evidence in support of the hypothesis of Continental Drift, first seriously proposed by Alfred Wegener (1880-1930) (Figure 6-8) in 1912 at a presentation before a scientific society and the publication of two scientific papers (de Queiroz, 2014), and later published by Wegener in his 1915 book, *The Origin of Continents and Oceans* (Wicander and Monroe, 2013; de Queiroz, 2014). However, the book was not translated into English until 1924 (de Queiroz, 2014).

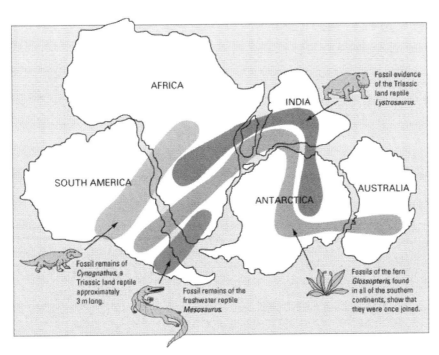

Figure 6-7. If the continents are rejoined as they were by the end of Permian time and the beginning of Triassic time, as suggested by Antonio Snider-Pellegrini (1802–1885), a French geographer, and Alfred Wegener (1880-1930), a meteorologist best known for advancing the hypothesis of Continental Drift, the locations of certain fossil plants and animals on present-day, widely separated continents would form definite patterns (shown by the bands representing distribution) (Kious and Tilling, 1996). (Source of Snider-Pellegrini Wegener fossil map: U. S. Geological Survey. This image has been modified from color to gray scale. This image is in the public domain.)

Alfred Wegener had a tremendous amount of factual evidence in support of the hypothesis that the continents were once assembled into a single supercontinent by about 250 million years ago (Ma), and that the supercontinent began to break-up about 200 Ma. But there had been several before Wegener that had noted the jigsaw puzzle match of the western edge of Africa with the eastern edge of South America and suggested that they had once been together with no ocean in between. Also before Wegener some scientists had noted the odd distribution of certain fossils around the globe. For example, Eduard Suess (1831-1914), an Austrian geologist and paleontologist, noted the similarities between Late Paleozoic (primarily

Pennsylvanian and Permian) aged plant fossils, the *Glossopteris* flora, and the evidence for Permian glaciation in the rock sequences of India, Australia, Africa, and

Figure 6-8. Photograph taken in 1930 of Alfred Wegener in Greenland during the German Expedition of 1930-31 to study the meteorology of Greenland and the thickness of the Greenland ice sheet. Wegener died in October of 1930, probably of heart failure due to overexertion, while trying to get food and supplies to some colleagues at a field station via dog sled. (Authors: Loewe, Fritz; Georgi, Johannes; Sorge, Ernst; Wegener, Alfred Lothar. Source: Archive of Alfred Wegener Institute via Wikimedia commons. This image is in the public domain.)

South America. He proposed the name *Gondwanaland* (now just called *Gondwana*) for a Late Paleozoic supercontinent composed of the above four continents (now, of course, Antarctica is added and also has plant fossils of the *Glossopteris* flora) (Tarbuck, Lutgens, and Tasa, 2013; Wicander and Monroe, 2013).

In 1910, the American geologist Frank Taylor (1860–1938), presented a hypothesis of continental drift that included the following, as summarized from Wicander and Monroe (2013):

- lateral movement of continents formed mountain ranges,
- a continent broke apart at the Mid-Atlantic Ridge to form the Atlantic Ocean,
- supposedly, tidal forces pulled formerly polar continents toward the equator when Earth captured the Moon about 100 million years ago.

Although Taylor was wrong about the mechanism of continental motion, his idea that the Mid-Atlantic Ridge was the linear position along which an ancient continent broke apart to form the Atlantic Ocean turned out to be correct (Wicander and Monroe, 2013; de Queiroz, 2014).

As we have already suggested above, it was Alfred Wegener, a German astronomer and meteorologist, who had assimilated the large amount of evidence available during the early 1900s and presented a strong argument for the hypothesis of continental drift (Kious and Tilling, 1996). Wegener proposed that all landmasses on Earth were originally united into a supercontinent that he named Pangea (may also be spelled as Pangaea) from the Greek meaning "all land" (de Queiroz, 2014). Wegener stated that Pangea began to split apart and fragment about 200 Ma (Tarbuck, Lutgens, and Tasa, 2013; Wicander and Monroe, 2013). Alexander du Toit (1878-1948), a South African geologist and Professor of Geology at Witwatersrand University, and one of Wegener's strongest supporters, further developed Wegener's ideas and gathered more geologic and paleontologic data in support of Wegener's continental drift hypothesis (Kious and Tilling, 1996; Wicander and Monroe, 2013). In 1937, du Toit published *Our Wandering Continents*, where he contrasted the coal deposits of the northern hemisphere continents with the glacial deposits of the same age in Gondwana, and suggested that the northern continents were closer to the equator during Late Paleozoic time, whereas the Gondwana continents straddled the South Pole. According to du Toit, Pangea first split apart to form two supercontinents, *Laurasia* situated in the northern hemisphere near the equator and the southern supercontinent referred to as Gondwanaland (or Gondwana, as we stated above) (Kious and Tilling, 1996). Then ultimately the break-up of Laurasia would form the northern continents of today, whereas the fragmentation of Gondwana would form the southern continents (Figure 6-9). According to Wegener, Pangea was surrounded by a universal ocean called

Panthalassa, allowing the fragmenting pieces of continental material from Pangea to "drift over" and/or "plow through" the denser material below (Kious and Tilling, 1996; Levin, 1999; Tarbuck, Lutgens, and Tasa, 2013; Wicander and Monroe, 2013). Wegener thought that the "bulldozing" leading edge of the slab of continental material may crumple to form mountain ranges such as the Andes (Levin, 1999). Although he amassed a tremendous amount of geologic, paleontologic, and climatologic evidence (especially with the work of du Toit and others), Wegener's ideas were not well received by the scientific community at the time, and by the late 1930s to early 1940s his hypothesis was in deep trouble. The main problem was that Wegener did not have a valid mechanism to explain how continents could plow through solid ocean crust. Geophysicists, such as Sir Harold Jeffries (1891-1989) of England, calculated that the ocean floor was far too rigid to allow continents to move over or plow through oceanic crust, regardless of the mechanism (Kious and Tilling, 1996; Levin, 1999; Wicander and Monroe, 2013).

Plate Tectonics

During the late 1960s, Plate Tectonic Theory became the new paradigm to explain past continental movement and represented a major revolution in the geosciences. Plate Tectonic Theory allowed for an adequate explanation of the movement of continents through time, the location of volcanic and earthquake belts on Earth, the location of major mountain systems, the location of various rock and mineral resources around the world, and the biogeographic and ancient biogeographic (paleobiogeographic) location and distribution of modern organisms and fossils. Plate tectonics (tectonics means large scale deformational movement) was a much more encompassing theory than continental drift. Plate movement is associated with Earth's rigid outer shell called the lithosphere. Earth's lithosphere consists of the crust and upper brittle portion of the mantle (Earth's structural layers consist of the crust, mantle, and core, [see any introductory physical geology text to review the origin, thickness, and composition of these]). The lithosphere averages about 100 Km in thickness (but continental lithosphere may be as thick as 250 Km and oceanic lithosphere is typically less than 100 Km thick) and is broken into about seven major lithospheric plates and many minor plates (Tarbuck, Lutgens, and Tasa, 2013; Wicander and Monroe, 2013). The largest plate is the Pacific Plate. Plates move slowly (about 2 cm to 15 cm per year) due to some type of convective heat flow within the mantle of Earth (Tarbuck, Lutgens, and Tasa, 2013; Wicander and Monroe, 2013).

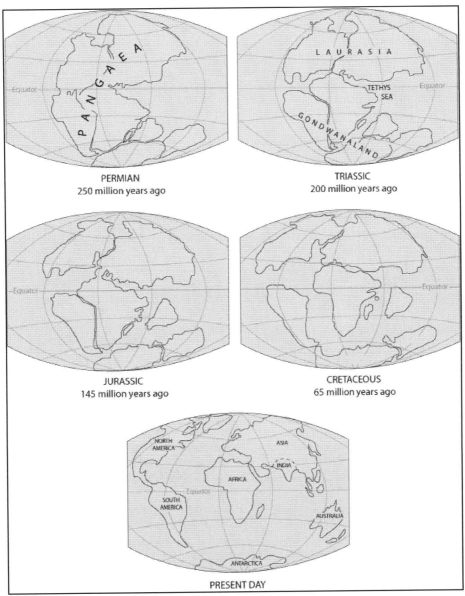

Figure 6-9. According to the continental drift hypothesis, the supercontinent Pangaea began to break up about 200 million years ago, at first forming the northern supercontinent Laurasia and the southern supercontinent Gondwanaland (or Gondwana); but eventually fragmenting into the continents as we know them today (Kious and Tilling, 1996). (Source of Break-up of Pangaea: U. S. Geological Survey. The image has been modified from color to gray scale. This image is in the public domain.)

Studies of continental drift waned until the1950s, when oceanographic research and studies of Earth's magnetic field revealed the very young age of oceanic crustal material; no oceanic crust was found that was older than about 180 Ma (Tarbuck, Lutgens, and Tasa, 2013; Wicander and Monroe, 2013). So, the modern oceans appear to have been formed due to the breakup of Pangea. But by what process did Pangea split apart and break up into the continents of today, with wide oceans separating the continents? By the late 1950s, a wealth of data was emerging from scientific studies of the ocean floor, which revived Wegener's ideas about continental drift and spurred an interest in conducting even more research into the nature of the ocean floor (Kious and Tilling, 1996). According to Kious and Tilling (1996):

> In particular, four major scientific developments spurred the formulation of the Plate Tectonic Theory: (1) demonstration of the ruggedness and youth of the ocean floor; (2) confirmation of repeated reversals of the Earth magnetic field in the geologic past; (3) emergence of the seafloor-spreading hypothesis and associated recycling of oceanic crust; and (4) precise documentation that the world's earthquake and volcanic activity is concentrated along oceanic trenches and submarine mountain ranges.

After World War I, echo sounding (basically a primitive form of sonar) greatly increased our knowledge of the topography of the ocean floor, and the ocean floor was turning out to be much more rugged than earlier thought. This echo sounding further demonstrated the ruggedness and continuity of the central Atlantic submarine mountain chain (later called the Mid-Atlantic Ridge) that had been suggested by earlier bathymetric measurements (Kious and Tilling, 1996). In addition, in the 1950s the pace of oceanographic surveys by many nations started picking up and more sophisticated instruments were allowing for more accurate data measurements. This revealed a worldwide submarine mountain chain that zigzagged mid-way between the continents, a mid-oceanic ridge system that rose 4500 meters above the average sea floor and was more than 50,000 kilometers in length (Kious and Tilling, 1996).

In addition to mapping mid-ocean ridges, ocean research prior to the 1960s had revealed magnetic anomalies on the sea floor. A magnetic anomaly is a deviation from the average strength of Earth's magnetic field at a particular location. It was determined that these magnetic anomalies of reversed polarity (opposite to the present polarity) and normal polarity (the same direction as present magnetic polarity) were parallel to mid-oceanic ridges, somewhat like strips on a zebra (Vine and Mathews, 1963; Vine and Wilson, 1965; Wilson, 1965; Kious and Tilling, 1996). It was known during the late 1950s and early 1960s that basaltic igneous activity occurred in the central rift valley on the crest of mid-oceanic ridges. Basalt is a fine-grained igneous rock that contains iron-rich silicate minerals (such as olivine [$(Mg,Fe)_2SiO_4$] and augite [$Ca(Mg,Fe,Al)(Al,Si)_2O_6$]), and also the iron-rich oxide, magnetite (Fe_3O_4). Thus, it was not a surprise to oceanographic scientists of the early 1960s that magnetic anomalies were present on the seafloor because the iron-rich minerals in basaltic volcanic flows along the crests of mid-oceanic ridges would align with their magnetic polarities parallel to Earth's magnetic field (and fixed or "frozen") as they cooled beyond a certain temperature (called the Curie temperature) (Vine and Wilson, 1965; Vine, 1966). This had already been determined on land for basaltic lava flows and it had also been shown that land volcanic basalt flows (that showed repetitive stacked lava layers) had a certain pattern of normal and reversed polarity with respect to absolute time. This, of course, would mean that Earth's magnetic field periodically reversed itself. Radiometric dating of land basaltic lava flows (that had cooled to form hard basaltic rock) resulted in absolute dates for particular magnetic polarity patterns. By matching these up with the patterns on the sea floor on opposite sides of mid-oceanic ridges, the age of the various magnetic bands could be determined. But why should there be a symmetrical alternating reversed and normal polarity magnetic band pattern on each side of mid-oceanic ridges? By matching up the land magnetic patterns with the patterns on each side of mid-oceanic ridges, it was determined that the basaltic crustal rocks just below the sediment on each side of mid-oceanic ridges were older the farther the rocks were from the mid-oceanic ridges. Thus, the oceanic crustal rocks are younger at the mid-oceanic ridges and get progressively older the farther they are from the mid-oceanic ridges. So, as basaltic lava oozes out of the rift zone on the crest of mid-oceanic ridges, it will cool to take on the magnetic polarity that is present at the time.

If Earth's magnetic polarity is as it is today, the cooled basaltic lava will take on the polarity of today's magnetic field, which is said to be normal (called a positive magnetic anomaly; this band would make a compass needle point towards the north magnetic pole). If the magnetic polarity of the cooled basaltic lava is the opposite of today's polarity (a negative magnetic anomaly), the cooled basaltic lava is said to have reversed polarity (Vine and Mathews, 1963; Vine and Wilson, 1965; Wilson, 1965; Vine, 1966; Menard, 1967; Kious and Tilling, 1996).

Based on oceanographic research of the 1950s, Harry Hess (1906-1969), a Princeton University Professor of Geology, and a Naval Rear Admiral during World War II, proposed the hypothesis (later accepted as a theory) of seafloor spreading in a landmark scientific paper in 1962 (Hess, 1962; Kious and Tilling, 1996; Levine, 1999; Wicander and Monroe, 2013). Hess (1962) inferred that continents did not plow their way through oceanic crust, but that most major plates consist of both continental and oceanic crust (except the Pacific Plate, which is primarily oceanic material) that moves over the mantle due to thermal convection cells within the mantle. Of course, we now know that the plates consist of lithosphere (that includes the crustal rocks and the very upper portion of the mantle [brittle portion of the mantle]) that moves over the asthenosphere (a hotter more plastic behaving layer of the upper mantle) (Kious and Tilling, 1996; Levine, 1999; Tarbuck, Lutgens, and Tasa, 2013; Wicander and Monroe, 2013).

Hess (1962) proposed that heat transfer within the mantle generated thermal convection cells that would rise towards the surface and would diverge to fracture and split the crust (and entire lithosphere) apart. The rising hot material would result in thermal expansion of the crust (and entire lithosphere) that would form elevated mid-oceanic ridges. As the hot mantle material diverged, fractures in the lithosphere would result in pressure-release partial melting of the mantle (asthenosphere) below, generating basaltic magma that would rise and flow out at the mid-oceanic ridge as basalt flows in the rift valley or intruded as basaltic or gabbroic (gabbro is a coarse grained igneous rock of basaltic composition) dikes (tabular discordant intrusions) into the deeper oceanic crust. As the lava or magma cooled it would form new oceanic crust (and lithosphere), that would then fracture apart and move to each side riding "piggy-back" on the mantle (asthenosphere) below. Thus, the seafloor would spread as new oceanic crust (and lithosphere) formed. As oceanic crust (and lithosphere) moved to each side of a mid-oceanic ridge it would be removed from the position of maximum thermal upwelling and would cool, contract, and sink to oceanic depths.

As more and more magnetic anomalies (as discussed above) were mapped across mid-oceanic ridges, this magnetic data was the most persuasive evidence in support of the theory of seafloor spreading (Hess, 1962; Vine and Mathews, 1963; Vine and Wilson, 1965; Wilson, 1965; Menard, 1967; Bonatti, 1968; Kious and Tilling, 1996; Levine, 1999; Tarbuck, Lutgens, and Tasa, 2013; Wicander and Monroe, 2013; de Queiroz, 2014). But in addition, further oceanographic studies from the late 1960s to the present, including three international ocean drilling programs (the Deep Sea Drilling Project [DSDP], 1966-1983; the Ocean Drilling Program [ODP], 1985-2003; and the Integrated Ocean Drilling Program [IODP], 2003-Present), have added a tremendous amount of data in support of seafloor spreading (Tarbuck, Lutgens, and Tasa, 2013; Wicander and Monroe, 2013). For example, studies across mid-oceanic ridges have shown that oceanic sediments are thin on the flanks of mid-oceanic ridges, but increase in thickness on the ocean floor with distance from mid-oceanic ridges (Levin, 1999; Tarbuck, Lutgens, and Tasa, 2013; Wicander and Monroe, 2013). Assuming a fairly constant rate of sedimentation across the oceans, this would imply an increasing age for oceanic sediments with distance from mid-oceanic ridges. Also from drill samples of oceanographic research ships, the age of sediment (based on microfossils) just above the oceanic basalt becomes older and older with distance from a mid-oceanic ridge (Tarbuck, Lutgens, and Tasa, 2013; Wicander and Monroe, 2013). (Note: We usually cannot radiometrically date the oceanic crustal basalt directly because it flowed into water and has been chemically altered such that radiometric dates are often not accurate. However, we can radiometrically date volcanic islands in the ocean [such as the Hawaiian Islands chain, which formed above a mantle plume or "hot spot" located in the mantle below the Pacific plate] because the upper portions of the islands are formed of subaerial volcanic rock of basaltic composition. For example, absolute dates of the Hawaiian Islands show that the islands get older to the northwest. Based on the distance of the islands apart and their absolute age dates, we can determine the past rate of seafloor spreading of this portion of the Pacific Plate.)

As we now know, based on more recent seismic research (the study of sound waves moving through Earth), the asthenosphere exists beneath the lithosphere (see Figure 6-10). It is hotter and weaker than the lithosphere and allows for motion of the lithosphere independent of the asthenosphere, but the lithosphere may ride over the asthenosphere due to convective motion within the asthenosphere (even though the asthenosphere is a plastic behaving solid).

All major interactions among lithospheric plates (or tectonic plates) occur along their boundaries. There are three major types of plate boundaries: 1) divergent plate boundaries, which are constructive boundaries that form new lithosphere where plates pull apart; 2) convergent plate boundaries, which are destructive boundaries that destroy or deform lithosphere where plates converge and one plate subducts below another or deformational lithospheric compression takes place; and 3) transform plate boundaries (or transform fault plate boundaries), where plates merely slide horizontal past each other. Earthquakes (caused by movement of rock material along faults [crustal fractures with displacement along them] in Earth's crust) occur at all plate boundaries. In fact, the boundaries of tectonic plates may be determined by mapping the epicenters of earthquakes for several years. However, volcanic activity only takes place at divergent and subducting convergent plate boundaries, usually not along transform plate boundaries (Levin, 1999; Tarbuck, Lutgens, and Tasa, 2013; Wicander and Monroe, 2013).

Most divergent plate boundaries are located along the mid-oceanic ridges (Figure 6-10). This is where seafloor spreading is taking place. As the two plates move apart (diverge) along mid-oceanic ridges, fractures form along the central rift valley (a valley or canyon along the crest of the ridge due to divergence) at the crest of the ridge, and below the crest in the deeper rock. This lowers the pressure on the asthenosphere, allowing pressure release and decompression partial melting to take place. As magma forms, it rises through fractures and cools, with some actually flowing into the rift valley (sometimes referred to as the rift zone) and cooling, to form new seafloor (i. e., mantle material upwells to create new seafloor). As this process continues, new seafloor and lithosphere are constructed (Levin, 1999; Tarbuck, Lutgens, and Tasa, 2013; Wicander and Monroe, 2013).

Divergent boundaries can also form on land. Obviously, this must be the case if the process of thermal convective mantle upwarping is responsible for splitting supercontinents (such as Pangea) apart to form smaller continents (and we strongly accept this process both for seafloor spreading and rifting continents apart) (Tarbuck, Lutgens, and Tasa, 2013; Wicander and Monroe, 2013). A prime example of a zone of divergence is the East African Rift Valley in eastern and southern Africa (see Wicander and Monroe, 2013, p. 49; or most college introductory geology textbooks). This is the first stage in the rifting of a continent apart. If this process of rifting continues the sea will flood into the rift valley and a linear sea will develop (Tarbuck, Lutgens, and Tasa, 2013; Wicander and Monroe, 2013). This is the current situation, for example, of the Red Sea separating Arabia from Africa where the continental crust has broken, the sea has flooded into the rift, and it appears that seafloor spreading has been initiated (Omar and Steckler, 1995). The Red Sea is a relatively recent example of a young divergent plate boundary that has been spreading since approximately 34 Ma during the late Eocene Epoch (Omar and Steckler, 1995). A mid-oceanic ridge system will develop and a full-fledged ocean will form, if seafloor spreading continues in a linear sea such as the Red Sea (Tarbuck, Lutgens, and Tasa, 2013; Wicander and Monroe, 2013).

If Earth is not getting larger in volume (and we know that it is not because we have been measuring the circumference of Earth since the time of the ancient Greek natural philosopher Eratosthenes in 240 B.C.E. [Bennett, et al, 2010]), there must be, in addition to divergent plate boundaries where new sea floor and oceanic lithosphere is formed, destructive plate boundaries where oceanic lithosphere is destroyed. This destruction occurs at convergent plate boundaries where oceanic lithosphere is destroyed along subduction zones (where the edges of oceanic lithosphere break and are pushed or pulled back into the interior) (Figure 6-10). There are three types of convergent plate boundaries: 1) oceanic-continental convergent plate boundaries, 2) oceanic-oceanic convergent plate boundaries, and 3) continental-continental convergent plate boundaries.

The type of convergent plate boundary shown in Figure 6-10 is an oceanic-continental convergent plate boundary. In this type of convergence, the denser oceanic lithospheric slab breaks at the edge of the continental lithospheric plate and subducts as it sinks into the asthenosphere. Partial melting of the subducting plate and the asthenosphere above the subducting plate (due to the addition of water from the subducting plate) generates pockets of magma that rise into the continental lithosphere. Most of this magma will cool before reaching the surface to thicken the continental crust, but some of the magma will reach the surface to form continental volcanic mountains. Examples include the Andes Mountains of western South America where the Nazca plate is subducted beneath the South American plate, and the Cascade Mountain Range of the northwest U.S. and southwest Canada where the Juan de Fuca plate is subducted beneath the North American plate (Tarbuck, Lutgens, and Tasa, 2013; Wicander and Monroe, 2013).

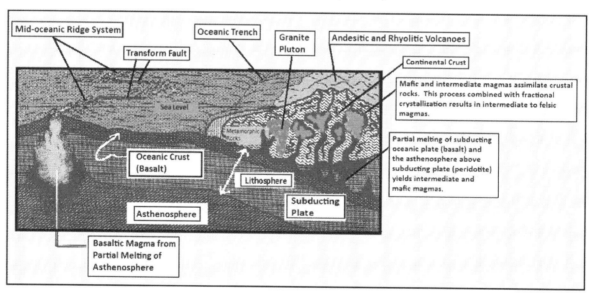

Figure 6-10. Diagram illustrating the origin of magma and igneous rocks at divergent plate boundaries along mid-oceanic ridges and along subduction zones associated with convergent plate boundaries. Note the oceanic trench along the boundary between the subducting oceanic lithospheric plate and the overriding continental lithospheric plate. The subducting lithospheric plate will eventually be destroyed and assimilated into the asthenosphere and/or lower mantle. Transform faults are also shown here that indicate offsets of the mid-oceanic spreading center (mid-oceanic ridge). (Sketch by E. L. Crisp, after several sources. The sketch is not to scale.)

Other examples of convergent plate boundaries are places where oceanic-oceanic convergence occurs. For this type of convergence, two oceanic plates converge and one plate descends beneath the other, forming an oceanic trench at the boundary between the two plates and a subduction zone beneath the overriding plate. Partial melting along the subducting plate forms rising magma bodies that often form volcanoes on the sea floor, and volcanic oceanic islands if the volcanoes rise above sea level. This forms what are called volcanic island arcs on the sea floor of the overriding plate. Examples of such volcanic island arcs are the Aleutian Islands off the coast of Alaska, the Japanese Islands, the Mariana Islands, and the Tonga Islands. These volcanic island arc systems occur along the margin of the Pacific Ocean above subduction zones (associated with oceanic trenches), as do other volcanic island arc and continental volcanic arc systems (such as the Andes Mountains). These volcanic arc systems around the Pacific Ocean basin are often referred to collectively as the "Ring of Fire" (Monroe and Wicander, 2005; Tarbuck, Lutgens, and Tasa, 2013).

The third type of convergent plate boundary is the continental-continental convergent plate boundary. Continental-continental convergence takes place when a subducting oceanic plate contains attached continental material, but the oceanic plate is subducting beneath different continental lithosphere. As the oceanic lithosphere portion of a lithospheric plate, that is also carrying continental material, completely subducts beneath a continent, two continental plates collide. Because of the tremendous forces involved in these plate collisions and because the frictional forces and the equal and low densities of these continental plates are too great for one continent to subduct beneath another continent, major deformational mountain systems are formed. A relatively new mountain range formed by this process is the Himalaya Mountains, today the mightiest mountain system in the world. This mountain system started forming about 40 to 50 million years ago as the Indian plate began to collide with the Asian plate (Wicander and Monroe, 2013). The two plates are still moving into each other and major earthquakes occur along the suture zone between these two plates as stresses build-up in the rocks and deformation occurs along the major fault systems parallel to the boundary between the two plates (Tarbuck, Lutgens, and Tasa, 2013; Wicander and Monroe, 2013). Other ancient collisional mountain systems are the Appalachians that formed during the Paleozoic

Era resulting from several orogenic pulses as Pangea formed, the Atlas Mountains of northwestern Africa that formed primarily during the late Paleozoic as Pangea was being built, the Alps of southern Europe that formed over several million years beginning in late Cretaceous as the Eurasian plate collided with the African plate, and the Ural Mountains of Russia that also formed during collisional orogenies as the formation of the supercontinent Pangea took place. (Note: Orogeny refers to a major mountain building episode.)

The final type of plate boundary is the transform plate boundary (also called transform fault plate boundary). At transform plate boundaries, plates slide horizontally past each other roughly parallel to the direction of plate movement. Lateral movement along this type of tear fault results in a zone of intensely shattered rock with numerous shallow earthquakes. The majority of transform faults connect two oceanic ridge segments and is marked by fracture zones (see Figure 6-10) (Tarbuck, Lutgens, and Tasa, 2013; Wicander and Monroe, 2013). One of the more famous transform plate boundaries is the San Andreas Fault of California, a right-lateral strike-slip fault that separates the Pacific Plate from the North American Plate (Schulz and Wallace, 2013). The Pacific Plate is moving to the northwest, whereas, relatively speaking, the North American Plate is moving to the southeast. This transform fault connects the East Pacific Rise oceanic ridge (spreading center) in the Gulf of California with the Juan de Fuca Ridge, which is the boundary between the Juan de Fuca and the Pacific plates (Wilson, 1965). Most of the earthquakes that take place in California result from movement along this transform fault (the San Andreas Fault; actually, a fault zone or system with many splinter faults coming off the main fault and/or trending approximately parallel to the main fault) (Schulz and Wallace, 2013; Tarbuck, Lutgens, and Tasa, 2013; Wicander and Monroe, 2013).

Summary of Biogeographic Evidence

Both Darwin and Wallace are considered to be biogeographic dispersalists, as are most biogeographers today. They both accepted that species have a center of origin and disperse outward from their center of origin, either by normal dispersal across corridors (easy, uninterrupted routes across continental or marine areas

[Kardong, 2005, 2008]) and filter bridges (connections that have restrictive barriers to certain organisms based on climate or ecology rather than geographic barriers [Kardong, 2005, 2008], such as the isthmus of Panama) or by long-distance dispersal across barriers (for example, oceans or mountain chains) (de Queiroz, 2014). Neither Darwin nor Wallace accepted continental movement or once present deep-water land bridges to explain the distribution of organisms across the major oceans, or to deep water oceanic islands (de Queiroz, 2014).

Long-distance dispersal, because of the low probability of this type of dispersal of organisms across a substantial barrier, is sometimes referred to as chance or sweepstakes dispersal (Quammen, 1996; Kardong, 2005; de Queiroz, 2014). But Charles Darwin (1859; 2005 [1845]), and certainly many others, felt that given the extremely old age for Earth, the probability of long-distance dispersal was much higher (Quammen, 1996; de Queiroz, 2014). Darwin, in fact, performed many experiments (particularly from 1854 to 1856) relative to long-distance dispersal across oceans (Darwin, 1859; de Queiroz, 2014; Costa, 2017). He put various types of seeds in containers filled with salt water for various lengths of weeks and months to determine if they would germinate afterwards (Quammen, 1996; de Queiroz, 2014; Costa, 2017). He did experiments to determine if freshwater snail hatchlings would attach to the feet of ducks, and thus have the potential to be transported with the duck across great oceanic distances (de Queiroz, 2014; Costa, 2017). Darwin also forced seeds into the stomachs of fish. He then feed the fish to eagles, storks, and pelicans. Afterwards he collected the droppings of the birds that contained the seeds and tried to germinate the seeds. Many of the seeds did germinate (Quammen, 1996; de Queiroz, 2014). He also did experiments with floating branches with fruits (that contained seeds) to determine how long they would float in salt water (de Queiroz, 2014). All these experiments (and more) convinced Darwin that long-distance dispersal across large expanses of ocean were actually quite likely given long episodes of time (Quammen, 1996; de Queiroz, 2014; Costa, 2017).

More modern researchers have noted that tropical storms and hurricanes can generate strong currents and winds. Strong winds may blow migrating birds off course and into new and different areas (Kardong, 2005), such as offshore islands, like the Galapagos Islands, about 900 kilometers (about 560 miles) off the coast of Ecuador. Also during storms, strong currents may uproot and transport trees and other vegetation to form rafts that may contain clinging land vertebrates and other terrestrial animals (Kardong, 2005, 2008). Thus, natural flotsam (floating debris) may be dislodged during storms that may be large and massive enough to carry growing plants, in addition to various types of animals, such as termite colonies, gecko eggs, snakes, and rats (Quammen, 1996). For example, de Queiroz (2014) discusses a situation that occurred during hurricanes in 1995 that dislodged a vegetation raft (trees and other floating vegetation) from Guadeloupe Island in the Caribbean. The raft contained about fifteen live green iguanas (*Iguana iguana*) (see Figure 6-11 for an example) on board. About a month later, the raft, with the live iguanas, washed onto shore on the distant island of Anguilla, about 280 kilometers (175 miles) to the northwest. These iguanas had not lived on the island of Anguilla prior to this (also see Censky, Hodge and Dudley, 1998; Lawrence, 1998; Kardong, 2005).

Figure 6-11. A green iguana (*Iguana iguana*) resting on a log in the middle of the Sarapiqui River near Chilamate, Costa Rica. (Photo by E. L. Crisp, July 2008)

This may be similar to how land iguanas reached the Galapagos Islands from the western coast of South America and eventually evolved into later species of land iguanas and marine iguanas there. However, the few initial iguanas coming from the mainland of South America would have a restricted gene pool (in terms of alleles) as compared to the large population on the mainland and thus the islands populations are based on limited variety of genetic traits (due to chance); this restricted gene pool in new and isolated locations is called the founder effect (Kardong, 2005, 2008).

Charles Darwin (in a letter to Charles Lyell in 1860) hypothesized that there was a common land iguana ancestor on the Galapagos Islands that evolved into the current Galapagos land iguanas (genus *Conolophus*, divided into two, or possibly three species) and the current Galapagos marine iguanas (*Amblyrhynchus cristatus*) (Grant and Estes, 2009). Marine iguanas are endemic only to the Galapagos Islands and feed on algae and seaweed. The fact that Galapagos land iguanas and marine iguanas do sometimes hybridize and produce live offspring with a mixture of characteristics between the two is evidence of fairly close genetic composition, even though the offspring are probably sterile (Charles Darwin Research Station Fact Sheet, 2006). Modern genetic studies indicate that the Galapagos land iguanas and marine iguanas diverged from a common ancestor over 10 million years ago, on what are now submerged volcanic islands. Most modern geneticists tend to favor Darwin's simpler hypothesis for divergence of the land iguanas and the marine iguanas on the islands, as opposed to divergence on the South American continent prior to dispersal to the Galapagos Islands (Grant and Estes, 2009).

Long distance dispersal across oceans certainly appears to have been important in the past to explain the distribution pattern of some organisms across these great barriers. However, the distribution of many Gondowanan fossils, such as the *Mesosaurus*, *Cynognathus*, *Lystrosaurus*, and *Glossopteris* genera that we mentioned earlier when discussing continental drift and plate tectonics, must be explained by land connections that have been severed. Thus, for example, as Africa and South America initially diverged due to plate tectonic activities during the Jurassic Period (with final separation about 110 Ma, during the Cretaceous Period), descendant populations of Gondwanan animals and plants must have been separated by the ever-widening southern Atlantic Ocean. This splitting of the continuous distributional range of a taxonomic group (taxon, plural taxa) into two or more parts by the formation of some sort of barrier to dispersal is referred to as vicariance biogeography (de Queiroz, 2014). Of course, the taxa that become isolated from each other by the vicariance separation will evolve in isolation from each other and ultimately give rise to different species in the separate areas.

Most long-distance separation of closely related organisms, from one continent to another continent across oceans or true oceanic islands (islands that form in the deep ocean due to volcanism and have never been connected to continental land bodies) offshore to mainland continental bodies (such as the Galapagos Islands and the Hawaiian Islands) must be explained by either long distance dispersal or vicariance dispersal. Vicariance biogeography explains the fossil distribution for many past organisms, especially Gondwanan relicts that have remained on southern continents after the break-up of Gondwana. But vicariance biogeography does not appear to be the best explanation for the endemic organisms that are present on true oceanic islands, such as the Hawaiian Islands or the Galapagos Islands. For example, molecular clock studies (based on the rate of mutations in organisms) indicate that almost all the Hawaiian taxa separated from continental relatives within the last 20 million years, most within the last 5 million years (if this were explained by a once present land connection, this number would have to be in excess of 70 million years) (de Queiroz, 2014). Thus, these taxa must have reached the Hawaiian Islands by long distance dispersal (de Queiroz, 2014). This appears to be the case for other deep oceanic islands, such as the Galapagos Islands.

But vicariance biogeography is important in explaining much of the distribution and diversity of life. For example, let's take the case of the formation of the Isthumus of Panama (including Panama and other Central American countries) to connect the North American Continent and the South American Continent. Prior to about 35 million years ago, North America and South America were completely separate continents where evolution had taken its course to yield unique animal and plant species endemic to each continent. However, because of the subduction of the Cocos tectonic plate beneath the Caribbean tectonic plate, a series of volcanic islands (as a volcanic island arc system) started emerging between North America and South America by about 15 million years ago. Between 2 to 7 million years ago (some geologists say about 3 to 5 million years ago) this volcanic/plate tectonic process had formed a continuous land bridge between North America and South America, the Isthumus of Panama (Wainwright, 2007; Wicander and Monroe, 2013; de Queiroz, 2014). This connection formed a filter bridge connecting the two continents, such that animal and plant species crossed the bridge in both directions, but at different rates. Thus, the distribution and diversity of species in the Americas underwent a rapid and profound change (Wainwright, 2007). The mammalian fauna was particularly affected. During most of the Cenozoic Era, South America was an island continent and had many different mammals than North America, including many marsupial (pouched) mammals. Once the connection was made to North America, migrants from North America replaced many of the species then endemic to South America, particularly many of the convergently similar marsupial mammals of South America. Only a few mammal groups from South America migrated across the filter bridge to establish themselves in North America, including the opossum, porcupine, and sloths (Kardong, 2005; Wicander and Monroe, 2013).

Once formed, the Isthumus of Panama served as a filter bridge between North and South America, but served as a barrier to migration between the Caribbean Sea and the Pacific Ocean. Prior to the formation of the Isthumus of Panama, a homogeneous community of bottom-dwelling invertebrates inhabited the shallow seas of the area, including the Caribbean Sea and the shallow Pacific continental shelf and near shore bays. After the barrier formed, different species evolved on opposite sides of the barrier because there were no longer common gene pools for species on opposite sides of the Isthumus of Panama (Wicander and Monroe, 2013).

CHAPTER 7

THE HISTORY AND DIVERSITY OF LIFE ON EARTH

INTRODUCTION

The story of life on Earth begins with the origin of the Universe. Obviously, the basic ingredients to make our Earth came from the Universe, and scientists are confident that the ingredients that make up life originally came from Earth. The most accepted theory for the origin of the Universe is the Big Bang Theory, which basically states that about 14 billion years ago an infinitely dense point of pure energy "exploded" and expanded outward to eventually evolve into our present Universe. The Universe was initially extremely hot (about 100 billion degrees Celsius or more), but as expansion continued the Universe cooled, matter formed and clumped together by gravity, and stars, planets, and galaxies formed. According to the Big Bang Theory, the Universe was initially concentrated into an extremely dense region billions of times smaller than a proton. The initial core exploded to send energy (and eventually matter) expanding outward in all directions (Arny, 2006; Hester et al, 2007; Bennett et al, 2010; Wicander and Monroe, 2010; Monroe and Wicander, 2011; Wicander and Monroe, 2013).

There are two main lines of evidence for the Big Bang Theory and an expanding Universe (these are the two main lines of support, there are several others):

- 1) Distant galaxies are moving away from Earth at high speed and the velocities of recession increase with distance from Earth. Thus, the Universe is expanding. This was first discovered by Edwin Hubble (1929).

He noted that the spectral lines (wavelengths of light) of the receding galaxies are shifted toward the red end of the spectrum, thus a shift towards longer wavelengths. We say that the light is redshifted. The Doppler Effect causes this redshift of the light from receding galaxies. This effect is the apparent shift of the wavelength of waves caused by movement of the source of the waves relative to an observer. The Doppler Effect is commonly observed by most people when moving vehicles approach and pass. The sound waves from the moving vehicle are shifted towards shorter wavelength as the vehicle approaches an observer (the observer hears a higher pitched sound) and is shifted towards longer wavelengths as the vehicle moves away from an observer (the observer hears a lower pitched sound). A source of light that is moving towards an observer will have an apparent shift of the wavelengths to shorter wavelengths (blueshift of the light waves), whereas a source of light moving away from an observer will appear to have the wavelengths shifted to longer wavelengths (redshift of the light waves). Light coming from receding stars or galaxies appears to be redder, that is, the wavelengths are shifted towards the red end of the spectrum. If the stars or galaxies were moving towards us, the wavelengths would be shifted towards the shorter wavelengths and the light would appear bluer. The wavelengths of hydrogen spectra emitted from receding galaxies are shifted to progressively longer wavelengths for galaxies at larger and larger distances from Earth. The velocity of recession of galaxies from each other is proportional to the distance between them, thus farther galaxies are receding with higher velocities. Based on the idea of an expanding Universe and the rate of recession of galaxies from each other, astronomers can use the expansion rate to calculate how long ago the galaxies were together at a single point. The estimate works out to about 14 Ga (13.77 Ga) (Jones, 2014); thus, astronomers accept that the Universe is about 14 billion years old (Arny, 2006; Hester et al, 2007; Bennett et al, 2010; Wicander and Monroe, 2013).

- 2) There is a cosmic background radiation permeating the Universe. This cosmic background radiation is thought by astronomers to be the "after glow" of the Big Bang. Arno Penzias and Robert Wilson received the Noble Prize in physics in 1978 for their discovery of the cosmic background radiation. In 1965, as astronomers with Bell Telephone Laboratories in New Jersey, they discovered a background radiation coming from outside our solar system that corresponds to a constant temperature of 2.7 kelvin (or 2.7°C above absolute zero). They postulated that this 2.7 kelvin temperature is the relic of the initial high temperature of the early Universe. This cosmic background radiation has a maximum wavelength of about 2 mm (in the microwave range of the electromagnetic spectrum). The radiation has all the characteristics of radiant heat and is the same kind of radiant heat as emitted by an object with a temperature of 2.7°C above absolute zero. So, the initial extremely hot radiant heat from the primordial fireball (of the "Big Bang") has cooled down to 2.7 Kelvin, and will eventually cool to absolute 0. This is direct evidence of the "Big Bang" (Arny, 2006; Hester et al, 2007; Bennett et al, 2010; Wicander and Monroe, 2013; PBS, Penzias and Wilson Discover Cosmic Microwave Radiation, accessed 10/03/2014; Wikipedia, Discovery of Cosmic Microwave Background Radiation, accessed 10/03/2014).

Astrophysicists can theoretically go back to a fraction of a second (10^{-43} seconds) following the "Big Bang". Before this, the Universe consisted of pure energy and the four major forces of nature that we recognize were not distinct entities (they may have been unified). Those four major forces of nature are 1) the gravitational force, 2) the electromagnetic force, 3) the strong nuclear force (holds nuclear particles together), and 4) the weak nuclear force (involved in radioactive decay). At the birth of the Universe, the temperature and density were extremely high and only pure energy could exist. Within the first second after the "Big Bang" the four forces of nature separated out, the Universe inflated rapidly, quarks and electrons started forming, as did antiquarks and antielectrons (antimatter), and matter and antimatter collided and annihilated each other, but fortunately the Universe was asymmetrical and had more matter than antimatter. Thus, a slight excess of matter was left over, which makes up our present Universe. Finally, quarks formed protons and neutrons (remember a proton is simply the nucleus of hydrogen atoms) and some of the neutrons were fused with protons to form helium nuclei and lesser amounts of lithium nuclei (Arny, 2006; Hester et al, 2007; Bennett et al, 2010; Wicander and Monroe, 2013).

By about 300,000 years after the "Big Bang" the temperature had cooled to about 3000 Kelvin, electrons combined with nuclei to form complete atoms of hydrogen and lesser amounts of helium and lithium, and photons of light separated from matter and light burst forth for the first time. By about 200 to 300 million years after the "Big Bang" matter began collecting into clouds of various sizes that eventually collapsed due to gravity to form stars and galaxies. In other words, the Universe became "clumpy" and formed stars that collected into galaxies by 13.5 billion years ago. The first stars and galaxies consisted mostly of hydrogen, with a small amount of helium and very rare lithium. Thus, the early Universe consisted primarily of 100% hydrogen and helium. Today the Universe has about 98% hydrogen and helium, with 2% heavier elements that formed inside of stars (Arny, 2006; Hester et al, 2007; Bennett et al, 2010; Wicander and Monroe, 2013).

After stars began forming, heavier elements formed in the cores of stars by a process known as nucleosynthesis. Stars are born when hydrogen in the core starts fusing into helium. Low mass stars (about the size of our Sun or a little more massive) will start fusing helium into carbon when they use up the hydrogen in the core. When the core has been converted to carbon, a low mass star cannot generate enough pressure by gravitational collapse (the core will collapse because of gravity if no energy is pushing out due to fusion) to fuse carbon into higher elements. Then the star will collapse to the size of the Earth and become extremely hot to form a white dwarf, but it will never fuse again so will eventually cool off to become a black dwarf (a burned-out chunk of carbon). High mass stars can generate enough pressure to fuse carbon into heavier elements. However, as a high mass star successively fuses elements higher than carbon, a problem occurs when the core is fused into iron. Iron absorbs the energy of gravitational collapse and will not fuse into higher elements. This results in rapid collapse (implosion) of the star and a rapid rebound (i.e., the star explodes). The explosion of a massive star is called a supernova. During a supernova explosion, nuclear particles are forced together at extremely high velocities. This process results in further fusion to form all the elements heavier than iron. This supernova remnant debris is then returned to space to eventually form recycled stars with a higher percentage of heavier elements. So, as a result of the explosion of stars, hydrogen, helium, and the heavier elements are returned to space to form clouds of dust and gases (nebula) that can then collapse to form additional stars. Our Sun is a second or third generation star that has abundant heavy elements (although it still is mostly hydrogen and helium) (Arny, 2006; Hester et al, 2007; Bennett et al, 2010).

Between about 7 to 10 billion years ago the Milky Way Galaxy had formed. By about 5 billion years ago a large nebular cloud of dust and gases in the Milky Way Galaxy began gravitationally contracting (perhaps due to a supernova explosion nearby that sent a pressure wave through space) and would ultimately form our Sun and Solar System. So, our Solar System started forming about 5 billion years ago. According to the Solar Nebula Theory, our Solar System formed as the result of the contraction of a diffuse, gaseous nebular cloud (consisting primarily of hydrogen and helium, but with some dust containing heavier elements). As the nebular cloud contracted it flattened into a disk-shaped structure with greater than 90% of the material going to the center (by gravity) to eventual form the Sun. Eddy currents in the accretion disk attracted matter that would eventually condense and coalesce to form the planets. Our Solar System was pretty much as we see it today by about 4.6 billion years ago (Arny, 2006; Hester et al, 2007; Bennett et al, 2010; Wicander and Monroe, 2013).

Any theory to explain the origin and evolution of the Solar System must explain the following observations about the Solar System:

- All planets revolve around the Sun on or near the plane of the ecliptic (exception: Mercury is about 7 degrees off the plane of the ecliptic).

• All the planets (except Venus and Uranus) and most of the moons in our solar system rotate on their axis in a counterclockwise direction (as viewed from above the north pole of the Sun and looking down on the Solar System). Venus rotates slowly to the right, possibly due to a catastrophic collision early in the evolution of the Solar system. Uranus is tilted on its side with the north pole pointing towards the Sun, again possibly due to a catastrophic collision early in the evolution of the Solar System.

• All the planets revolve around the Sun in a counterclockwise direction.

• All the planets (except Uranus) have a rotational axis that is near perpendicular to the plane of the ecliptic.

• The difference in chemical composition and characteristics of the inner Terrestrial planets as compared to the outer Jovian planets. The Terrestrial planets have high densities (about 4 to 5.5 g/cm^3), no or few moons, and are primarily composed of metals and silicate rocks. The Jovian planets have low densities (about 0.7 to 1.76 g/cm^3), many moons, and are primarily composed of gases (hydrogen, helium, methane, ammonia, and water vapor) and frozen compounds rich in hydrogen.

• Asteroids, comets, and interplanetary debris and dust must be explained.

• The relatively slow axial rotation of the Sun must be explained.

As stated previously, the most accepted theory for the formation of the Solar System (which explains all of the above observations) is the Solar Nebula Theory. The Solar Nebula Theory states the following:

• About 5 billion years ago a portion of a large nebular cloud of interstellar material in a spiral arm of the Milky Way Galaxy began to collapse due to gravity. The initiation of gravitational collapse may have been due to a pressure wave from a nearby supernova explosion (which pushed matter closer together in the cloud and caused a stronger gravitational force within the cloud - remember Newton's inverse square law of gravitation).

• As the interstellar material collapsed it began to rotate in a counterclockwise direction with greater than 90% of the material collapsing to the center of rotation. Due to the increasing rate of spin as the material collapsed (because of conservation of angular momentum), the material flattened into a disk-shaped structure (a solar nebula).

• As the solar nebula continued to concentrate material in the center of the disk, the embryonic Sun formed with a rotating, turbulent cloud of material around it.

• Gases and solids began to condense in localized turbulent eddies to form particles that then collided because of gravitational attraction to form planetesimals (from a few meters to a few tens of meters in size). Dust from the original interstellar material may have served as nuclei for condensation.

• Planetesimals collided with each other to accrete into protoplanets and eventually into the planets we know today. This process to form the planets probably took about a million years or so.

• The inner Terrestrial planets (Mercury, Venus, Earth, and Mars), being near the evolving Sun (which was growing very hot due to gravitational collapse), could only form from condensation of metals and silicates (refractory elements) due to the high temperature. So not much gas could accumulate in the inner Solar System.

• The outer Jovian planets (Jupiter, Saturn, Uranus, and Neptune), which formed far from the embryonic Sun, in the outer much cooler region of the Solar System were formed from condensation of metals, silicates, and gases (because all could condense at the cooler temperatures). However, because gases (primarily hydrogen and helium) were much more abundant in the original interstellar material than refractory elements, the outer planets are primarily gaseous giants, but do have rocky, metallic cores.

• Eventually the core of the embryonic Sun grew so hot and was under such high pressure that hydrogen was fused into helium. This provided an outward pressure from the core that halted the gravitational collapse. Thus, our modern Sun came into existence as a nuclear fusion "furnace". When our Sun "turned on", solar radiation and particles (protons, neutrons, and electrons) began streaming outward from the Sun into the Solar System. This is referred to as the solar wind. The solar wind blew any additional gas and dust out of the inner part of the Solar System (most being captured by the Jovian planets). Our Sun has been fusing hydrogen into helium for about 5 billion years and has enough hydrogen to continue this process for about another 5 billion years. When the hydrogen is used up in the core, our Sun will start to die, and as we learned earlier will become a white dwarf, and finally a black dwarf (burned-out chunk of carbon).

• Planetesimals between Mars and Jupiter were affected by the gravitational fields of the two planets (particularly Jupiter) and could not accrete into a planet, thus the asteroid belt was formed.

• Icy planetesimals that formed in the outer solar system were attracted to the large Jovian planets, if they missed collision with them they were gravitational flung into other planets or a spherical region (50,000 to 100,000 astronomical units from the Sun) surrounding the solar system known as the Oort Cloud. These icy planetesimals are incipient comets and perhaps are occasionally pulled into the inner solar system as another star comes close to our Sun (maybe we have a sister star that revolves around our Sun). Comets often hit planets and moons.

• A small amount of dust and gases still remain in interplanetary space that was never accreted into larger bodies. Collisions of asteroids with each other generate fragmentary material (meteoroids) that move out of the asteroid belt, sometimes impacting planets and moons as meteorite collisions.

• The slow rotation of the Sun on its axis is due to the magnetic field lines of the Sun interacting with ionized gases that were in the solar nebula. This is referred to as magnetic braking.

• The formation of our Solar System was completed by about 4.6 billion years ago.

So, the Solar Nebula Theory adequately explains all the observations (evidence) relative to the origin of our solar system (Arny, 2006; Hester et al, 2007; Bennett et al, 2010; Wicander and Monroe, 2013).

How do we know the age of formation of the Solar System? Radiometric dating of meteorites that impact Earth gives us the age of the Solar System. Most meteorites are thought to be composed of primeval material left over from the formation of the Solar System that has not evolved since formation. Much of the meteorite bombardment of planets and moons in the early history of our Solar System was from material that had not accreted into a planet or moon. Many meteorites that hit planets and moons today are the pieces of asteroids that have collided and fragmented (most asteroids have also not evolved much since the formation of the Solar System) (Arny, 2006; Hester et al, 2007; Bennett et al, 2010; Wicander and Monroe, 2013). (Note: Chunks of material moving through interplanetary space are called meteoroids; when a meteoroid enters an atmosphere of a planet and heats up it is called a meteor [a "shooting star"], and when a meteoroid or meteor hits a planet or moon it is called a meteorite.)

There are basically three categories of meteorites, these being the following:

• Stones - of all meteorites found on Earth, 93% of them are stones. Stones are of two types: Chondrites and Achondrites.

◦ Chondrites are composed of olivine and pyroxene (iron, magnesium silicate minerals). Olivine and pyroxene compose the chondrules, small spherical grains that make up chondritic meteorites. The chondrules represent the original material from rapid cooling and condensation to form the meteorites. Most stony meteorites are these ordinary chondrites; however, some chondrites contain a small percentage of organic compounds, including amino acids. The chondrites that contain organic compounds are called carbonaceous chondrites. The organic compounds found in these meteorites are not of biologic origin, but may represent precursors of biologic organisms.

◦ Achondrites contain no chondrules and have a composition much like terrestrial basalts. They are interpreted to be pieces of large differentiated (density layered due to melting and subsequent recrystallization) asteroids that have fragmented due to collision. These meteorites have been altered after their origin.

• Irons - make up about 6% of all meteorites found on Earth. They are composed of large crystals of iron and nickel alloys and most likely formed in the interior of large differentiated asteroids where relatively slow cooling took place.

• Stoney-irons - make up only about 1% of all meteorites that impact the Earth. They are composed of about equal amounts of iron/nickel and silicates and probably represent broken fragments from the zone between the metallic and rocky portions of large differentiated asteroids.

• Another class of meteorites that are found on Earth (but are relatively rare) are chunks of rock that (based on their composition) represent surface material from the Moon or Mars. These meteorites formed when Mars or the Moon were hit by meteorites with a glancing blow that shattered surface rock with such force as to knock some of the material into space, where it eventually hit Earth.

Radiometric dating of most meteorites (in particular, the chondrites) yields an age of 4.6 billion years, thus this is thought to represent the age of the Solar System, and also the age of Earth (Arny, 2006; Hester et al, 2007; Bennett et al, 2010; Wicander and Monroe, 2013).

Shortly after accretion from within the disk of the solar nebula (a rotating cloud and accretion disk of dust and gases surrounding the evolving Sun) about 4.6 billion years ago (4.6 Ga), Earth was a rapidly rotating, hot, barren, waterless planet, bombarded by comets and meteorites, with no continents, intense cosmic radiation and widespread volcanism (Levin, 1999; Wicander and Monroe, 2010). This concept for the relatively simultaneous formations of Earth and the other planets, the moons, asteroids and comets, and of course our star (the Sun), is called (as you now know) the Nebular Theory or Solar Nebula Theory (Arny, 2006; Hester et al, 2007; Bennett et al, 2010; Wicander and Monroe, 2013).

From 4.6 Ga to about 4.0 Ga (now referred to as the Hadean Eon,) Earth was barren, hot, and by 4.0 Ga had perhaps cooled enough to allow a thin dark igneous crust (similar in composition to the rock peridotite, which today makes up Earth's mantle) to form, but may have been completely molten soon after formation. There was lots of primordial heat from gravitational collapse of formation and radioactive decay (Earth was much more radioactive than today and additional short-lived radioisotopes were present) (Wicander and Monroe, 2013). Additional heat was added to Earth soon after its formation by the impact of a Mars-sized planetesimal that spewed out vaporized material that took up orbit around Earth to eventually condense to form our Moon (Arny, 2006; Hester et al, 2007; Bennett et al, 2010; Wicander and Monroe, 2013). The Homogeneous Accretion Theory to explain Earth's heterogeneous internal stratification (Earth's differentiation) states that even though Earth was internally homogeneous upon initial formation, it became molten or semi-molten soon after formation and the more dense elements, such as iron and nickel, sank to the center to eventually form a crystalline metallic inner iron-nickel core surrounded by a liquid iron-nickel outer core. Lighter elements, such as silicon, oxygen, magnesium, etc., floated upward and eventually formed a mantle and crust rich in silicate minerals (Wicander and Monroe, 2013).

By about 4.0 Ga (the beginning of the Archean Eon), the atmosphere consisted primarily of CO_2, H_2O vapor, N_2, H_2, CH_4, NH_3, SO_2, H_2S, and CO from very active volcanic outgassing, with hardly any molecular oxygen (O_2). Because of the lack of atmospheric oxygen, there certainly was no ozone (O_3) layer, so there was no filtering of ultraviolet radiation by the atmosphere (Wicander and Monroe, 2013). Earth rotated on its axis in about 10 hours. The moon was much closer and caused a huge tidal effect. Earth eventually cooled, and by 3.8 Ga water condensed from the atmosphere and began to accumulate in the ocean basins (Miller and Levine, 2008; Wicander and Monroe, 2013). Meteorite and comet bombardment slowed and was probably very slow by 3.9 billion years ago. By 3.8 billion years ago small areas of continental crust may have formed (Wicander and Monroe, 2013). Life had definitely emerged by about 3.5 billion years ago (probably by as early as 3.8 billion years ago) (Kardong, 2008; Miller and Levine, 2008; Tarbuck, Lutgens, and Tasa, 2013; Wicander and Monroe, 2013).

THE EVOLUTION OF THE ATMOSPHERE AND OCEANS

The very first atmosphere on Earth consisted of H and He, the two most abundant gases present in the solar nebula. These gases boiled off and were swept to the outer solar system by the solar wind once the Sun "turned on" (began fusing hydrogen into helium in its core to become a true star) (Arny, 2006; Hester et al, 2007; Bennett et al, 2010). Soon afterwards, a primeval secondary atmosphere formed via volcanic outgassing from Earth's interior. Modern volcanoes outgas water vapor (H_2O), carbon dioxide (CO_2), carbon monoxide (CO), and molecular nitrogen (N_2), with lesser amounts of sulfur dioxide (SO_2), hydrogen sulfide (H_2S), chlorine gas (Cl_2), and molecular hydrogen (H_2) (Bada and Lazcano, 2009; Tarbuck, Lutgens, and Tasa, 2013; Wicander and Monroe, 2013). Volcanoes probably emitted these gases during Hadean and Archean time also. These gases would have given a mildly reducing atmosphere (Bada and Lazcano, 2009). From chemical reactions in the atmosphere, the gases methane (CH_4) and ammonia (NH_3) probably formed during late Hadean and Archean time and were fairly abundant in the atmosphere (Tarbuck, Lutgens, and Tasa, 2013; Wicander and Monroe, 2013), although some scientists think that these two gases would have been nearly absent or present only near volcanoes or hydrothermal vents (Bada and Lazcano, 2009).

Notice that molecular oxygen (O_2) is not a product of volcanic outgassing today, and most likely was not during Hadean and Archean time either. Thus, we are confident that no (or very little) molecular oxygen (O_2) was present in the atmosphere at the beginning of Archean time. However, by late Archean (about 2.0 -2.5 Ga) photochemical dissociation of water vapor (by incoming ultraviolet radiation) had generated some molecular oxygen (this may have eventually supplied up to 2% of our present atmospheric oxygen; at 2% moleuclar oxygen, ozone forms to cut-off photochemical dissociation) (Wicander and Monroe, 2010). By the end of Archean time (about 2.5 Ga) maybe the atmosphere had about 1% or less of our present atmospheric oxygen (Wicander and Monroe, 2010). The other major source of atmospheric oxygen is photosynthesis, which may have started by about 3.5 billion years ago in early Archean by cyanobacteria (sometimes informally and erroneously called blue-green algae) (Levin, 1996; Ward, 2006; Wicander and Monroe, 2010; Tarbuck, Lutgens, and Tasa, 2013). Oxygenic photosynthesis is a process in which carbon dioxide and water combine into organic molecules and oxygen is released as a waste product, as the following equation illustrates:

$$6CO_2 + 6H_2O + sunlight + chlorophyll ==> C_6H_{12}O_6 + 6O_2.$$

Although cyanobacteria may have started photosynthesizing about 3.5 Ga, molecular oxygen (as a byproduct of the photosynthesis) did not start accumulating significantly in the atmosphere until about 1.8 Ga (in the Proterozoic Eon). At first, the molecular oxygen reacted with dissolved iron in the shallow oceans to form iron oxide compounds (such as hematite [Fe_2O_3] and magnetite [Fe_3O_4]) and with silica to form chert (microcrystalline quartz [SiO_2]) resulting in banded-iron formations (BIFs). The thickest accumulations of BIFs formed from about 2.1 Ga to 1.9 Ga in early Proterozoic time (mid-Precambrian) (Levin, 1996; Ward, 2006; Wicander and Monroe, 2010; Wicander and Monroe, 2013; Tarbuck, Lutgens, and Tasa, 2013; Canfield, 2014). BIFs around the world are the chief sources of iron ore today. After most of the iron from the oceans had been oxidized, molecular oxygen began to accumulate in ocean water and the atmosphere. The first extensive red beds (hematite cemented sandstones, siltstones, and shales) started forming on land (from river deposits) by about 1.8 Ga (early Proterozoic Eon), indicating a significant component of molecular oxygen in the atmosphere (Levin, 1996; Ward, 2006;

Wicander and Monroe, 2010; Wicander and Monroe, 2013; Tarbuck, Lutgens, and Tasa, 2013). Once free oxygen was present in significant quantity, an ozone layer formed that blocked most of the incoming ultraviolet radiation (Levin, 1996; Wicander and Monroe, 2010; Tarbuck, Lutgens, and Tasa, 2013).

THE ORIGIN OF LIFE ON EARTH

Despite the fact that creationists often speak of Darwin's ideas about the origin of life on Earth, as they imply he presented in *On the Origin of Species*, Darwin did not deal with the topic of the origin of life in *On the Origin of Species*. In his book, *On the Origin of Species*, Darwin only dealt with organisms that had already come into existence and that then evolved because they had variations that natural selection could work on. Darwin, in fact, did not deal with this topic at all in his various publications, except to indicate in a letter to botanist Joseph Hooker (one of Darwin's closest friends and colleagues – the other closest friend and colleague was geologist Charles Lyell) in February 1871 that life may have originated in a "warm little pond". In an essay entitled *On the Origin and Evolution of Life on a Frozen Earth* by John C. Priscu for the website called *Evolution of Evolution – 150 Years of Darwin's "On the Origin of Species"* sponsored by the National Science Foundation, Priscu (accessed 2014) presents the following quote from Darwin's letter to Hooker:

> It is often said that all the conditions for the first production of a living organism are present, which could ever have been present. But if (and Oh! what a big if!) we could conceive in some warm little pond, with all sorts of ammonia and phosphoric salts, light, heat, electricity, etc., present, that a protein compound was chemically formed ready to undergo still more complex changes, at the present day such matter would be instantly devoured or absorbed, which would not have been the case before living creatures were formed.

In the essay, Priscu (accessed 2014) was presenting an argument for the origin of life on Earth under very cold conditions in glaciers, as opposed to warm or hot conditions that have been proposed by many other researchers. He says the following relative to this idea:

> Freezing concentrates molecules, allowing for a high probability of self-organization into more complex molecules, while at the same time, reducing the potential to degrade the molecules. The mineral surfaces within ice veins,

and inclusions associated with impurities also provide a scaffolding to assist with the synthesis and assembly of complex molecules. Recently, experiments have shown that simple monomeric molecules concentrated in ice veins for almost 30 years can produce precursors for nucleic acid bases.

He goes on to further state that according to the "Snowball Earth Hypothesis" that about 600 million years ago during the Proterozoic Eon (Neoproterozoic Era) the Earth experienced an ice age so severe that Earth would have been completely frozen over (even the tropics) for 10 million years or more with ice thicknesses exceeding one kilometer. This ice would have served as a refuge for only the hardiest microbes and as he states in the following:

> The concentration of microbes within ice veins in this frozen environment would favor intense chemical and biological interactions between species, which would entice the development of symbiotic associations, and perhaps influence the development of more complex life-forms through evolutionary time. As such, these ice-bound habitats provided opportunities for microbial evolution, and the acquired biological innovations may have triggered the Cambrian explosion, or the seemingly sudden appearance of most major groups of complex organisms, which occurred immediately after this snowball Earth event.

Well, whether life originated in a warm shallow sea (or in a "warm little pond"), in deep sea hot springs along mid-oceanic ridges, or as discussed above in very cold glaciated environments, the origin of life is still hotly (no pun intended) debated. However, we are very confident that life started as quite simple organisms, basically as a cell membrane enclosing some cytoplasm (cell fluid) and DNA. So, how did life first come into existence? To originate by natural processes via abiogenesis, (i.e., from non-living material) life must have passed through prebiotic stages that mimicked life (like replication), but was not really what we would think of as alive because of not being able to sustain and maintain life functions. Well, what is life? Are viruses alive? We need to define what we mean by life.

There are two basic features that identify life, one being the ability to replicate and the other to maintain metabolism (Kardong, 2008). Thus, some biologists do not consider viruses to represent living organisms, but some do (Griffin, 2015, personal communication); however, they are right on the line between living and non-living. Viruses can only perform these two tasks when they invade a true living cell and hijack the replication and metabolic processes of the cell to make more copies of themselves and use the energy of the invaded cell for metabolic processes.

So, how did the first living cells appear? Many scientists today hypothesize that the first living cells may have appeared through four mains steps (Urry, et al, 2014). According to Urry, et al, 2014, the first step would have involved the abiotic synthesis of monomers or small organic molecules such as amino acids and nitrogenous bases (such as the nucleotides that are essential for the make up of RNA and DNA). The second step in this process would have been the polymerization of these monomers (small molecules) into larger macromolecules (polymers), such as proteins and nucleic acids (RNA and DNA) (Wicander and Monroe, 2013; Urry, et al, 2014). Next would be the separation of these molecules from the surroundings by a semipermeable membrane that would allow further chemical reactions within the membrane partitioned area separate from the surroundings (Wicander and Monroe, 2013; Urry, et al, 2014). Finally, the fourth step would be the origin of self-replicating molecules that would allow hereditary information to be passed from one generation to the next (Wicander and Monroe, 2013; Urry, et al, 2014). In addition to these four steps, an energy source would have needed to be present to supply energy for chemical reactions. This source may have been supplied by ultraviolet radiation hitting Earth (remember there was no ozone layer present to filter out the bulk of uv radiation prior to the origin of life) and/or lightening in the primitive atmosphere.

The Miller-Urey Experiment

In 1953 Stanley Miller, a graduate student of Harold Urey at the University of Chicago, published the results of his experiment to test the Oparin-Haldane Hypothesis (proposed independently in the 1920s by the Russian chemist A. I. Oparin and the British scientist J. B. S. Haldane) of an ancient Earth with a reducing atmosphere consisting primarily of hydrogen, methane, ammonia, and water vapor that produced the organic compounds necessary for the origin of life (Miller, 1953; Miller and Urey, 1959; Bada and Lazcano, 2009; Wicander and Monroe, 2013; Urry, et al, 2014). Basically, the Oparin-Haldane Hypothesis for the origin of life was similar to Darwin's "warm little pond" hypothesis that we talked about previously. Miller's experiment tested the Oparin-Haldane Hypothesis by setting up laboratory conditions that simulated a reducing atmosphere on an early Earth that resulted in the chemical formation of organic compounds in the ancient ocean that were the building blocks for the origin of life on Earth (Miller, 1953; Miller and Urey, 1959; Bada and Lazcano, 2009; Wicander and Monroe, 2013; Urry, et al, 2014). Miller's laboratory set-up heated water from a half-filled flask (representing the ancient ocean) that circulated water vapor to an upper flask that was filled with the gases thought to represent the ancient atmosphere. Electrodes in the upper flask simulated lightening to supply energy for chemical reactions. The gas then passed through a condenser resulting in liquid water, with compounds, that then passed downward to a trap at the bottom of the set-up that could be sampled for analysis. Miller's experiment yielded several amino acids found in organisms, along with other organic compounds (Miller, 1953; Miller and Urey, 1959; Bada and Lazcano, 2009; Wicander and Monroe, 2013; Urry, et al, 2014). Later analyses by Miller of vials produced by his 1953 experiment yielded several more amino acids in lower concentrations and other scientists using different gaseous mixtures to simulate the ancient Earth atmosphere also produced amino acids and other organic compounds. Some scientists recently have even undertaken Miller-Urey-type experiments with gaseous mixtures of carbon dioxide (CO_2), water vapor (H_2O), and molecular nitrogen (N_2) that were only mildly reducing or even neutral and have also produced amino acids and other organic molecules (Bada and Lazcano, 2009). In fact, more recent experiments have produced many of the 20 essential amino acids required by life on Earth (Wicander and Monroe, 2013; Urry, et al, 2014). According to Urry, et al, 2014, these types of experiments show that the synthesis of organic molecules (such as amino acids, the building blocks of proteins) is possible under various assumptions about the exact composition of Earth's ancient atmosphere.

Polymerization to Form Macromolecules

Monomers, such as amino acids and nitrogenous bases, are the basic building block units of life, but more complex macromolecules are needed for life, such as nucleic acids and proteins. How did this polymerization of monomers occur to yield more complex macromolecules? Some recent researchers have shown that the abiotic synthesis of RNA monomers can form spontaneously from simpler molecules, and further, solutions containing RNA nucleotides and amino acids dripped onto hot sand, clay, or rock have formed polymers without the aid of enzymes or ribosomes (Urry, et al, 2014).

Water usually causes depolymerization, however, according to Wicander and Monroe (2013), researchers have recently synthesized molecules known as proteinoids when heating dehydrated concentrated amino acids. Some of these proteinoids consist of more than 200 linked amino acids. These concentrated amino acids spontaneously polymerized to form proteinoids. Perhaps similar conditions for polymerization existed on early Earth, but the proteinoids needed to be protected by an outer membrane or they would break down (Monroe and Wicander, 2013).

Cell-like Structures and Protocells

Experiments show that proteinoids and other organic polymers (such as lipids) may spontaneously aggregate into microspheres (Monroe and Wicander, 2013) or vesicles (Urry, et al, 2014) when added to water. These microspheres or vesicles are bounded by cell-like membranes and grow and divide much as bacteria do (Wicander and Monroe, 2013; Urry, et al, 2014). According to Urry, et al, 2014, these abiotically produced vesicles can reproduce and increase in size ("grow"). These vesicles may also have selectively permeable membranes and can absorb montmorillonite (a type of clay) particles with RNA attached to them. They further say that recent experiments have shown that some vesicles can perform metabolic functions (reactions) using an external source of reagents. Wicander and Monroe (2013) and Miller and Levine (2008) discuss a similar argument, based on research by scientists, for cell-like structures called proteinoid microspheres. Proteinoids (or thermal proteins) consisting of more than 200 linked amino acids have been synthesized by researchers (Wicander and Monroe, 2013). These proteinoids have been shown in experiments to spontaneously form microspheres with a cell-like outer covering and that, somewhat similar to bacteria, grow and divide (Wicander and Monroe, 2013). However, these microspheres divide in a nonbiologic way (Wicander and Monroe, 2013).

The big question is, of course, how did DNA and RNA originate? As we know, the characteristics of organisms are stored in DNA that is then transcribed into RNA and ultimately translated into proteins (Miller and Levine, 2008). Also, as we know, during reproduction of cells (and organisms), DNA is duplicated and passed on to the next generation. But how did these processes come into existence? In modern organisms, DNA (or in some rare cases RNA) is necessary for reproduction, but these nucleic acids cannot replicate without enzymes (which are proteins). However, enzymes are not typically made without the involvement of nucleic acids (Wicander and Monroe, 2013). Yet in recent experiments simulating ancient conditions on Earth, researchers have shown that some small RNA molecules can self-replicate without the help of protein enzymes (Kardong, 2008; Miller and Levine, 2008; Wicander and Monroe, 2013). Thus, a RNA-based form of life may have preceded the DNA-directed protein synthesis form of life that exists today (Kardong, 2008; Miller and Levine, 2008, Bada and Lazcano, 2009). The actual steps involved in the origin of life still remain elusive, but this is an active area of research today.

Over time, however, life on Earth has emerged and evolved into an extremely diverse array of organisms that belong to three domains, the prokaryotes (possess the small, simple, non-nucleated prokaryotic cell), Bacteria and Archaea, and the eukaryotes (possess the larger, more complex, nucleated cell), Eukarya (or Eukaryota) (Figure 7-1). Starting as simple prokaryotic single-celled organisms sometime between about 4 billion years ago and about 3.5 billion years ago, that were no doubt heterotrophic (had to get energy resources from the exterior environment) and anaerobic, life has evolved diverse metabolic mechanisms and survival strategies, from simple one-celled organisms into complex multicelled organisms. By 3.5 billion years ago, prokaryotes similar to today's cyanobacteria had evolved into photoautotrophs that were capable of making their own food by photosynthesis and expelling molecular oxygen as a waste byproduct (Tarbuck, Lutgens, and Tasa, 2013; Wicander and Monroe, 2013). By about 2 billion years ago, prokaryotes evolved symbiotic relationships that ultimately led to the eukaryotic cell (Margulis, 1998; Kardong, 2008; Tarbuck, Lutgens, and Tasa, 2013; Wicander and Monroe, 2013). With the larger eukaryotic cells also came the ability to reproduce sexually and thus allow for more variation in the offspring, thus creating variants that natural selection could work on more effectively to evolve new species. Sometime around 1.2 billion years ago, some colonial eukaryotes were experimenting with multicellularity to form various types of algae and by approximately 600 million years ago primitive multicelled animals were present on Earth. With multicellularity, three basic and different lifestyles evolved to gain nutrients from the environment, plants photosynthesize to make their own food, fungi absorb nutrients from their environment, and animals ingest nutrients (usually other organisms) (Kardong, 2008).

THE PRECAMBRIAN EARTH AND LIFE HISTORY

All time prior to the beginning of the Cambrian Period (about 541 million years ago), of the Paleozoic Era, of the Phanerozoic Eon, is informally referred to as Precambrian time. Precambrian time lasted more than 4 billion years and represents about 88% of geologic time, with the Phanerozoic Eon representing about 12% of geologic time (Wicander and Monroe, 2013). The Precambrian episode of time is divided into three eons, the Hadean Eon, the Archean Eon, and the Proterozoic Eon (see Figure 3-1 and the Geological Society of America Geologic Time Scale at the following website: http://www.geosociety.org/science/timescale/).

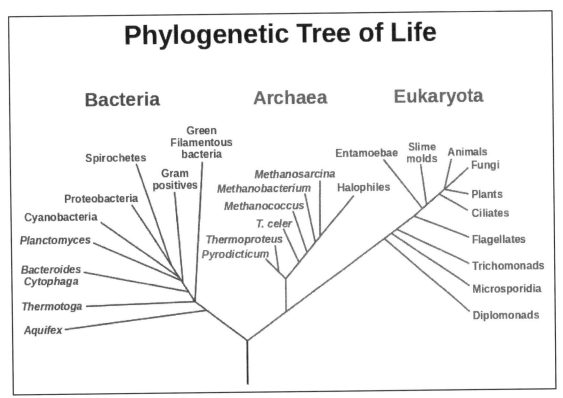

Figure 7-1. A phylogenetic tree of life on Earth based on ribosomal RNA sequence comparisons showing the three domains of life as proposed by Carl R. Woese, Otto Kandler, and Mark L. Wheelis (1990). The detailed relationships of the three domains as well as the position of the root of the tree are still being studied and debated. However, presently the evidence does support that Archaea is more closely related to Eukarya (Eukaryota) than to Bacteria (Woese, Kandler, and Wheelis, 1990). The above diagram has been slightly modified from that presented by Woese, Kandler, and Wheelis (1990) and also they spelled the domain on the right (with eukaryotic cells) as Eucarya. (The author of this vector version as uploaded to Wikimedia commons is Eric Kaba. The source of this figure is NASA Astrobiology Institute [found in an article] via Wikimedia commons. This image is in the public domain.)

Each continent has Precambrian Shield areas with exposed (outcropping) Precambrian rocks. The Shield areas and buried areas of basement rock (crystalline rocks, such as igneous or high rank metamorphic rocks below the younger sedimentary cover on the continents), called stable plateforms, make up the nuclei of continents (Tarbuck, Lutgens, and Tasa, 2013; Wicander and Monroe, 2013). In North America, the Canadian Shield covers much of Canada, part of Greenland, and parts of the northern U.S. The shield areas have subdued topography with numerous lakes and poor surface drainage. Precambrian rocks also exist in uplifts, mountains (see Figures 7-2 and 7-3), and below Phanerozoic sediments under most of the North American Continent (and other continents also). The Precambrian nuclei of the continents formed by early plate tectonic processes (Tarbuck, Lutgens, and Tasa, 2013; Wicander and Monroe, 2013).

Figure 7-2. Precambrian (Archean Eon) rocks exposed in the Teton Range of Grand Teton National Park in Wyoming. The rocks are gneiss, schist, and grantite that are 2.8 to 2.5 billion years old (2.8 to 2.5 Ga). This photo is of the Cathedral Group that towers over Jenny Lake. The highest peak in the photo is Grand Teton at 13,770 ft. (4,197 m). The Teton Range formed less than 10 million years ago (Love, Reed, and Pierce, 2007; Wicander and Monroe, 2013). (Photo by E. L. Crisp, August 2012.)

Figure 7-3. A cobble of the metamorphic rock gneiss (with a quartz vein cutting across it) found on the shore of Jenny Lake in Grand Teton National Park, Wyoming. This cobble was probably eroded from the Teton Mountain Peaks shown in Figure 7-2 and is Precambrian (probably Archean) in age. (Photo by E. L. Crisp, 2012.)

The Hadean Eon

The Hadean Eon is represented by very few rocks on Earth and represents the time episode from the origin of Earth 4.6 Ga to 4.0 Ga. Remember from Chapter 3, that the oldest rocks on Earth are about 4.0 Ga, so these rocks are right at the age of the boundary between the Hadean Eon and the Archean Eon. Some continental crust may have evolved by late Hadean time, or earlier. Sedimentary rocks in Australia contain detrital zircons ($ZrSiO_4$) dated at 4.4 billion years old (Wicander and Monroe, 2013), so source rocks at least that old must have existed during Hadean time. These rocks indicate that some kind of Hadean crust may have been present, but its distribution is unknown. The oceans and atmosphere had most likely started forming by late Hadean time, but we have very few rocks that yield any evidence to their state during Hadean time.

The Archean Eon

Most of our previous discussion of the evolution of the oceans and atmosphere described events that were taking place during the Archean Eon. As we stated previously, the Archean atmosphere was rich in water vapor (H_2O), carbon dioxide (CO_2), and molecular nitrogen (N_2); these gases reacting in this early atmosphere with ultraviolet radiation probably formed ammonia (NH_3) and methane (CH_4). This early atmosphere persisted throughout the Archean and was probably chemically reducing (rather than oxidizing), because oxygen was not a significant component of the atmosphere during Archean time (Miller, 1953; Miller and Levine, 2008; Wicander and Monroe, 2013). Some of the evidence for this is the abundance of reduced minerals, such as pyrite (FeS_2), in Archean detrital sedimentary rocks (Wicander and Monroe, 2013).

Remember that the gases present in Earth's ancient atmosphere primarily came from volcanic outgassing from Earth's interior. Oxygen is not a product of volcanic outgassing. Two processes account for introducing free oxygen into the atmosphere, both of which began during the early Archean time. Probably the first process that generated oxygen in Earth's early atmosphere was photochemical dissociation of water vapor (H_2O) due to bombardment by ultraviolet radiation in the upper atmosphere. This type of radiation breaks-up water molecules and releases their oxygen and hydrogen, ultimately resulting in the addition of molecular oxygen (O_2) and hydrogen (H_2) into the atmosphere. Most of the lighter hydrogen eventually escaped Earth's gravitational pull or reacted with other molecules in the atmosphere, but the molecular oxygen accumulated in the atmosphere. During Archean time (and also today), the activities of organisms that practiced photosynthesis were more important in generating molecular oxygen (O_2). As we previously discussed, cyanobacteria had started the process of photosynthesis by at least 3.5 Ga, releasing molecular oxygen as a waste byproduct of that photosynthesis (Schopf, 2011).

So, life emerged in early Archean time in the interval from about 4.0 Ga to 3.5 Ga, probably about 3.8 Ga (Kardong, 2008; Wicander and Monroe, 2013). In fact, carbon isotopic data (ratios of carbon-13 to carbon-12, living organisms prefer carbon-12) from 3.8 Ga metasedimentary rocks (sedimentary rocks that have been mildly metamorphosed) from Isua, Greenland indicate that these rocks contain organic carbon, thus indicating that life existed on Earth by 3.8 Ga. The earliest organisms (which emerged prior to 3.5 billion years ago) were probably anaerobic, heterotrophic prokaryotes in the form of bacteria and/or archaea. However, the oldest known fossils are thought to be 3.5 Ga cyanobacteria from western Australia (Schopf, 2011). These are photosynthesizing (thus photoautotrophic), anaerobic (because there was little, if any, molecular oxygen in the atmosphere or oceans at this time), prokaryotic bacteria. Cyanobacteria form layered mound-like (or reef-like) structures called stromatolites. The oldest definite stromatolites are in 3 Ga rocks of South Africa (Wicander and Monroe, 2013). However, probable stromatolites are present in the 3.5 Ga rocks of the Apex Chert of the Warrawoona Group from Western Australia (Schopf, 2011; Wicander and Monroe, 2013). In addition, actual remains of what appear to be filamentous cyanobacteria (or some form of photosynthesizing filamentous bacteria or filamentous cyanobacterium-like organisms) have been found in stromatolitic layers of the 3.5 Ga Apex Chert in western Australia (see Lane, 2009, p. 74-77; see Schopf, 2011, p. 94-95; see Wicander and Monroe, 2013, p. 164-165). Although some have challenged whether these remains in western Australia are really cyanobacteria or photosynthesizing filamentous bacteria, or truly fossils at all (Lane, 2009), the discoverer of these supposed fossils, William Schopf (professor of paleobiology at the University of California, Los Angeles), still argues forcefully that they really were some form of filamentous bacteria that were capable of oxygenated photosynthesis (they broke down water to yield molecular oxygen as a waste byproduct) (Schopf, 2011). The following statement by Lane (2009, p. 76) illustrates this point: "All in all, said Schopf, cyanobacteria, or something that looked very much like them, had already evolved on the earth by 3,500 million years ago, just a few hundred million years after the end of the great asteroid bombardment that marked the earliest years of our planet, so soon after the formation of the solar system itself."

Many Precambrian sedimentary and metasedimentary rocks have abundant stromatolites. In fact, stromatolites are the most abundant Precambrian fossils; however, they did not become abundant until about 2 Ga during the Proterozoic Eon (Tarbuck, Lutgens, and Tasa, 2013; Wicander and Monroe, 2013). In the early 1900s, paleontologist Charles Doolittle Walcott (1850-1927) (became director of the U.S. Geological Survey in 1894, elected to the National Academy of Sciences in 1896, became secretary of the Smithsonian Institute in 1907, and served as president of the American Association for the Advancement of Science in 1923) described structures, now called stromatolites, in chert from the early Proterozoic Gunflint Iron Formation of Ontario, Canada that he proposed represented reefs constructed by algae (Wicander and Monroe, 2013; Wikipedia, Charles Doolittle Walcott, accessed 11/04/2014). However, it was not until 1954 that paleontologists demonstrated that stromatolites are of organic origin (Wicander and Monroe, 2013). Today, we know that stromatolites form in hypersaline marine lagoons as irregular layered structures, isolated reef-like columns or bioherms, and columns linked by irregular layered structures. The stromatolites are the result of mats of cyanobacteria that trap sediment grains and crystals on their sticky (due to mucilaginous secretions from the cyanobacteria) surfaces and build-up millimeter-sized layers as the cyanobacteria continue to grow upward to receive light for photosynthesis (Schopf, 2011; Tarbuck, Lutgens, and Tasa, 2013; Wicander and Monroe, 2013; Canfield,

2014).   Modern carbonate cyanobacterial stromatolites form in extreme environments (such as hypersaline lagoons) that lack grazing and burrowing metazoans (such as snails) that would eat and/or disrupt the microbial mats (Schopf, 2011; Tarbuck, Lutgens, and Tasa, 2013; Wicander and Monroe, 2013; Canfield, 2014). This is why they are not abundant in Phanerozoic rocks, but were abundant during most of the Precambrian, even in shallow normal marine carbonate environments, prior to the evolution of grazing and burrowing metazoans (Schopf, 2011; Canfield, 2014). Modern and ancient cyanobacteria form mats that grow in intertidal and subtidal carbonate environments (but extreme salinities for the modern forms) relatively free of detrital sedimentation. The cyanobacteria secrete a mucus-like material that forms a sticky surface. As carbon dioxide is used by the cyanobacteria in photosynthesis and removed from the water, the water becomes supersaturated with calcium carbonate. This results in the precipitation of calcium carbonate crystals that fall upon the sticky surface of the cyanobacterial mat and become trapped. Eventually this blocks the sunlight for the cyanobacteria, thus they grow through the sediment layer and form another layer of the mat above the sediment layer; this process is repeated over and over to form the millimeter-scale laminated carbonate (usually $CaCO_3$ as aragonite or calcite) structure of stromatolites (Schopf, 2011; Canfield, 2014). Typically, the recrystallization of carbonate crystals in stromatolites destroys most of the structure of the cyanobacteria cells (or other stromatolite-forming microorganisms), leaving only a thin film of amorphous carbonaceous matter (Schopf, 2011). However, if, during early diagenesis (change of sediment after deposition), silica (SiO_2 as microcrystalline quartz, chert) replaces the original carbonate, intact cells of cyanobacteria (and/or other microbes) may be preserved (Schopf, 2011).

The Proterozoic Eon

As we have already discussed, the Archean atmosphere contained very little free oxygen, thus the atmosphere was not strongly oxidizing as it is today. The amount of molecular oxygen (O_2) present at the beginning of the Proterozoic Eon (2.5 Ga to 541 million years ago [541 Ma]) was probably no more than 1% of the modern value (molecular oxygen is 21% of the gases making up our modern atmosphere). Even by the end of the Proterozoic, molecular oxygen may not have exceeded 10% of the present level (Wicander and Monroe, 2013).

Cyanobacteria and the structures that they formed, stromatolites, as we already know, were present during the Archean Eon, but stromatolites did not become common until about 2.3 billion years ago (Wicander and Monroe, 2013). These photosynthesizing cyanobacteria and to a lesser extent photochemical dissociation of water by ultraviolet radiation, resulted in the addition of molecular oxygen to the evolving atmosphere (Ward, 2006; Wicander and Monroe, 2013). Geologic evidence indicates that prior to about 2.3 Ga there was no, or very little, free oxygen in the atmosphere and little oxygen dissolved in seawater (Ward, 2006). But molecular oxygen started increasing dramatically by 2.3 billion years ago, no doubt due to the increased activity of photosynthesizing cyanobacteria (Ward, 2006). This increase in oxygen in the shallow oceans and the atmosphere at about 2.3 to 2.4 Ga is referred to as the Great Oxidation Event (GOE) (Holland, 2006; Canfield, 2014). This increase in atmospheric oxygen eventually led to the formation of an ozone layer in the stratosphere that filtered out most of the harmful (to life) ultraviolet radiation (Ward, 2006).

Remember that during the latter part of the Archean and early Proterozoic, molecular oxygen released by photosynthesizing cyanobacteria was oxidized to form banded iron formations (BIFs). As discussed earlier, there was a major episode of BIF deposition from about 2.1 Ga to 1.9 Ga (Canfield, 2014), but 92% of all BIFs formed in the interval from about 2.5 Ga to 2.0 Ga (Wicander and Monroe, 2013). This certainly appears to be related to the increase in cyanobacterial oxygenic photosynthesis at about 2.3 Ga. The increased deposition of BIFs continued until about 1.9 Ga, and then ceased. These BIFs form the bulk of our modern iron ore (Wicander and Monroe, 2013). Extensive continental red beds started forming about 1.8 Ga, although some red beds are known as far back as about 2.2 Ga (Ausich and Lane, 1999; Canfield, 2014), indicating a significant component of molecular oxygen had started moving from the oceans into the atmosphere. This suggests that most of the reduced iron in the oceans had been oxidized (Holland, 2006; Wicander and Monroe, 2013).

Fossils of the Archean consist only of several types of bacteria and archaea and early Proterozic fossils are restricted to bacteria and stromatolites. Although these organisms were most likely abundant during the latter part of the Archean and early Proterozoic, very little evolutionary diversification had taken place since the origin of the first living organisms. All organisms of the Archean and early Proterozoic consisted of one-celled prokaryotes with relatively low diversity. These organisms reproduce asexually by binary fission (replication of DNA and splitting of the cell into two daughter cells), thus evolution proceeds at a very slow pace due to the limited amount of variation in these prokaryotic populations. However, the increase in oxygen in the shallow oceans that began with the GOE made possible the evolution of a more complex form of life with larger cells that undergo sexual reproduction, the Eukarya, consisting of eukaryotic cells. Because eukaryotic cells undergo sexual reproduction, the reshuffling of genes allows for more allelic diversity from generation to generation. Eukaryotic cells are much larger than prokaryotic cells and require molecular oxygen to metabolize, thus they could not have evolved before free oxygen was available in seawater and the atmosphere.

Let's compare prokaryotic cells with eukaryotic cells. Prokaryotic cells are very small, 1 to 10 micrometers (µm), whereas typical eukaryotic cells range from 10 µm to 100 µm. Prokaryotic cells have a single circular loop of DNA in the cytoplasm, but eukaryotic cells have their DNA arranged in linear chromosomes that are surrounded by a membrane-bounded nucleus. Prokaryotes do not have membrane-bounded organelles within the cells, as opposed to eukaryotes that do, such as mitochondria (where cellular metabolism takes place) in all eukaryote cells and chloroplasts (plastids that contain chlorophyll and allow photosynthesis to take place) in the cells of some eukaryote cells (various types of algae and in plants). As stated previously, prokaryotic cells primarily reproduce asexually by binary fission, however, a type of sexual conjugation does allow horizontal transfer of genetic material (typically a circular plasmid containing DNA) in most bacteria (Urry, et al,

2014). However, eukaryotic cells primarily reproduce sexually via mitosis and meiosis.

Eukaryotic organisms are all aerobic, so they depend on free oxygen to perform metabolism. Thus, they could not have evolved before some free molecular oxygen was present in the seas and atmosphere. Fossils of eukaryotic cells have been found in rocks dated to 1.8 billion years old (Erwin and Valentine, 2013; Urry, et al, 2014). However, it appears likely that eukaryotic cells had evolved by about 2 Ga (Sagan, 1967; Margulis, 1998; Kardong, 2008; Lane, 2009; Erwin and Valentine, 2013; Wicander and Monroe, 2013; Urry, et al, 2014).

Remember, molecular oxygen started increasing in the atmosphere, and thus free oxygen was also dissolved in shallow seawater, beginning with the GOE about 2.3 Ga. By 1.8 Ga, extensive red beds were abundant, thus significant free molecular oxygen was in the atmosphere and dissolved in shallow seawater. Oxygen is a strong reactant, and thus would be toxic (destructive) to living cells. So, prokaryotes living after the GOE would have to have evolved mechanisms to deal with the toxicity of molecular oxygen. Some forms of bacteria and archaea went extinct as molecular oxygen increased or they migrated into anaerobic environments, such as in deep anoxic ocean waters or anoxic bottom sediments of the oceans or bottom waters of deep lakes, and in swamps and bogs. Certainly, any mutations that would allow some bacteria to use oxygen to break down food (aerobic metabolism) would be selected for. Also, as oxygen increased in the atmosphere, abiogenic organic matter would have been quickly oxidized (Sagan, 1967), thus making all life at the time dependent, directly or indirectly, on cellular photosynthesis (Sagan, 1967). Thus, anaerobic heterotrophic prokaryotes, in order to live and grow, had to consume organic matter produced by photosynthesizing or chemoautotrophic prokaryotes (Sagan, 1967). Therefore, the first step in the evolution of eukaryotes from prokaryotes was the ingestion of an aerobic heterotrophic prokaryotic microbe into the cytoplasm of an anaerobic heterotrophic prokaryotic microbe (Sagan, 1967). This relationship would have been mutually beneficial as an endosymbiotic relationship, with the host heterotrophic prokaryote receiving energy resulting from the metabolism of organic material by the ingested aerobic prokaryote, and the aerobic prokaryote would be protected and supplied with organic material (also ingested from outside the host cell) to metabolize for energy.   Eventually, this mutualistic endosymbiosis became obligate, with the engulfed aerobic prokaryote becoming a mitochondrion (Sagan, 1967).   This endosymbiotic relationship, associated with other changes, such as the infolding of the plasma membrane to form the endoplasmic reticulum and the nuclear membrane, resulted in the first aerobic heterotrophic eukaryote (Urry, et al, 2014).   Some of these aerobic heterotrophs formed similar symbiotic relationships with photosynthesizing autotrophic bacteria (cyanobacteria) to form ancestral photosynthetic eukaryotes (Sagan, 1967; Margulis, 1998; Urry, et al, 2014).

The concept outlined above was first proposed in 1967 by Lynn Sagan (later known as Lynn Margulis) in an article in the March issue of the *Journal of Theoretical Biology* entitled "On the Origin of Mitosing Cells." Later, she would refer to this hypothesis as the Serial Endosymbiosis Theory (SET) (see Margulis, 1998). Others today refer to this explanation for the evolution of the first eukaryotic cells as the Serial Endosymbiosis Hypothesis (Urry, et al, 2014).   But some refer to this explanation simply as the Endosymbiotic Theory for the origin of eukayotes (Levin, 1999; Wicander and Monroe, 2013).   At any rate, there is a significant amount of evidence in support of this explanation for the origin of Eukarya and the origin of mitochondria and chloroplasts (Urry, et al, 2014).   Urry, et al, 2014 state that studies of DNA sequence data of eukaryotic cells suggest that these cells have some of their genetic and cellular characteristics derived from archaea, and other characteristics are derived from bacteria.   They also list the following characteristics as evidence supporting the endosymbiotic origin of mitochondria and plastids (in particular chloroplasts) (here I am paraphrasing):

- Surrounding membranes of both mitochondria and chloroplasts have similar enzymes and transport systems that are homologous to the plasma membranes of living prokaryotes.
- These organelles replicate by dividing in a similar way as certain prokaryotes (in particular certain bacteria).

- Mitochondria and chloroplasts (as well as other plastids) contain one circular molecule of DNA that very much resembles bacterial DNA.

- These organelles also contain ribosomes and the cellular machinery to transcribe their DNA into mRNA and to translate that message into proteins. This would be expected if these organelles were derived from free-living prokaryotes.

- The ribosomes of mitochondria and chloroplasts (and other plastids) are more similar to ribosomes of prokaryotes than to eukaryotic cytoplasmic ribosomes.

Molecular studies of the comparisons of the entire genomes of mitochondrial DNA of animals, plants, fungi, and protists to each other and to the DNA sequences of major groups of bacteria and archaea indicate that mitochondria arose from a specific proterobacterium, and thus infer that mitochondria descended from a single common ancestor (Urry, et al, 2014). Similar studies indicate that plastids (including chloroplasts) arose only once from an ingested cyanobacteium (Urry, et al, 2014). Genomic studies also indicate that the host cell came from the archaea domain (Lane, 2009; Urry, et al, 2014). Of course, all this suggests that the ancestral eukaryote cell arose only once and the progeny of this cell gave rise to algae, fungi, plants, and animals (Lane, 2009; Urry, et al, 2014), with the last common ancestor (LCA) of all modern Eukarya arising about 2 Ga (Kardong, 2008; Lane, 2009).

Although the scenario that is presented above is reasonable based on the evidence that we have, there are still unanswered questions that remain. Was the original host cell for the eukaryotic cell really of archaean origin, or bacterial origin? How did the nuclear membrane really form, and why? What allowed the host cell to engulf other prokaryotes; was the host cell a phagocyte? Was the evolution of the eukaryotic cell based on vertical and gradual Darwinian evolution, or was it primarily due to lateral transfer of DNA between different prokaryotic cells and with more rapid chimeric evolution. These are very complex questions and these and other questions about the origin of the eukaryotic cell are an active area of modern research (see the discussion in Lane, 2009, p. 88-117, about the complex cell).

So, by about 2 Ga the eukaryotic cell is on the scene and starts diversifying and adaptively radiating into various niches as potentially sexually reproducing single-celled forms. The appearance of eukaryotic cells is a major milestone in the evolution of life on Earth, almost (but perhaps not quite) as important as the evolution of oxygenic photosynthesis by cyanobacteria during the Archean. However, we certainly would not be here if eukaryotic cells had not evolved.

By the beginning of the late Mesoproterozoic Era, of the Proterozoic Eon, about 1.2 Ga (see Geological Society of America geologic time scale at http://www.geosociety.org/science/timescale/), multicelled algae had evolved. In fact, fossils of multicellular algae date to about 1.2 Ga (Erwin and Valentine, 2013). Multicelled animals had evolved by at least about 600 Ma. Choanoflagellates, solitary and colonial unicellular eukaryotes that feed on bacteria, probably evolved into sponges, the first animals, prior to 600 Ma (Erwin and Valentine, 2013). Fossil impressions of other soft-bodied animals have been found in rocks of the Ediacara Hills of South Australia and at other places around the world (Levin, 1999; Erwin and Valentine, 2013; Wicander and Monroe, 2013; Fortey, 2012). These soft-bodied fossils are now referred to as the Ediacaran Fauna and represent complex precursors to the animal phyla of the great Cambrian diversification, some may even be related to later phyla, such as Cnidaria, Mollusca, Annelida, and Arthropoda (Levin, 1999; Wicander and Monroe, 2013) that follow in the Cambrian Period of the Phanerozoic Eon. The last period (from about 635 to 541 Ma) of the Neoproterozoic Era is now referred to as the Ediacaran Period. Based on molecular clock studies, Erwin and Valentine (2013) in their book, *The Cambrain Explosion: The Construction of Animal Biodiversity*, suggest that the metazoan (multicelled animal) stem ancestor lived earlier than 780 Ma (Note: the beginning of the Cambrian Period of the Phanerozoic Eon begins about 541 Ma) in the Cryogenian Period (from 850 to 635 Ma) of the Neoproterozoic Era. They also state that the stem bilaterian (bilaterally symmetrical) animal lived between 700-670 Ma and the last common ancestor of protostome and deuterostome animals lived approximately 670 Ma.

During the last 500 to 600 million years of the Precambrian, plate tectonics, with a style similar to modern plate tectonics, was shaping the isolated continental masses into two major supercontinents, one after the other. The first, that geologist have significant evidence to recognize with some certainty, was Rodinia, which was assembled between 1.3 to 1.0 Ga (Wicander and Monroe, 2013). Rodinia started fragmenting about 750 Ma, with the fragmented pieces of continental lithosphere reassembled to form the supercontinent Pannotia by 650 Ma (Wicander and Monroe, 2013). By 550 Ma, near the end of the Neoproterozoic Era, Pannotia fragmented to give rise to the continental configuration that existed at the beginning of the Phanerozoic Eon.

During the Proterozoic Eon, large changes were taking place in Earth's atmosphere. As we have discussed, oxygen was increasing in the atmosphere as the Proterozoic progressed past the Great Oxidation Event (GOE) at about 2.3 Ga. As we also discussed, Earth's ancient atmosphere during the Hadean and Archean consisted of very little oxygen in the atmosphere, but carbon dioxide was one of the most abundant gases in the atmosphere. It is quite evident from geochemical studies of Earth rocks that as oxygen increased in the atmosphere during the Proterozoic; carbon dioxide was decreasing (Wicander and Monroe, 2013; Canfield, 2014). This also was the case during the Phanerozoic Eon. Why is this?

Certainly, oxygenic photosynthesis removes carbon dioxide from the atmosphere in the form of organic carbon, which becomes locked in molecules like carbohydrates and sugars (CH_2O, for example $C_6H_{12}O_6$), and adds molecular oxygen (O_2) to the atmosphere. So, as oxygenic photosynthesis increased in the oceans and atmosphere, more organic carbon formed to remove carbon dioxide from the atmosphere. Thus, organic carbon production liberated oxygen to the atmosphere at the expense of carbon dioxide which was fixed in the organic carbon that resulted from cyanobacterial photosynthesis, plant growth in forests, swamps, and bogs and phytoplankton growth in oceans with the subsequent organic material buried in ocean sediments. Eventually much of this carbon was locked up in fossil fuels, such as coal, crude oil, and natural gas. As we burn these fossil fuels for energy, we liberate carbon dioxide back into the atmosphere (of course, carbon dioxide is a greenhouse gas that absorbs energy from the Sun to make the atmosphere hold heat and thus induces global warming). In addition, large amounts of carbon dioxide are dissolved in ocean waters (and have been during past times); the following equations will illustrate how this takes place (assuming reaction between oceanic surface waters and the atmosphere and also rainwater reacting with atmospheric carbon dioxide and then the resulting water draining into the oceans):

$$H_2O + CO_2 \longleftrightarrow H_2CO_3$$

$$H_2CO_3 \longleftrightarrow H^+ + HCO_3^-$$

$$HCO_3^- \longleftrightarrow H^+ + CO_3^{-2}.$$

Today (and most likely during past time periods), in warm shallow marine environments on continental shelves and epicontinental seas (eperic seas), the bicarbonate ion (HCO_3^-) is (and was) the most abundant. However, in deep, cold oceanic depths, CO_2 is (and was) the most abundant chemical species. Thus, much of the carbon dioxide in the atmosphere becomes locked-up in the oceans by the above process.

In addition, both organic activity and inorganic chemical activity may result in the precipitation of calcium carbonate oceanic sediments, at least above the carbonate compensation depth (CCD) of between 4300 to 5000 meters in modern oceans, depending on temperature and pressure. In fact, today, warm shallow tropical and subtropical marine waters around the world are supersaturated with respect to calcium carbonate (Berner, 1971) and thus are precipitating calcium carbonate as calcite or aragonite, either by organisms (primarily) that make their skeletons of calcium carbonate (such as algae and other phytoplankton, corals, molluscs, etc.) or by inorganic precipitation (such as oolite formation in shallow, agitated marine waters – for example, surrounding the Bahamian Islands). This situation has been the case on Earth for billions of years, starting significantly in the Proterozoic Eon about 2.3 Ga (Levin, 1999). The following chemical equation illustrates this process:

$$Ca^{+2} + 2HCO_3^- \longleftrightarrow CaCO_3 + H_2O + CO_2.$$

So, calcium carbonate sediments that accumulate above the CCD will eventually be lithified into limestone or dolostone (if magnesium ions are available to react with the limestone). Limestones and dolostones are referred to as carbonates (see Chapter 2). Carbonates have been forming on Earth for billions of years and thus have served as sinks for much of the carbon that was once in the atmosphere as carbon dioxide. Thus, much of the original carbon dioxide has been removed from the atmosphere by this process.

THE PHANEROZOIC EON

The Paleozoic Era

The Cambrian "Explosion" of Animals

Charles Darwin, in *On the Origin of Species*, was greatly concerned about the then perceived absence of fossils in Precambrian rocks (then referred to as Azoic [meaning without life] rocks) below the Cambrian System of rocks (in Darwin's time, referred to as Silurian rocks) and also the sudden appearance of the major groups of animals in the lower strata of the Cambrian (or Silurian in Darwin's time). Today we often refer to this sudden appearance of the major animal phyla as recorded in the lower and middle portion of the Cambrian System of rocks as representing the Cambrian "explosion" of animal life. Now we know that this rapid diversification of animal phyla to produce the major phyla of today occurred over a span of about 20 million years (or less), from about 541 Ma to 521 Ma (primarily during the Terrenuvian Epoch of the Cambrian Period) based on the current fossil record (Lee, Soubrier, and Edgecombe, 2013). Still, this is a relatively short time span for the rapid morphologic disparity of animal phyla observed in the so called "explosion" fossils of the Lower to Middle Cambrian. We will return to this discussion, but for now let's return to Darwin's concern, often referred to as Darwin's dilemma (Conway Morris, 2006). During Darwin's time the fossil record of the Precambrian, as we know it today, was absent. So, based on his theory of slow, gradual evolution via natural selection, how was he to reconcile the sudden appearance of advanced animal life with his evolutionary theory? The following statement from Darwin (1859, p. 306-308) illustrates his dilemma:

On the sudden appearance of groups of Allied Species in the lowest known fossiliferous strata. — There is another and allied difficulty, which is much graver. I allude to the manner in which numbers of species of the same group, suddenly appear in the lowest known fossiliferous rocks. Most of the arguments which have convinced me that all the existing species of the same group have descended from one progenitor, apply with nearly equal force to the earliest known species. For instance, I cannot doubt that all the Silurian trilobites have descended from some one crustacean, which must have lived long before the Silurian age, and which probably differed greatly from any known animal. Some of the most ancient Silurian animals, as the Nautilus, Lingula, &c., do not differ much from living species; and it cannot on my theory be supposed, that these old species were the progenitors of all the species of the orders to which they belong, for they do not present characters in any degree intermediate between them. If, moreover, they had been the progenitors of these orders, they would almost certainly have been long ago supplanted and exterminated by their numerous and improved descendants.

Consequently, if my theory be true, it is indisputable that before the lowest Silurian stratum was deposited, long periods elapsed, as long as, or probably far longer than, the whole interval from the Silurian age to the present day; and that during these vast, yet quite unknown, periods of time, the world swarmed with living creatures.

To the question why we do not find records of these vast primordial periods, I can give no satisfactory answer. Several of the most eminent geologists, with Sir R. Murchison at their head, are convinced that we see in the organic remains of the lowest Silurian stratum the dawn of life on this planet. Other highly competent judges, as Lyell and the late E. Forbes, dispute this conclusion. We should not forget that only a small portion of the world is known with accuracy. M. Barrande has lately added another and lower stage to the Silurian system, abounding with new and peculiar species. Traces of life have been detected in the Longmynd beds beneath Barrande's so-called primordial zone. The presence of phosphatic nodules and bituminous matter in some of the lowest azoic rocks, probably indicates the former existence of life at these periods. But the difficulty of understanding the absence of vast piles of fossiliferous strata, which on my theory no doubt were somewhere accumulated before the Silurian epoch, is very great. If these most ancient beds had been wholly worn away by denudation, or obliterated by metamorphic action, we ought to find only small remnants of the formations next succeeding them in age, and these ought to be very generally in a metamorphosed condition. But the descriptions which we now possess of the Silurian deposits over immense territories in Russia and in North America, do not support the view, that the older a formation is, the more it has suffered the extremity of denudation and metamorphism.

The case at present must remain inexplicable; and may be truly urged as a valid argument against the views here entertained.

We will return to Darwin's concern shortly. But today we know that by the early Cambrian Period of the Paleozoic Era of the Phanerozoic Eon, which began about 541 million years ago, several types of invertebrate organisms had evolved hard mineralized exoskeletons; thus, fossilization of animals became much more abundant and widespread near the beginning of the Phanerozoic. However, the actual base of the Cambrian System of rocks is defined as the first occurrence of the trace fossil (ichnofossil) *Treptichnus pedum* (Peng, Babcock, and Cooper, 2012; Erwin and Valentine, 2013) (also referred to as *Trichophycus pedum*, *Phycodes pedum*, or *Manykodes pedum*) [Peng, Babcock, and Cooper, 2012]). These trace fossils indicate both horizontal and vertical burrowing (for nutrients), whereas Ediacaran trace fossils show only horizontal grazing because the sea floor was covered with microbial (primarily cyanobacterial) mats (Peng, Babcock, and Cooper, 2012; Erwin and Valentine, 2013) (also see Martin, 2012). Thus, *Treptichnus pedum* indicates a significant increase in bioturbation of the marine substrate at the beginning of Cambrian time as compared to the end of Ediacaran time.

Fossils of animals with hard mineralized exoskeletons are first encountered near the base of the Cambrian System and are known as the small shelly fauna or small shelly fossils (SSFs) (Gould, 1989; Levin, 1999; Budd, 2003; Peng, Babcock, and Cooper, 2012; Erwin and Valentine, 2013; Wicander and Monroe, 2013). SSFs are usually very small, typically less than 1 mm to 2 mm (Erwin and Valentine, 2013). However, although some SSFs represent complete shells of forerunners to such shelled animals as brachiopods and molluscs, and others of unknown affinities, some represent parts (referred to as sclerites) of larger animals (Erwin and Valentine, 2013). But most all of the SSFs have been preserved by a secondary phosphatization process (Porter, 2004), even though some of the fossils may have originally had calcium carbonate or silica compositions. The SSFs, because of the phosphatization process, often have very fine detail preserved and this process is considered to be a form of exceptional preservation (Budd, 2003; Porter, 2004). The SSFs are present in Cambrian rocks that date from about 535 Ma to about 515 Ma (see Erwin and Valentine, 2013, Figure 6.6, p. 155) (also see Peng, Babcock, and Cooper, 2012, Figure 19.15, p. 476).

Fossils with more typical mineralized shells, such as trilobites with a dorsal exoskeleton and hypostome (a plate surrounding the mouth on the ventral surface), composed primarily of calcium carbonate, originally as low magnesium calcite (Wilmot and Fallick, 1989), do not appear in the Cambrian rock record until about 521 Ma (Erwin and Valentine, 2013). Once present, trilobites adaptively radiate during the remainder of Cambrian time to form one of the most abundant and diverse groups of the Cambrian Period, with over 600 genera (Levin, 1999; Wicander and Monroe, 2013). Trilobites belong to the phylum Arthropoda, class Trilobita, and were swimming and bottom crawling marine scavengers that typically lived in warm, shallow seas. The name of this group is based on the fact that the trilobite body is divided into three longitudinal lobes, a median axial lobe with a pleural (side) lobe on each side of the axial lobe (Moore, Lalicker, and Fischer, 1952). The body is also divided transversely into three distinct sections, the cephalon (head region), the thorax (abdomen), and the pygidium (tail region). Trilobites, like all arthropods, are segmented animals. Each segment usually bears a pair of jointed appendages, which serve as antennae and feeding appendages on the cephalon and walking or swimming appendages on the thorax and pygidium (Moore, Lalicker, and Fischer, 1952). As with many arthropods, trilobite walking and swimming appendages are typically biramous, or two-pronged, with a gill branch (for respiration) above and a leg branch below (Gould, 1989). As with all arthropods, at various stages of growth the exoskeleton was shed to allow for an increase in size. This process of shedding the exoskeleton is called ecdysis (Moore, Lalicker, and Fischer, 1952). Many fossil impressions of trilobites (and other arthropods) are simply the disregarded remains of the animals' previous exoskeletons. Thus, it is easy to see that one individual trilobite could form several fossil remains of itself. Many trilobite fossils are found in a ball-shaped, rolled-up position. Some paleontologists interpret this as the result of a defensive mechanism to protect the soft underside of the animals, which was not covered by the exoskeleton. When a predator approached, the trilobite would just roll-up into a ball. Perhaps the old age trilobite, when near death, rolled itself up into a ball and waited to die. One Upper

Ordovician (Cincinnatian Series) trilobite, the genus *Flexicalymene*, is very often found in a rolled-up position (see Figures 3-23 and 3-24).

Trilobites remained very abundant and diverse until the end of the Ordovician Period, when they suffered a great loss of diversity during the end Ordovician mass extinction. They hung on until the end of the Paleozoic Era, but finally went extinct at the great mass extinction event at the end of the Permian Period. For the entire Paleozoic Era, over 15,000 species of trilobites have been described (Wicander and Monroe, 2013). Figures 7-4 through 7-7 show some typical trilobites of the early to middle Paleozoic Era.

Figure 7-4. A photograph of a slab of Middle Cambrian Wheeler Formation shale of Utah containing one of the most familiar and abundant trilobites, *Elrathia kingii* (Meek, 1870) (the larger trilobite on the right) (see Gaines and Droser, 2003) and the small blind agnostid trilobite *Itagnostus interstrictus* (White, 1874) (formerly known as *Peronopsis interstricta* [White, 1874]) (see Wikipedia at http://en.wikipedia.org/wiki/Itagnostus, accessed 12/01/2014). The *Elrathia* specimen is 3.5 cm in length and the *Itagnostus* specimen is 9 mm in length. (Photo by E. L. Crisp. The fossil specimens are owned by E. L. Crisp.)

Figure 7-5. A photograph of the large (23 cm in length) Cambrian trilobite *Cambropallas telesto* Geyer, 1993 from Morocco. (Photo by E. L. Crisp. The fossil specimen is owned by E. L. Crisp.)

Figure 7-6. A photograph of the Upper Ordovician (Cincinnatian Series) trilobite *Flexicalymene meeki* (Foerste, 1910). The trilobite is in a matrix of limy (calcareous) shale and is 2.9 cm in length. The pygidium is partially folded downward. (Photo by E. L. Crisp. The fossil specimen is owned by E. L. Crisp.)

Figure 7-7. A photograph of two rolled specimens of the middle Devonian trilobite *Eldredgeops rana* (Green, 1832) [formerly known as *Phacops rana* (Green, 1832)], in limestone matrix. The larger rolled specimen is about 1.7 cm in width. Note the large compound eyes in the larger specimen. There has been a recent technical change in the name to *Eldredgeops rana* (Green, 1832) (as indicated above), but many refuse to use or are unaware of the name change (particularly amateur paleontologists). (Photo by E. L. Crisp. The fossil specimens are owned by E. L. Crisp.)

In addition to sponges (carryovers from the Proterozic Eon), trilobites, and other mineralized invertebrate groups (such as brachiopods and molluscs) represented by SSFs, another wave of evolution appears (perhaps this appearance is due to preservational bias) to have begun about 521 Ma during Age 3 of Cambrian time (Stage 3 of Cambrian rocks) (see Erwin and Valentine, 2013, Figure 6.6, p. 155) as represented in the fossil record by exceptionally preserved lagerstätten faunas of Cambrian Stage 3 through Stage 5 (see Erwin and Valentine, 2013, Figure 6.6, p. 155, or the GSA Geologic Time Scale at

http://www.geosociety.org/science/timescale/). The evidence for this rapid (some say explosive) appearance of many new animal body plans is primarily based on original organic carbon films of soft-bodied animals preserved by exceptional processes related to rapid anoxic burial conditions (Gould, 1989; Erwin and Valentine, 2013). One of the older and richest of these fossil assemblages is from the Chengjiang fauna of Yunnan Province, China that appears in the lower portion of Cambrian Stage 3 (Erwin and Valentine, 2013). Many new body plans for life had evolved by this time and there was a tremendous morphologic disparity of body plans with low taxonomic diversity within each major group of animals (Erwin and Valentine, 2013). Also, the lagerstätte in Middle Cambrian rocks of the Burgess Shale in Canada show many of these strange new body plans (both soft bodied and with hard exoskeletons). Charles D. Walcott (remember, we talked about Walcott earlier) discovered the Burgess Shale Fauna in 1909 in British Columbia, Canada (Gould, 1989; Sepkoski, 1993; Wicander and Monroe, 2013). These fossils are representative of soft bodied organisms (and some mineralized skeletons, about 15% of the deposit [Erwin and Valentine, 2013]) that lived during middle Cambrian time, between about 505 to 509 Ma, during Cambrian Stage 5 (Peng, Babcock, and Cooper, 2012; Erwin and Valentine, 2013). The Burgess Shale fauna is interpreted to have lived in a shallow sea on mud banks on the steep face of a reef wall (Sepkoski, 1993). Mudslides on the steep reef escarpment carried these shallow water animals into deeper marine basin anoxic (without oxygen) waters, where they were covered by muddy sediment and preserved as fine carbon films (Sepkoski, 1993). As Erwin and Valentine (2013) point out, although the Chengjiang fauna and the Burgess Shale fauna have received the most publicity from the popular press and paleontologists, there are several other Lower to Middle Cambrian lagerstätten deposits of animal fossils in the strata between the Chengjiang fauna and the Burgess Shale fauna that contain exceptionally preserved "explosion" fauna. Some notably examples of these lagerstätten are "the Sirius Passet fauna of Greenland, the Emu Bay Shale in Australia, the Sinsk biota in Russia, and the Kaili and Guanshan biotas in China (fig. 6.6)" (Erwin and Valentine, 2013).

Remarkable morphologic diversity is present within the fossils of these soft bodied organisms. Most of the species found in the Chengjiang and Burgess Shale fauna are arthropods or arthropod allies and the majority of these forms can be placed into living phyla (Erwin and Valentine, 2013). Further, it appears that all of the major phyla (of invertebrates and Chordata) of today appear to be present in the Chengjiang fauna, the Burgess Shale fauna and the other Cambrian lagerstätten as mentioned above (Erwin and Valentine, 2013). In fact, Stephen Jay Gould in his 1989 book *Wonderful Life: The Burgess Shale and the Nature of History* states that all the major phyla were present in the Burgess Shale and that no new major phyla have evolved since.

A link to the ancestry of vertebrates may be present within the Burgess Shale fossil accumulation. Very well preserved fossils (114 specimens) of *Pikaia gracilens* (Walcott) have been found in the Burgess Shale (Conway Morris and Caron, 2012). *Pikaia* is thought to have been a worm-like segmented organism that appears to have a notochord (a stiff cartilaginous rod running dorsally down the back that supports the dorsal nerve cord), thus it would be a primitive member of the phylum Chordata (which includes all vertebrates) (Gould, 1989; Sepkoski, 1993; Conway Morris and Caron, 2012; Erwin and Valentine, 2013). The fossils of *Pikaia* from the Burgess Shale are about 4 to 5 cm in length and appear to display what are thought to represent V-shaped bundles of muscles (seriated myotomes or myomeres, about 100) along its flanks (Gould, 1989; Sepkoski, 1993; Conway Morris and Caron, 2012; Erwin and Valentine, 2013). This type of muscle pattern is only known in chordates. Based on the features observed, particularly a notochord, V-shaped muscle bundles, and gills; *Pikaia* appears to be a member of the subphylum Cephalochordata, similar to living lancelets (Gould, 1989; Sepkoski, 1993). In addition to *Pikaia*, reputed vertebrate (subphylum Vertebrata) fossils have been found in Stage 3 rocks close to the Chengjiang localities (Erwin and Valentine, 2013); the earliest example of such a vertebrate is the genus *Myllokunmingia* (see Erwin and Valentine, 2013, p. 177).

The Paleozoic After the Cambrian Period

Early Paleozoic encompasses the Cambrian, Ordovician, and Silurian. By the end of the early Paleozoic, the southern continent of Gondwanaland had formed. The northern continents were somewhat scattered. North America was a barren lowland. The seas transgressed onto the North American continent and regressed several times during the early Paleozoic. During the Silurian Period shallow marine basins on the continent evaporated leaving rock salt and gypsum deposits. The Taconic Orogeny, the first mountain building episode that would eventually form the

Appalachian Mountains, affected eastern North America at the end of Ordovician time.

During the Early Paleozoic, life, except perhaps for bacteria and archaea, was restricted to the marine environment (oceans and seas). As we have already discussed, the first animals with hard parts, such as shells (perhaps for protection from predators), evolved during the Cambrian Period. Of course, this greatly increased the probability of organisms being preserved in the rock record and accounts for the abundance of fossils during the Phanerozoic Eon. The word Phanerozoic is derived from the root words phaneros (meaning to see) and zoic (meaning or referring to animals, or as usually taken, life in general).

Animal life in the seas consisted primarily of several invertebrate groups, including sponges, corals, brachiopods, trilobites, and mollusks (in the form of bivalves, gastropods, cephalopods). As we have already mentioned, vertebrates (in the form of primitive fishes) had evolved by Cambrian time. These first fish were agnathians (jawless fish). Some agnathians, referred to as ostracoderms, had bony plates covering their heads and thoracic regions (most likely for protection from predators). Even land plants had evolved from green algae by Late Ordovician to Early Silurian.

The Late Paleozoic consists of the Devonian, Mississippian, Pennsylvanian, and Permian Periods. The supercontinent of Pangea had formed by the end of the Permian Period. Several Mountain belts formed as continents collided with each other. The world's climates become more seasonal. The worst mass extinction event in the history of life on Earth occurred at the end of the Paleozoic Era; life was almost wiped out on Earth at the end of the Permian Period.

During the Late Paleozoic, organisms diversified dramatically. By Devonian time, fishes had evolved into two groups of jawed bony fishes – the ray-finned fishes (like bass, catfish, etc.) and lobe-finned fishes (like modern lungfish). Insects invaded the land in the Devonian Period. Amphibians (the first tetrapods) evolved from lobe-fined fishes by Late Devonian time and during the remainder of Paleozoic time diversified rapidly. Present-day lobe-finned fish (such as lung fish) are characterized by muscular pectoral (chest or shoulder region) and pelvic fins. The fins have articulating bones rather than radiating bones or cartilage (as in ray-fined fish), with the fin attached to the body by a fleshy shaft (Wicander and Monroe, 2013). Three orders of lobe-finned fish are recognized: 1) coelacanths, such as the so called "living fossil" *Latimeria* (a rare modern genus of Sarcopterygii [lobe-finned fishes plus terrestrial vertebrates]), thought to have succumbed to extinction at the end of the Cretaceous Period, but first found alive in 1938 off the coast of Africa; 2) lungfish (including both fossil forms and modern lungfish of Africa, South America, and Australia); and 3) crossopterygians. The crossopterygians are an important group of lobe-finned fish because amphibians evolved from them. The crossopterygians that gave rise to tetrapods (four-limbed terrestrial vertebrates, including amphibians) reached over 2 m in length and were the dominant freshwater predators during the Late Paleozoic (Wicander and Monroe, 2013).

The Devonian genus *Eusthenopteron*, which lived about 385 million years ago, is a good example of the type of crossopterygian lobe-finned fish that gave rise to tetrapods (Zimmer, 2010; Zimmer and Emlin, 2013). This large fish had an elongated body that enabled it to move swiftly in the water and paired muscular fins that could be used for very limited locomotion on land (but eventually modified by natural selection in descendants for better locomotion on land). The morphologic and structural similarity between crossopterygian fish and the earliest amphibians (such as *Ichthyostega* and *Acanthostega*) is striking, representing one of the better documented transitions from one major group to another (Wicander and Monroe, 2013; see Zimmer, 2010, Figure 4.3, p. 67). (Also see our discussion of "*Tiktaalik roseae*, An Intermediate Fossil in the Rise of Tetrapods from Fish", beginning on page 159.) The oldest tetrapod (in this case, amphibian) fossils are found in the Upper Devonian Old Red Sandstone of eastern Greenland in rocks that are about 360 to 365 million years old (Benton, 1993; Dawkins, 2004; Clack, 2006; Clack, 2012; Steyer, 2012; Wicander and Monroe, 2013; Zimmer, 1998; Zimmer, 2010; Zimmer and Emlen, 2013). These amphibians, which belong to genera such as *Acanthostega* and *Ichthyostega*, possessed streamlined bodies, long tails, and fins. In addition, they had four legs with digits at the end of the forelimbs and hindlimbs (but more than five), a strong backbone, a rib cage, and pelvic and pectoral girdles;

all of which were structural adaptations for maneuvering through shallow water filled with weeds and water-logged vegetation within tropical deltaic and coastal plain wetlands, and in some cases for walking on land (perhaps to escape a predator) (Benton, 1993; Dawkins, 2004; Clack, 2006; Clack, 2012; Steyer, 2012; Wicander and Monroe, 2013; Zimmer, 1998; Zimmer, 2010; Zimmer and Emlen, 2013). The older, now unfavored, Romer theory (named for Alfred S. Romer [1894-1973], a vertebrate paleontologist and biologist), that tetrapods evolved from lobe-finned fishes during the Late Devonian because, according to Romer, it was very seasonally arid all over the world during this time and lobe-finned fish living in ponds and rivers that were drying up used their fleshy lobe-fins to drag themseleves to a pond or river that still had water in it. Of course, in this theory, since lobe-fins had lungs they could survive by breathing the air while out of the water. Also, as natural selection worked on these populations during the later Devonian, any mutations towards limbs with digits (and other tetrapod characteristics) would result in increased success at moving onto land and would be selected for, thus tetrapods evolved from these lobe-finned fishes (Dawkins, 2004; Zimmer, 1998; Zimmer, 2010; Zimmer and Emlen, 2013). As Romer (1962, p. 57) states this himself:

> "The amphibians appear to have evolved from crossopterygian ancestors toward the close of the Devonian, an age during which seasonal droughts were, it seems, common over much of the earth. Lungs, already present in the ancestral fishes, are an excellent adaptation for use under stagnant water conditions. But when a stream or pool dries up completely, a typical fish is rendered immobile and dies. Some further development of the fleshy fins already present in crossopterygians would give their fortunate possessor the chance of crawling up or down the stream bed (albeit with considerable pain and effort at first) and enable him to reach some surviving water body where he could resume a normal piscine existence."

Because amphibians did not evolve until the Late Devonian, they were a minor element of the Devonian terrestrial ecosystem. In fact, insects evolved during the earlier Devonian and were the first animals to make the transition to the terrestrial environment. Like other groups that moved into new and previously unoccupied niches, amphibians underwent rapid adaptive radiation and became abundant during the Mississippian, Pennsylvanian, and early Permian Periods (Wicander and Monroe, 2013).

The next major evolutionary innovation of the tetrapods was the amniote egg. This allowed tetrapods to invade dry land because they did not have to return to the water to lay gelatinous eggs and did not have to have a free-swimming larval stage (as is the case with amphibians). The amnion is a membrane that surrounds the amniotic cavity. The amniotic cavity is filled with the amniotic fluid that baths the embryo. In most amniotes, there is a leathery or mineralized outer shell, inside of the outer shell is another membrane around the entire egg (most mammals have lost the outer shell because they evolved live birth). The amniote egg also had a yolk for nutrition and a special bladder (the allantois) for waste from the developing embryo.

Amniotes are divided into two major groups, Synapsida, with only one opening in the skull behind the eye socket for jaw muscle attachment, and Reptilia. Modern reptiles include both anapsid reptiles, with the skull completely roofed over by bone (such as turtles and their kin), and diapsid reptiles, with two openings in the skull behind the eye socket for muscle attachment. It is perhaps best to think of the first amniotes of Late Mississippian age as just primitive amniotes that gave rise to two great groups (clades) of amniotes, Synapsida and Reptilia.

The synapsids are one of two great lineages of the amniote tetrapods. All mammals are synapsids. Some extinct forms of synapsids are called "mammal-like reptiles" by some, but this is not the best name for them. As we will see later, synapsids (even primitive ones) are not reptiles at all (according to the modern definition of Reptilia that includes birds). The split between synapsids and reptiles from more primitive amniotes occurred about 310-320 million years ago (Fastovsky and Weishampel, 2005). All synapsids have a distinctive skull type. Their skull is a departure from the primitive tetrapod skull. In the primitive condition for a tetrapod skull there is a sheet of interlocking bones covering the brain case. Synapsids have a single opening in the skull below the postorbital and squamosal bones of the skull (behind the eye orbit). This opening is called the Lower Temporal Fenestra (or Infratemporal Fenestra). Jaw muscles pass through the opening to attach to the upper part of the skull roof. The primitive synapsids *Dimetrodon* and *Edaphosaurus*, finbacks or sailbacks of the Pennsylvannian and Permian Periods, belong to this group of amniotes.

The synapsids radiated during the latter part of the Paleozoic and by medial Triassic were the dominant terrestrial vertebrates, with a worldwide distribution. They had diversified into many herbivorous and carnivorous forms. The synapsids suffered greatly during the late Triassic mass extinction event and by late Jurassic were reduced to a clade of tiny, scrappy, furry night dwellers called mammals (mammals first arose in late Triassic time from therapsid ancestors).

Reptilia are the other great clade of amniotes. The clade Reptilia includes the anapsid and diapsid reptiles. Modern representatives of Reptilia are turtles, snakes, lizards, crocodiles (and alligators), the Tuatara, and the birds (notice here that if we do not include the birds we would have a paraphyletic group). From the past, Reptilia include the dinosaurs, pterosaurs, plesiosaurs, mosasaurs, and icthyosaurs. Today there are about 15,000 species of Reptilia (of course this number includes the birds) (Fastovsky and Weishampel, 2005).

In diapsid reptiles the openings in the temporal region provided space for bulging jaw muscles and more space for their attachment to the skull. There are two major clades of diapsid reptiles: 1) Lepidosauromorpha – snakes, lizards, and the Tuatara of New Zealand, and 2) Archosauromorpha, which includes the clade Archosauria. The archosaurs are the crocodiles, pterosaurs, dinosaurs, and birds (and some primitive archosaurs of the Triassic Period, certain ones of which were in the ancestral line to the dinosaurs).

Land plants make their appearance in the early Paleozoic, but evolve rapidly during the later Paleozoic. Fossils of the oldest land plants are found in rocks of Upper Ordovician age. Most paleobontanists agree that the ancestral stock of land plants first evolved in the sea and then a branch of this ancestral stock adapted to the freshwater environment, eventually some freshwater forms evolved into land plants. The first land plants of Late Ordovician and Early Silurian were probably nonvascualar land plants. These bryophytes (liverworts, hornworts, and mosses) do not have vascular tissue with specialized cells to transport nutrients and water, and to aid in support of the plant. Nonvascular plants live in moist areas and are usually small and low to the ground (Wicander and Monroe, 2013).

Vascular plants, which make up the bulk of land plants today, had evolved by middle Silurian time and a major diversification of vascular plants took place during Early Devonian time. Extensive tropical coal swamps formed during the Late Paleozoic, particularly during the Carboniferous Period, usually referred to as the Mississippian Period and the Pennsylvanian Period in North America.

The Permian mass extinction occurred about 251 million years ago and was the greatest mass extinction ever recorded in Earth history; even larger than the Ordovician, Devonian, and Triassic crises, as well as the better known Cretaceous/Paleogene (K/Pg) mass extinction that wiped-out the non-bird (nonavian) dinosaurs. About 96% percent of marine species were eliminated as a result of this Permian event (Benton, 1993; Benton, 2003; Steyer, 2012). About 75% of families of land animals went extinct. Land plants were also hit hard during this mass extinction event.

Perhaps we should digress a little here to define mass extinction events and some of their possible causes. What is "extinction"? Extinction is the total disappearance of a species or higher taxon, so that it no longer exists anywhere. Extinction was not accepted by most naturalists until Baron Georges Cuvier (1769-1832) presented a paper in 1796 to the French Institute proving beyond doubt that mammoths and mastodons are extinct relatives of the elephant.

Estimates of the number of species that have lived on the earth range from about 5 to 50 billion. It is estimated that there are approximately 50 million species extant today. Given this, then over 99.9% of all species that have ever lived are extinct! An average extinction rate between 2 and 4.6 families per million years was reported by Raup and Sepkoski in 1982. There are 5 intervals of "mass extinction" that stand out against this "background extinction" rate. Survivorship curves, illustrate that long-lived species have no greater probability of survival than short-lived species. Species do not become better or worse at avoiding extinction as they persist in time; old species have the same chance of extinction as young ones! Remember, we discussed the Red Queen Hypothesis (Van Valen, 1973) earlier. In *Alice through the Looking Glass*, the Red Queen told Alice that she must keep running and running as hard as she can to stay in the same place. Evolution of species is like this, so evolution is a zero-sum game: species compete for limited resources; natural selection constantly improves organisms to keep up with competing species; and each species' environment deteriorates as competitors evolve new adaptations. Thus, species must constantly improve themselves to avoid extinction. Species are constantly going extinct, but new species are constantly evolving into existence. Therefore, there is a background extinction rate. Mass extinction events occur when the extinction rate exceeds the background extinction rate.

As we mentioned above, there have been five major extinction events (referred to as "The Big Five") during the Phanerozoic Eon, those being the following: 1) the Late Ordovician event, 2) the Late Devonian event, 3) the Permian – Triassic (P/Tr) event, 4) the Late Triassic event, and 5) the Cretaceous – Paleogene (K/Pg) (Note: K is for Kreta, meaning chalk – for the extensive chalk deposits in rocks of Cretaceous age.) event. The Permian – Triassic extinction event that occurred at the end of the Paleozoic Era (and the beginning of the Mesozoic Era) is considered to be the mother of all extinctions for its severity. Life was almost extinguished during this time, about 251 million years ago. This event was long and protracted. It occupied the last 2 stages of the Permian Period, a time span of about 8 million years. It took over 100 million years for species diversity to recover from this extinction event (Benton, 2003).

The Mesozoic Era

The Mesozoic Era lasted from 251 Ma to 66 Ma and consists of the Triassic, Jurassic, and Cretaceous Periods. The Mesozoic begins with the supercontinent Pangea still intact and much of the world's land above sea level. The breakup of Pangea begins during the Early Jurassic. The North American plate began to override the old Farralon Plate forming a convergent plate boundary along the western portion of NA. The mountains of western North America began forming along this convergent boundary (and associated subduction zone that initiated volcanism in these mountains).

Survivors of the Great Permian Extinction diversify and adaptively radiate into the available niches. Gymnosperms become the dominant land plants. Amniotes (the first true terrestrial animals), including reptiles and more mammal-like amniotes (and mammals by late Triassic) adapt to the dry Mesozoic climate. These amniotes had shell-covered or leathery-covered eggs that could be laid on land (they did not have to return to water to lay their eggs). Dinosaurs had evolved by Late Triassic and are beginning to dominate the land. One group of meat-eating dinosaurs (theropods) evolved into the birds during late Jurassic time (see our discussion of the first bird, *Archaeopteryx*). Many reptile groups, including the non-bird dinosaurs and many other groups of animals, become extinct at the close of the Mesozoic, 66 million years ago, during the Cretaceous/Paleogene (K/Pg) extinction event (formerly referred to as the K/T extinction event or Cretaceous/Tertiary extinction event).

Life of the Triassic Period

The organisms of the Triassic included the holdovers from the Permian extinction (club mosses, ferns, gymnosperms, reptiles, more mammal-like amniotes, etc.), as well as the appearance of icthyosaurs (marine reptiles), pterosaurs (flying reptiles), dinosaurs, and modern conifers. Rapid evolutionary adaptive radiation was taking place by the end of the Triassic (Tarbuck, Lutgens, and Tasa, 2013; Wicander and Monroe, 2013).

Life of the Jurassic Period

Large herbivorous (vegetarian) dinosaurs roamed Earth, such as *Brachiosaurus*, *Apatosaurus*, *Camarasaurus*, *Diplodocus*, and *Stegosaurus*. These plant-eating dinosaurs feed on lush growths of ferns and palm-like cycads (sometimes the Jurassic Period is called the Age of Cycads). Smaller but vicious carnivorous dinosaurs, including *Allosaurus*, stalked and fed upon the great herbivores. Oceans were full of fish, squid, and coiled ammonites, along with great ichthyosaurs and long-necked plesiosaurs. As mentioned above, one group of small theropod (meat-eating) dinosaurs evolved into the first bird, *Archaeopteryx*, during Late Jurassic time about 150 Ma (Tarbuck, Lutgens, and Tasa, 2013; Wicander and Monroe, 2013).

Life of the Cretaceous Period

During the Cretaceous the first horned (for example, *Triceratops*), bone-headed (such as *Pachycephalosaurus*), and duck-billed dinosaurs (like the hadrosaurids, *Edmontosaurus* and *Parasaurolophus*) appeared. The mighty carnivore, *Tyrannosaurus rex*, lived the last 2 million years of the Cretaceous, hunting vegetarians like the ones mentioned in the previous sentence. We find the first fossils of many insect groups, modern mammal and bird groups, and the first flowering plants. However, the end of the Cretaceous, during the Cretaceous/Paleogene (K/Pg) (formerly referred to as the Cretaceous/Tertiary [K/T]) extinction event, brought the end of many previously successful groups of organisms, such as the non-bird (nonavian) dinosaurs and ammonite cephalopods. (Note: The Tertiary Period is no longer preferred by geologists for the first period of the Cenozoic Era, rather the Paleogene is preferred. The Tertiary Period includes both the Paleogene and Neogene Periods of the Cenozoic Era.)

The Cretaceous/Paleogene (K/Pg) Mass Extinction Event

Two major hypotheses have been proposed for the K/Pg mass extinction event, these being 1) a bolide impact from outer space and 2) volcanism resulting in ecological collapse. The bolide impact hypothesis basically states that Earth was hit by a large (10 Km) asteroid, meteor, or comet about 66 million years ago at the end of Cretaceous time. This idea suggests a sudden catastrophic extinction of many forms of life. The volcanism and ecological collapse hypothesis proposes that changes in topography (due to recession of Cretaceous epicontinental seaways [eperic seas]), and climate, vegetation, and/or animal life changes resulted from volcanism. Thus, an ecological collapse occurred that resulted in major extinctions of organisms. This idea also suggests a gradual extinction. Many geologists and paleontologists support the hypothesis that increased volcanism during Late Cretaceous time resulted in climatic changes that stressed the biota and caused the gradual extinction of many species of organisms. Some evidence suggests that the climate was becoming cooler during the latest part of Late Cretaceous (this may be related to Plate Tectonic related volcanic activity and the withdrawal of the eperic seas from the continents (which in itself could cause more seasonality at mid-latitudes, but also would decimate many shallow water invertebrates).

The bolide impact hypothesis states that the impact caused a huge dust cloud in the atmosphere that circled the world and blocked solar radiation to cause rapid global cooling, similar to what has been proposed for a nuclear winter due to an all-out nuclear war on Earth. The "smoking gun" for this hypothesis is the Chicxulub Crater on the Yucatan peninsula, an approximately 150 Km crater that is mostly covered with sediment and was discovered by a petroleum geologist during the 1970s. The bolide impact idea was first proposed in a 1977 paper by Luis and Walter Alvarez based on the discovery of a one inch Gubio, Italy clay layer that represents the fallout of dust from the atmosphere due to an asteroid or comet impact with Earth. The clay layer at the K/Pg boundary was found to have an unusually high content of iridium (a rare element). Crustal rocks on Earth do not have high iridium content, but meteorites do. Many other iridium rich clay layers have since been found around the world at K/Pg boundary (Alvarez, 1997).

Proponents of the volcanism theory also believe the climate got colder during the latter part of the Cretaceous (last few million years), but because of increased volcanism during Late Cretaceous time. They propose that volcanoes blew huge amounts of volcanic ash and sulfur compounds into the atmosphere, which blocked solar radiation and thus caused the climate change and the gradual mass extinctions. The gradualistic volcanism proponents do not think the volcanism occurred suddenly, but occurred over several million years during latest Cretaceous time and the climate gradually got colder. However, some proponents of volcanism also propose a tremendous greenhouse warming right at the end of Cretaceous time due to the increased output of carbon dioxide from the Deccan Traps volcanism in India, which is radiometrically dated to the last few million years of the Cretaceous period (Tarbuck, Lutgens, and Tasa, 2013; Wicander and Monroe, 2013).

The Cenozoic Era

The Cenozoic Era is often called the Age of Mammals, but could also be called the Age of Angiosperms (flowering plants). This era represents a smaller fraction of time than either the Paleozoic or the Mesozoic. In North America, most of the continent was above sea level throughout the Cenozoic. There was significant mountain building, volcanism, and earthquake activity in western North America during the Cenozoic. The Rocky Mountains were completed during the Early Cenozoic.

The Basin and Range Province formed in western North America as upthrown and downthrown faults that formed basins (in the downthrown blocks) and mountain ranges (in the upthrown blocks). This tectonic (large scale deformation) movement resulted from the overriding of the old Farallon oceanic plate by the North American continental plate. Because of this, the entire western interior was uplifted during the Cenozoic. This resulted in reelevation of the Rockies and caused many of the western streams to vigorously downcut, creating spectacular gorges and canyons (such as the Grand Canyon of the Colorado River and the Black Canyon of the Gunnison River). Extensive basaltic volcanism in Washington and Oregon resulted in the Columbia River Plateau basalts. Also, due to subduction along the northwest margin of North America, the volcanic Cascade Range formed. Later the coast ranges of California and Oregon formed (Tarbuck, Lutgens, and Tasa, 2013; Wicander and Monroe, 2013).

Eastern North America was stable during the Cenozoic, with extensive sedimentation on the continental shelves. The eroded Appalachian Mountains were uplifted by isostatic adjustments due to the removal of sediment by erosion. During the Miocene Epoch, this resulted in the streams cutting down to form the present rugged terrain of the Appalachian Mountains. The Pleistocene "Ice Age" (the Pleistocene Epoch) has dominated the last 2.6 million years of the Cenozoic, except for about the last 10,000 years of the Holocene Epoch (Tarbuck, Lutgens, and Tasa, 2013; Wicander and Monroe, 2013).

CHAPTER 8

MECHANISMS OF EVOLUTION

INTRODUCTION

As we have continued to study the evolution of organisms over time and with new and sophisticated tools, we have come to realize that evolution is relentless, constantly acting on biological communities (Thompson, 2013). This is the case even for physical environments that are relatively stable and have not changed much recently, to which we would think that organisms have adapted. However, organisms in these environments continue to "adapt, speciate, and go extinct" as "continually changing webs of interacting species" (Thompson, 2013). Darwin (1876, p. 429) may have had this in mind when he imagined the competition and coevolution of species on a tangled bank of a stream, as stated in a portion of the last paragraph of the sixth edition (final and definitive edition) of *On the Origin of Species*:

> It is interesting to contemplate a tangled bank, clothed with many plants of many kinds, with birds singing on the bushes, with various insects flitting about, and with worms crawling through the damp earth, and to reflect that these elaborately constructed forms, so different from each other, and dependent upon each other in so complex a manner, have all been produced by laws acting around us. These laws, taken in the largest sense, being Growth with Reproduction; Inheritance which is almost implied by reproduction; Variability from the indirect and direct action of the conditions of life, and from use and disuse: a Ratio of Increase so high as to lead to a Struggle for Life, and as a consequence to Natural Selection, entailing Divergence of Character and the Extinction of less-improved forms. Thus, from the war of nature, from famine and death, the most exalted object that we are capable of conceiving, namely, the production of the higher animals directly follows.

Evolutionary success is not only dependent on survival, but evolutionary fitness is known to be associated with reproductive success. The more fit an organism is to its environment the better its adaptation to that environment. Organisms do not become better adapted to their environment because of a "need" to adapt in order to survive (this idea of "need" to adapt is a Lamarckian concept and is not supported by evidence). The more offspring that an individual contributes to the next generation, the more fit the individual. At the level of the gene, fitness is measured by the success of one genotype (or individual gene - allele) over another genotype (or individual gene - allele). However, the environment (both the biotic and physical environment) applies selection pressure to the phenotype (observable characteristics or traits of an organism), not the genotype (inherited instructions carried in the DNA of an organism's genes). Thus, the environmental factors working on the phenotype will result in certain phenotypes having greater reproductive success than other phenotypes; but because the phenotypes are determined by their genetic make-up, the genotype ultimately changes. Therefore, the gene pool of the population changes, as a result evolution will occur as a consequence of environmental selection pressures. Populations of a species evolve, but this evolutionary adaptation is a result of environmental selection pressure that works on the phenotypes of individuals in a population, eventually changing the gene frequencies within a population of organisms. This nonrandom process is referred to as natural selection (Darwin, 1859; Futuyma, 2005; Kardong, 2008; Futuyma, 2009; University of California Museum of Paleontology, accessed 2010). Although natural selection works on individuals, it only has meaning when we speak of the fitness of groups or genotypes within a population so that the statistical average of reproductive success occurs for a particular genotype (Futuyma, 2009, p. 283). By the way, "Survival of the fittest" is not an adequate substitute phrase for natural selection because, according to Futuyma (2005), it is wrong and misleading. People often get the wrong idea of what this means and as pointed out by Futuyma (2005), this term should be discarded, abolished. Herbert Spencer, a contemporary of Charles Darwin, first coined this phrase (see the following link at Wikipedia which discusses the history of this phrase:

http://en.wikipedia.org/wiki/Survival_of_the_fittest).

Charles Darwin (1859) discussed Variation Under Domestication in the first chapter of *On the Origin of Species*. He stresses how man has used inherent variations in domesticated organisms and selective breeding of those animals and plants over time to produce extraordinary variants of the original forms. Of course, humans have performed this selective breeding in order to produce breeds that are of more value or use to them, such as cows that give more milk, corn that yields more (or less) oil, or homing pigeons that were faster and more accurate. But sometimes breeders just fancied certain types of pigeons (certain leg feathers or chest postures) or dogs and cats. Kardong (2008, p. 124-128) gives a good discussion of various examples of the products of artificial selection (or selective breeding) of animals and plants. He defines artificial selection as "The weeding out of organisms by humans for human purposes" (Kardong, 2008, p. 124). We all know the power of artificial selection when we compare the various breeds of dogs that arose by selective breeding, such as a Chihuahua compared to a Great Dane or English Bulldog. Dogs first diverged from wolves (*Canus lupus*) about 15,000 years ago in East Asia and dog breeders have since used artificial selection to produce many breeds of dogs (Kardong, 2008).

In chapters 2 and 3 of *On the Origin of Species*, Darwin discusses variations in nature and the struggle for existence. Eventually, in chapter 3, he compares the superior power of natural selection to the weaker power of artificial selection in the following statement (Darwin, 1859, p. 61):

> Owing to this struggle for life, any variation, however slight and from whatever cause proceeding, if it be in any degree profitable to an individual of any species, in its infinitely complex relations to other organic beings and to external nature, will tend to the preservation of that individual, and will generally be inherited by its offspring. The offspring, also, will thus have a better chance of surviving, for, of the many individuals of any species which are periodically born, but a small number can survive. I have called this principle, by which each slight variation, if useful, is preserved, by the term of Natural Selection, in order to mark its relation to man's power of selection. We have seen that man by selection can certainly produce great results, and can adapt organic beings to his own uses, through the accumulation of slight but useful variations, given to him by the hand of Nature. But Natural Selection, as we shall hereafter see, is a power incessantly ready for action, and is as

immeasurably superior to man's feeble efforts, as the works of Nature are to those of Art.

NATURAL SELECTION

As stated above, natural selection is a process that works on the phenotypes in the population, ultimately resulting in the survival and reproductive success of phenotypes that are most fit for their environment. Thus natural selection leads to a change in the genotypic frequencies in a population over time.

Seemingly, by our standards of time, natural selection is a very slow process, acting over hundreds of generations. However, viruses and bacteria (because of very rapid generation reproduction) may evolve significantly in human lifetimes; some examples are HIV (Human immunodeficiency virus), antibiotic resistance in certain strains of bacteria, and resistance of agricultural pests to pesticides (Thompson, 2013).

There are also many examples of more complex organisms that studies have shown to have recently evolved by natural selection. For example, studies (for a period of over 40 years) by Peter and Rosemary Grant of the finches in the Galapagos Islands have shown that ground finches on a small island there vary in beak thickness and beak depth based on extended environmental fluctuations (droughts and wet periods), with the beaks adapting to become deeper and thicker during droughts in order to eat larger and harder seeds (Weiner, 1994; Futuyma, 2005; Grant and Grant, 2008; Grant and Grant, 2014).

Miller and Harley (2007) present an interesting and dramatic case study concerning the evolution of (primarily via natural selection) and then the rapid ecological collapse of the cichlid population of Lake Victoria in southeast Africa. Cichlids have shown rapid speciation in southeast African lakes after having invaded Lake Kivu approximately 100,000 years ago. This rapid adaptive radiation (adaptive evolution by natural selection of many species from a common ancestor) resulted from the cichlids invading rivers and finding their way to other lakes along the East African Rift Zone. This allowed isolation of populations of cichlids from each other and thus reduced the size of the gene pool in each isolated population permitting natural selection to take a slightly different route in individual populations. In addition, many cichlid species occupy rocky bottoms within the lakes that are separated by lighter colored sandy bottoms. They tend to remain rather isolated and breed within like subpopulations in the rocky bottom areas because the lighter sandy bottomed areas allow predators to detect them more easily.

Further complicating the evolutionary history of the Lake Victoria cichlids was the isolation of Lake Victoria populations about 40,000 years ago by volcanic eruptions. There is also strong evidence that Lake Victoria dried out after 40,000 years ago, forming many isolated smaller ponds and lakes, but refilled about 14,700 years ago. This event, no doubt, caused the extinction of some species, but the isolation in smaller ponds and lakes may have resulted in further speciation (or at least sub-speciation).

Until the 1950s, Lake Victoria's 500 species of cichlids were a prime example of a diverse and abundant population of these fish. But then the giant Nile perch was introduced into Lake Victoria to increase the fishery. This proved disastrous to the cichlid population. The exotic Nile perch, a voracious predator, decimated the cichlid population, decreasing the numbers from 99% of the total fish population to only 1%. Further, because many species of cichlids are algae eaters, the algae grew uncontrolled in the lake and the decay of algal organic material depleted the oxygen within the lake waters, further reducing the cichlid population. In addition to all this, because the Nile perch is very oily, fisherman deforested the area around the shores of Lake Victoria to get wood to burn and dry the fish. The deforestation allowed runoff of sediment laden water into the lake to increase the turbidity. The increased turbidity decreased the visibility of the water such that the brightly colored cichlids could not recognize their own species for mating. Also mouth brooding cichlids could not see their eggs or young in the water. Needless to say, there has been severe extinction of cichlid species such that the number of remaining species has been drastically reduced (also the abundance within each species has been greatly reduced).

This scenario is a prime example of the rapid evolution of a group via adaptive radiation, followed by an ecological collapse of the group due to the uninformed introduction of a foreign species into what was a relatively stable ecosystem. Of course, there are many other examples of man's disruption of world ecosystems by uneducated actions similar to this example.

Another example of the rapid natural selection of a trait (color) by a species over a relatively short historical time span (about thirty years to evolve from dominantly one color pattern to the other) is the evolution of black (melanic) varieties of the peppered moth (*Biston betularia*) during the latter part of the nineteenth century and first half of the twentieth century in Great Britain (Kettlewell, 1955; Ridley, 1996; Grant, 1999; Grant, 2004; Rudge, 2005; Kardong, 2008; Futuyma, 2009; Cook, et al, 2012; Thompson, 2013). *Biston betularia* typically occurs in two color varieties, one very dark (melanic) and the other light (peppered, light grey to white with black flecks). The black form (*Biston betularia* form *carbonaria*) of this trait for the moth is dominant over the peppered form (*B. betularia* form *typica*) (Kettlewell, 1955; Grant, 1999; Majerus, 2009). However, there are intermediate phenotypes between the peppered form (*typica*) and the black form (*carbonaria*) that have color patterns that range between the two extremes; these are collectively referred to as *B. betularia* form *insularia* (Kettlewell, 1955; Kettlewell, 1958; Kettlewell, 1965; Lees and Creed, 1977; Grant, 1999; Grant, 2004). Lees and Creed (1977) studied the genetics of the *insularia* forms relative to the *carbonaria* and *typica* forms and concluded that the several phenotypes occur as a result of multiple alleles at a single locus, with the *carbonaria* allele dominant to all *insularia* alleles (at least three) and to the *typica* allele. However, the *insularia* form of the moth is not common in areas where industrial melanism has been studied (Grant, 1999), so we will only discuss here the case for the *carbonaria* (melanic) form and the *typica* (peppered) form. Peppered moths (*Biston betularia*) reproduce one generation per year and the Mendelian inheritance of black color (melanism) for the scales on the wings and body has been strongly established (Grant, 2004). Many independent genetic researchers have studied the various populations of the peppered moth, as well as the several phenotypes, in Britain (and America) from the mid-1800s to the present, "...but the volume of data demonstrating the heritability of *carbonaria* is beyond dispute" (Grant, 2004, p. 101).

The first melanic forms (*carbonaria*) of *B. betularia* were found near Manchester, Great Britain, but represented less than 1% of the *B. betularia* population in Britain. However, by 1895, 98% of this species of moth were of the *carbonaria* form (melanic) near industrial centers (such as Manchester) across Great Britain. So, it appeared that peppered moths were evolving rapidly towards the *carbonaria* form during the latter part of the nineteenth century as a result of pollution due to the rapid advance of the industrial revolution across Britain (Grant, 1999; Majerus, 2009). The pollution given off by the burning of coal in Britain's industrial complexes involved the release of sulfur dioxide, that kills lichens (which are light colored), and airborne soot that darkens the surfaces that it lands on (such as tree bark on the trunks and limbs of trees). Thus, trees in the forests around Manchester and other industrial centers of Britain and Europe became darkened by pollution (and also in the United States, where industrial melanism also occurred somewhat later on the same species of moth [*Biston betularia*]) (Grant, 1999; Grant, 2004; Majerus, 2009). Since the melanic form of the moth is dark, it was better camouflaged than the peppered lighter form of the moth, and therefore the dark form was less likely to fall prey to predation by birds (thought to be the major predator for moths) on the trunks and branches of trees. Thus, the dark melanic forms of the moth increased rapidly in polluted regions due to industrial melanism, whereas the lighter peppered form was dominant in the less polluted areas (Grant, 1999; Majerus, 2009).

The above example of industrial melanism in the moth *Biston betularia* has been praised as one of the best teaching models of natural selection, and one of the most rapid cases of the change in a trait by natural selection that has been studied in nature. In fact, Kettlewell (1955), in one of his early articles relative to his studies of industrial melanism in the Lepidoptera, which was concentrated at first primarily on *B. betularia*, states (p. 323) that "The industrial melanism of the Lepidoptera is the most striking evolutionary change ever actually witnessed in any organism, animal or plant." In this article, he goes on to describe a mark-release-recapture experiment that he (and his assistants) performed that indicated that birds were the primary predators of the moths (which rested during the day and were active at night) and selected the most conspicuous moths resting on tree trunks and limbs to eat, thus supplying the selection pressure for the form of the moth that survived. Kettlewell's studies indicated that in polluted areas near the industrial area of Birmingham, where the *carbonaria* form represented about 85% of the moth population present, the black form (*carbonaria*) were recovered (recaptured) in significantly greater numbers than *typica* (Kettlewell, 1955; Kettlewell, 1958). This pattern was also the case in other polluted areas of Great Britain (Kettlewell, 1955; Kettlewell, 1958). Kettlewell (1955; 1958) also performed his experiments in areas that were not polluted, where the light (*typica*) form was recovered in significantly greater numbers than *carbonaria*.

Kettlewell's 1950s exemplary studies of rapid Darwinian natural selection of the melanic form of the peppered moth due to industrial melanism were so convincing that major textbooks during the last half of the 20th century used the peppered moth experiments and observations of Kettlewell (and others) as one of their major models of evolution in action (Majerus, 2009). And as pointed out by Rudge (2005, p. 373), "...Kettlewell's experiments were successful at providing among the first direct observation of a proximal mechanism of natural selection ever recorded." But Rudge (2005) goes on to explain that the reason the industrial melanism model for *Biston betularia* is so popular in biology textbooks is several factors that make it especially conducive for the teaching of biology (and in particular, evolution by natural selection). Rudge (2005, p. 373) further states the following relative to using industrial melanism as a teaching model (both for high school and college biology classes, but also for laypeople that are not strong in math and science: "These include the relative simplicity of the phenomenon - that is, it illustrates the simplest form of selection (directional), the advantage is conferred by one effect of a single gene, and the mechanism is intuitively obvious (birds will have the same difficulty spotting inconspicuous moths as humans do)".

However, during the late 1990s and the very early 21st century, this excellent example of natural selection was (and still is) attacked and misrepresented by advocates of Special (Divine) Creationism and Intelligent Design Creationism (Rudge, 2005; Majerus, 2009) (both of these are forms of creationism that do not accept modern evolutionary theory [except for some creationists who accept limited microevolution below the species level]; we will say more about creationism later in Chapter 12). (Note: This attack was at a time that was building up to the Tammy Kitzmiller et al. v. Dover Area School District et al. federal trial in Harrisburg, Pennsylvania during the fall of 2005. The federal judge, John E. Jones III, ruled that the Dover school board was advocating the teaching of a form of creationism [intelligent design creationism] when it required that the school present intelligent design as an alternate to evolutionary theory and that this requirement by the school board violated the Establishment Clause of the First Amendment to the Constitution of the United States.)

Although there may have been better ways to devise the experiments that Kettlewell performed to test the hypothesis that birds selectively chose *B. betularia* moths that are most conspicuous on their backgrounds, many of the experimental inadequacies that Kettlewell has been accused of by creationists (most of whom do not understand how natural selection works, or even science for that matter – and certainly not the genetics and ecology of the peppered moth) have been positively tested by others since the time of Kettlewell's work (see reviews and summaries by Grant, 1999; Rudge, 2005; Majerus, 2009; and Cook, et al, 2012). Most certainly Kettlewell knew what he was doing and did not try to deceive his readers (or commit scientific fraud) about the conclusions that he reached concerning industrial melanism relative to the peppered moth (as suggested by several of the creationist reviewers of his work) (again, see reviews by Grant, 1999; Rudge, 2005; Majerus, 2009; and Cook, et al, 2012). In fact, continued study of *Biston betularia* moths by several investigators to more recent times (latter half of the 20th century and the first part of the 21st century) has shown a reversal of the abundance of the melanic moth with respect to the more typical light colored moth (both in industrial areas of Europe and North America) resulting from changes in frequencies of melanic versus light colored moths because of the enactment of clean air laws, further illustrating the evolution of the peppered moth via Darwinian natural selection (see Grant, 2004; Majerus, 2009 and Thompson, 2013). Of course, the reason that creationists have attacked the peppered moth model of Darwinian evolution is because it is easy and straightforward to teach and for students and lay people to understand (Majerus,

2009). So, stated in the words of Majerus (2009, p. 72) (who had studied this model of Darwinian evolution for over a decade [with numerous publications] and had done extensive recent experimentation himself that bears on this topic), "The antievolution lobby is worried that if the peppered moth story is allowed to stand, too many people will be able to understand." Thus, as Majerus stated in the last sentence of his 2009 review article (p. 72), "Consequently, the peppered moth may become not only one of the best teaching examples of Darwinian evolution in action that we have but also a bedrock example of the difference between science and nonscience."

Because of attacks from creationists over Kettlewell's work on industrial melanism in the peppered moth (and even doubts by some biologists about the robustness of his experiments), Michael Majerus did experiments himself in his garden near Cambridge, England between 2001 and 2007 to test the validity of Kettlewell's conclusions. According to Jerry Coyne in an essay entitled "The Peppered Moth Story is Solid," appearing on his blog web page *Why Evolution is True* (see Coyne, accessed 06/24/2015), Majerus adequately tested Kettlewell's bird predation hypothesis for natural selection in peppered moth coloration (Majerus found in his studies that the birds preferently ate the darker [melanic] forms because the trees are now not polluted and thus the lighter colored moths are more camouflaged). Sadly, Michael Majerus died in January 2009 prior to publishing the results of his experiments on peppered moths (Coyne, accessed 06/24/2015; Wikipedia, Michael Majerus, accessed 06/24/2015). However, a group of four biologists headed by L. M. Cook summarized the work of Michael Majerus and presented his data in a 2012 article in *Biology Letters* (a peer reviewed journal) (Cook, et al, 2012; Coyne, accessed 06/24/2015; Wikipedia, Michael Majerus, accessed 06/24/2015). Cook, et al (2012) concluded that the work of Majerus certainly vindicated the work of Kettlewell and several others relative to industrial melanism in the peppered moth. They further presented the following summary and conclusion statement in the last sentence of the article: "The new data, coupled with the weight of previously existing data convincingly show that 'industrial melanism in the peppered moth is still one of the clearest and most easily understood examples of Darwinian evolution in action' [21]." (Note: [21] is a reference to Majerus, 2009, as also referenced in this book.)

Natural selection may act on a population to affect variable traits in three ways: 1) directional selection, moves the trait in one direction or the other in a changing environment away from a disadvantageous extreme for the trait, i. e., one extreme phenotype for the trait is fittest and therefore selected; 2) stabilizing selection (also referred to as normalizing selection) is selection against disadvantageous extremes for a trait, such that intermediate phenotypes for the trait are favored by natural selection; and 3) disruptive selection (also called diversifying selection), a type of selection where the more extreme values of trait distribution are favored over the intermediate phenotypes for the trait, i. e., two (or more) phenotypes are favored (fitter) than the intermediate phenotypes (if selection is strong for two extreme phenotypes, the population may split into two main phenotypes for this trait) (Miller and Harley, 1994; Campbell, 1996; Ridley, 1996; Campbell, Reece, and Mitchell, 1999; Freeman and Herron, 2007; Kardong, 2008; Miller and Devine, 2008; Futuyma, 2009; Thompson, 2013; Zimmer and Emlen, 2013; Urry, et al, 2014).

One way that we can express these different modes of natural selection is with a normal (or Gaussian) curve that plots the value of the phenotypic trait on the abscissa (x-axis) versus the frequency of occurrence (as raw numbers or percentages of the random sample that represents the population) on the ordinate (y-axis) (see Figure 8-1). Using these curves, we can compare the modes of selection that we discussed above assuming a heritable quantitative (or continuously varying) trait and we can compare the changes in the trait distributions in the population as time progresses (see Figure 8-2).

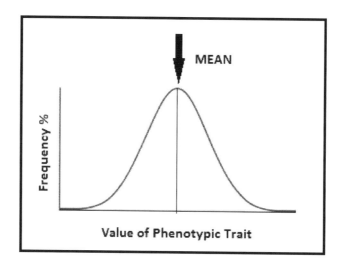

Figure 8-1. A diagram illustrating a normal (or Gaussian) statistical probability curve (sometimes referred to as a bell-shaped curve) that represents a hypothetical heritable, continuously varying, phenotypic trait, in a large random sample of a population for a specific organism. The line down the center of the curve dividing it into two equal parts represents the mean value for the phenotypic trait and is also shown by the position of the arrow. For example, a curve similar to this could depict the pattern of height in humans based on a random sample of 1000 people (of the same sex).

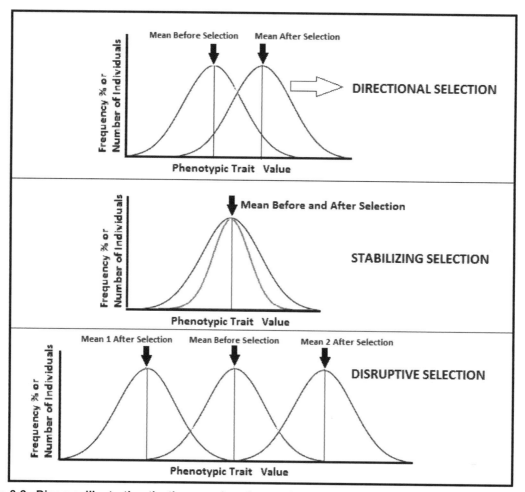

Figure 8-2. Diagram illustrating the three modes of natural selection.

In a changing environment for a particular trait, directional selection is typically the more common of the three modes of natural selection. According to Thompson (2013, p. 35), "The published studies suggest that directional selection may be more common than stabilizing or disruptive selection, but, ..., these patterns must be viewed cautiously." Melanic industrialism in the moth *Biston betularia* is an example of directional selection, as is the evolution of beak size and shape in Darwin's finches, both of which we have discussed previously. I have discussed in some detail the melanic industrialism example, but let's look a little closer at the work (lasting about 40 years) of Peter and Rosemary Grant of Princeton University relative to changes in beak size of Darwin's finches on Daphne Major, a small volcanic island in the Galápagos archipelago. For brevity and to illustrate a case of directional selection, I will only discuss one example for one of Darwin's finches, the medium ground finch (*Geospiza fortis*), of the change in beak depth (determined by the Grants to be most associated with survival during stressful times) both before and after the severe drought on Daphne Major during 1977. The Grants had already determined that beak depth is a highly heritable trait (Boag, 1983; Weiner, 1994; Grant and Grant, 2008; Grant and Grant, 2014).

The Grants arrived on Daphne Major to begin studies of the finches there in 1973. Daphne Major was small enough that they (and their colleagues and graduate assistants) could capture, band, and measure most of the birds on the island. From 1973 through 1976 rain was plentiful on Daphne Major, so the food (both small, soft seeds and larger, hard and tough seeds) was plentiful to supply the larger finches, with large, deep beaks (primarily the medium ground finch, *Geospiza fortis* during the 1970s) and the smaller finches, with smaller, less deep beaks (some of the *G. fortis*, but also the cactus finch, *G. scandens*, and the smallest of the finch species on Daphne Major, the small ground finch, *G. fuliginosa*). (Note: the large ground finch was not a significant component of the finch population during the 1970s, but did immigrate in significant numbers to the island during the 1980s.) But 1977 brought a severe drought, only 24 mm of rain fell on the island from December through June (the wet season), compared to 127 mm and 137 mm during the wet periods for 1976 and 1978 (Boag and Grant, 1981).

Throughout the drought of 1977 the total mass of seeds on the island went down and the average size and hardness of the seeds increased (Weiner, 1994). The smaller, softer seeds were consumed first by all the finches, but when they were gone only the larger of the medium ground finches (*Geospiza fortis*) with the deeper beaks could eat the hard seeds that remained (Weiner, 1994; Boag and Grant, 1981; Grant and Grant, 2008; Grant and Grant, 2014). As the food supply decreased the total number of finches on Daphne Major decreased from 1,400 in March 1976, to 1,300 in January 1977, to fewer than 300 in December 1977 (Weiner, 1994). *G. fortis* numbers decreased from 1,200 individuals at the beginning of 1977 to less than 300 birds in December 1977, an 85% decrease (Weiner, 1994; Boag and Grant, 1981). Certainly, this was a bottleneck effect that reduced the overall genetic variation in the surviving members of the *G. fortis* population on Daphne Major. According to Boag and Grant (1981, p. 84) relative to their study of the very stressful directional selection of *G. fortis* during the 1977 drought: "Our results are consistent with the growing opinion among evolutionary ecologists that the trajectory of even well-buffered vertebrate species is largely determined by occasional 'bottlenecks' of intense selection during a small portion of their history…Occasional strong selection of heritable characters in a variable environment may be one of the keys to explaining the apparently rapid adaptive radiation of the Geospizinae in the Galapagos…" (Note: Geospizinae is the subfamily to which the 14 species of Darwin's finches are classified.)

For *G. fortis*, only the largest birds with the deepest beaks that could crack open the hardest seeds left on the island, survived. The surviving *G. fortis* birds averaged 5 to 6% larger than the birds that died (Weiner, 1994). The average beak size of the *G. fortis* breeding (parental) birds in 1976 prior to the drought (before selection) was 10.68 mm (millimeters) long and 9.42 mm deep, whereas the average beak size of the *G. fortis* birds that survived the 1977 drought (after selection) was 11.07 mm in length and 9.96 mm in depth (Weiner, 1994). Beak depth is the most important factor in determining the ability of *G. fortis* to crack larger seeds. So, a difference of just over 0.5 mm in beak depth for these birds was the difference between life and death.

The above change of the breeding stock beak depth of 1976 (before selection) to the beak depth of the surviving *G. fortis* birds after the 1977 drought (after selection) "was the most intense episode of natural selection ever documented in nature" (Weiner, 1994, p. 78). As stated by Boag and Grant (1981, p. 82) for the strong and rapid directional natural selection indicated above: "Selection intensities, ..., are the highest yet recorded for a vertebrate population."

But, even though intense and very rapid natural selection had taken place during the 1977 drought for the medium ground finch (*G. fortis*), had this translated into any evolutionary change? In order to determine if an evolutionary change had occurred, the beak depths of the fully grown offspring hatched in 1978, of the 1977 drought survivors, had to be compared to the fully grown offspring hatched in 1976 (before selection). This was completed by Peter and Rosemary Grant and the results were published in 1995 (Grant and Grant, p. 244, Table 2) and 2003 (Grant and Grant, p. 969, Figure 5). In the 2003 article in the journal *Bioscience*, the Grants show histograms of beak depth plotted versus numbers of finches for the 1976 offspring as compared to the 1978 offspring (see Grant and Grant, 2003, p. 969, Figure 5; also see Zimmer and Emlen, 2013, p. 222, Figure 8.4). The two histograms show an obvious directional shift in the mean of beak depths to deeper beak depths in the 1978 offspring, with the beak depths for the 1976 offspring averaging about 9.2 mm as compared to the beak depths of the 1978 offspring averaging about 9.7 mm (Grant and Grant, 2003, p. 969, Figure 5; Zimmer and Emlen, 2013, p. 222, Figure 8.4). So, yes, a statistically significant evolutionary change had taken place in beak depth for the *G. fortis* finches from 1976 to 1978 due to natural selection resulting from the 1977 drought (Weiner, 1994; Grant and Grant, 2003; Grant and Grant, 2008; Zimmer and Emlen, 2013; Grant and Grant, 2014). Grant and Grant (2003, p. 969) state this well in the last sentence of their caption for Figure 5: "Evolutionary change between generations is measured by the difference in mean between the 1976 population before selection and the birds hatched in 1978."

So, a significant and rapid evolutionary increase in beak depth occurred as a result of the 1977 drought. But just a few years later, 1983, a most dramatic ecological change occurred that affected natural selection in *G. fortis*. This was the year of an unusually strong El Niño event (the most severe in 400 years) (Grant and Grant, 2008) and the associated heavy rains. The plants responsible for the large, hard seeds (primarily *Tribulus cistoides*) that were critical to the survival of deep beaked *G. fortis* during the 1977 drought were grown over and smothered by the rapid and extensive growth of *Merremia aegyptica* vines (Grant and Grant, 2008). The heavy rains (1359 mm in 1983) and rapid growth of *M. aegyptica* vines continued for eight months and the effects of the El Niño were evident for the next few years (Grant and Grant, 2008; Grant and Grant, 2014). The *M. aegyptica* vines and other plants that grew quickly during the El Niño year primarily supplied small seeds that favored small-beaked *G. fortis* finches with more pointed beaks (Grant and Grant, 2008). Thus, by 1985, when the next drought began, "the direction of evolution had reversed" (Grant and Grant, 2008, p. 55) to favor *G. fortis* birds with smaller, more pointed beaks.

Oscillating directional selection in beak depth (and other beak traits and body size) has occurred in *G. fortis* over the 40 years that the Grants and their colleagues have been studying Darwin's finches on Daphne Major. Thus, *G. fortis* has changed in morphology over the 40 years of study (Grant and Grant, 2014) and "is smaller in average body size and has a more pointed bill than it did in 1973, ..." (Grant and Grant, 2008, p.57).

If a species, like *G. fortis*, experiences oscillating directional selection for a specific trait over an extended period of time, the net result might not be much net evolutionary change. Thus, if we only saw the beak depth for *G. fortis* in 1975 (before the drought of 1977, which selected for larger, deeper beaks) and then again in the late 1980s (after the El Niño event of 1983 that resulted in selection for smaller, more pointed beaks), we would conclude that not much evolutionary change had occurred in beak size and shape. When Charles Darwin looked at the fossil record he saw what appeared to be very slow evolutionary change, so he reasoned that natural selection must occur at a very slow pace also. In the fossil record it is rare to get successive generations of a species preserved and typically there may be thousands or sometimes even a few million years between fossil occurrences for a particular species. The following statement by Peter Grant, when interviewed by

Jonathan Weiner for his 1994 award-winning book *The Beak of the Finch* (see p. 111) will make this clearer:

> 'There's quite a bit of wobble in the fossil record too', says Peter Grant. 'But you don't normally see it. When you look at a series of fossils, the usual practice is to take 3,000 or 5,000 or 10,000 years as the unit. You take the average position at the beginning and the end, and average the rate of change in between.
> Even then, you see wobbles in the record. It's one in a thousand wobbles, if you like. You rarely get successive generations preserved in fossils. It's like looking at specimens of Darwin's finches from 1874, 1932, and 1987. You may get some wobbles there – but you miss all sorts of action in between.'

How do we measure rates of evolution? The British naturalized Indian geneticist and evolutionary biologist J. B. S. Haldane (1949) was the first to devise a method for measuring the rate of evolution (Haldane, 1949; Weiner, 1994; Thompson, 2013). Haldane wanted to answer the question of how much a measured character in a population has evolved (changed) in a certain direction over a specific time interval (Thompson, 2013). Haldane (1949) called the unit of evolution the darwin (named in honor of Charles Darwin). The parameters needed to determine the rate of evolution in darwins is the mean of a group of measurements of a trait of a species at a certain time (x_1), the mean of a group of measurements of a trait of a species at another time (x_2), and the amount of time between the two means of the measurements ($t_2 - t_1 = \Delta t$) (Thompson, 2013). Haldane (1949), also a mathematician, devised the following formula to determine the rate of evolution in darwins if time is measured in millions of years (Thompson, 2013):

$$d = \frac{\ln x_2 - \ln x_1}{\Delta t}$$

A darwin measures the amount of change in a trait over the time interval of one million years as the change in *e* (the natural logarithm base, approximately 2.718), so 1*e*/million years equals 1 darwin. This measure of evolutionary rate can only be used for length measurements (not area or volume measures of biologic traits).

This method has been used in the past by several workers (Haldane, 1949; Gingerich, 1983; MacFadden, 1992; Gingerich, 1993; Gingerich, 2009; and others) to determine the rate of morphologic evolution in fossil animal lineages. Gingerich

(1983, p. 159) determined that this method might not give accurate results when comparing "narrow ranges of morphological variation over a wide range of time intervals."   He found an inverse relationship to the interval of time over which measurements were taken when comparing morphological evolution in lab selection experiments, historical colonization events, and the fossil record. Gingerich (1983) found very high rates for lab selection experiments (12,000 darwins to 200,000 darwins, with a mean of 58,700 darwins), moderate values for historical colonization events (0 to 79,700 darwins, with a mean of 370 darwins), and low values for fossils (0 to 26.2 darwins, with a mean of 0.08 darwins). But the method seems to work well if similar time ranges for morphological change are compared, but the times must be well calibrated. According to MacFadden (1992, p.191), relative to studies of morphologic change in fossil horse dental parameters during the Tertiary and Quaternary (Paleogene, Neogene, and Quaternary), the following is the case:

> Calculation of darwins can be used to study both microevolutionary and macroevolutionary changes, for example, within populations of chronoclinal species or between ancestral and descendant species... However, in any of these cases, an obvious requirement is that the ages of the populations or species being studied are well calibrated. This method, whether it is within or between species, does not account for evolutionary reversals in individual character trends between measured endpoints, so darwins actually provide an assessment of minimum rates of morphological change.

(Note: Microevolution is evolution below the species level, whereas macroevolution is evolution above the species level [such as speciation, or the evolution of higher taxonomic levels]. Chronoclinal species [also called chronospecies or paleospecies] are lineages represented in the fossil record by almost continuous series with no obvious breaks in morphologic traits such that, in order to divide the lineage into species, arbitrary breaks must be chosen [Gingerich, 1979; Kardong, 2008]. Of course, this has to be the result of phyletic gradualism [slow, gradual evolution] [Gingerich, 1979]).

If we were to calculate the rate of evolution in darwins of beak depth for *Geospiza fortis* before and after the drought year of 1977 (adult offspring in 1976 compared to adult offspring in 1978), we would get a value of about 26,500 darwins. Of course, the El Niño event of 1983 resulted in evolution of smaller values for beak depth and resulted in a reversal of the evolutionary rate due to the 1977 drought. Thus, the net evolutionary rate from 1976 through about 1990 was much less.

Another measure of the rate of evolutionary change that Haldane (1949) suggested may be better than the darwin unit; this is called the haldane (Gingerich, 1993; Gingerich, 2009; Thompson, 2013). According to Thompson (2013, p. 17):

> As Haldane wrote, variation within a population is the raw material available for evolution. He therefore based his alternative measure on change in standard deviations per generation. Calculation of an evolutionary rate in haldanes therefore requires a solid assessment of variation in a trait in addition to the average value. Haldanes are also more intuitive when thinking about the relentlessness of evolution, because they measure change per generation, whereas darwins are measured in change per million years. In recent years, evolutionary rates have become commonly measured in haldanes, although studies also commonly report rates in darwins.

The following formula may be used to calculate evolutionary rate in haldanes:

$$\mathbf{h} = \frac{\left[(\ln x_2 - \ln x_1)S_p\right]}{\mathbf{g}}$$

The equation determines the evolutionary rate of change in a trait in standard deviations per generation (g). The two means are the same measurements as for darwins, but in addition the pooled standard deviation (S_p) of the two samples is required (Thompson, 2013).

As opposed to directional selection, where one or the other of the extreme values of variation of a trait is selected for, or disruptive selection, where both ends of the normal distribution of an initial population are selected for, stabilizing selection reduces the variation in a population (reduces the standard deviation around the mean) for a particular trait such that the population becomes more similar for that specific trait. In stable environmental settings, this may be the favored mode of natural selection because the population has adjusted to local conditions (Kardong, 2008). A classic example of stabilizing selection is for human birth masses (or weights in pounds) of 3 to 4 kilograms (6.6 to 8.8 pounds), as compared to larger and smaller newborns; i. e., newborns of 3 to 4 kilograms are favored for survival over larger or smaller infants (Campbell, Reese, and Mitchell, 1999; Kardong, 2008;

Coyne, 2009). Smaller newborn babies have higher heat loss, get sick more often, often are premature with developmental problems, and have higher infant mortality rates than average sized newborns (Campbell, Reese, and Mitchell, 1999). Larger, heavier babies experience more difficult births because of problems getting through the pelvic birth canal and may be injured or die in the birthing process. Larger newborns also have higher infant mortality rates (Campbell, Reese, and Mitchell, 1999).

Another clear example of stabilizing selection is the changes in body size of Atlantic cod (*Gadus morhua*) off the coast of Norway as measured in juveniles of the coastal population using a standardized seine over the past ninety years (Thompson, 2013, p. 36-37). The mean body length of the fish has not changed over this time period, but the standard deviation has decreased significantly, indicating stabilizing selection (Thompson, 2013, p. 37). These studies have been supplemented more recently with mark-recapture studies. As pointed out by Thompson (2013, p. 37), "These studies show that selection over the past century has acted against the fish with extreme traits: large and fast-growing juvenile cod at the one extreme, and small and slow growing juveniles at the other extreme. Together, the survey and mark-recapture studies suggest strong selection that restricts variation in juvenile cod size and growth rate to a narrow range near the average for the population."

More rarely than directional selection and stabilizing selection (although not as studied as those modes of selection), disruptive selection may occur in a population due to phenotypic polymorphism. Polymorphism is said to be present in a population when two or more morphologies (termed morphs) are sufficiently abundant to be easily distinguished (Campbell, Reece, and Mitchell, 1999). For disruptive selection to be the result of polymorphism, natural selection must favor variation at some gene loci. The ability of natural selection to preserve diversity in a population is referred to as balanced polymorphism (Campbell, Reece, and Mitchell, 1999, Futuyma, 2009). Environmental patchiness may result in selection for different phenotypes (and thus genotypes if the traits in question are highly heritable) in different portions of the general habitat environment (microhabitats) or different niches (such as slightly different food resources, an example of multiple-niche polymorphism) (Campbell, Reece, and Mitchell, 1999; Futuyma, 2009). For example, studies of the African finch called the black-bellied seedcracker (*Pyrenestes ostrinus*) in south-central Cameroon by Thomas Bates Smith during the late 1980s and early 1990s illustrate an example of balanced polymorphism that has led to disruptive (diversifying) selection (Smith, 1990; Smith, 1993; Campbell, Reece, and Mitchell, 1999; Freeman and Heron, 2007; Futuyma, 2009). This is an example of multiple-niche polymorphism (Futuyma, 2009). These finches have a bimodal distribution of bill width (Smith, 1990; Smith, 1993; Futuyma, 2009) that appears to be due to a single allele difference between the wide beaks and the narrow beaks (Futuyma, 2009). The two different morphs of beak size (primarily beak width) for these finches result in different feeding efficiencies for processing their primary food, seeds of different species of sedges. The narrow-beaked finches process soft seeds more efficiently, whereas, the wide-beaked finches crack the hard seeds more efficiently. Smith (1990, 1993) determined that finches that have intermediate beak widths have lower fitness for survival on the major food source than the wide- or narrow-beaked finches. Unlike in the Galapagos Islands, where climate variations from year to year affected the directional selection of Darwin's finches, south-central Cameroon has a wet and dry season, but there is little variation from year to year in total rainfall (Smith, 1990). Thus, selection of beak width primarily occurs during the dry season when sedge seeds are sparser (Smith, 1990). According to Futuyma (2009, p. 317-318): "Smith banded more than 2700 juvenile birds and found that survival to adulthood was lower for birds with intermediate bills than for wide- or narrow-billed birds. Thus, diversifying selection, arising from the superior fitness of different genotypes on different resources, appears to maintain the polymorphism."

To summarize, although natural selection is the most important concept in the biological sciences (and one of the most important concepts in the history of thought) to explain the common descent and diversity of organisms on Earth (Futuyma, 2005), it is basically a very simple process. As pointed out by Shaw, et al (2005), Darwin's (and Wallace's) idea of natural selection may be reduced to the following: 1) individuals vary in the traits they possess (thus, variations exist for traits within a population), 2) most of these traits can be genetically inherited, and 3) that some individuals survive and reproduce better than others because they possess certain traits (i.e., some individuals are better adapted to their environment than others within the population and thus are more fit in terms of survival and reproducing their inheritable traits in their offspring).

SEXUAL SELECTION

Sexual selection is a form of natural selection, primarily in animals, in which males compete for mating access to the females and females choose their mating partners (sometimes these roles are reversed). Darwin was the first to propose this concept in his 1871 *The Descent of Man and Selection in Relation to Sex* (Darwin, 1871; Shaw, et al, 2005). Sexual selection happens when some individuals within a population are more successful at attracting mates and mating than others in the population because of the traits that they possess, and thus pass on more of their traits to the next generation (i.e. more reproductively successful). Many males have strong sexually dimorphic characteristics to attract females, such as the large antlers of the Wapiti elk to attract females and fight with competing males (Futuyma, 2009), the elaborate peacock's tail to attract peahens, or the male red winged blackbird that flashes his colorful shoulder feathers at the edge of his territory during courtship and territorial defense (Kardong, 2008). In most cases, the female is much drabber or less elaborate in coloration or other sexually dimorphic features than the male, but she gets to choose the male that she is most attracted to for mating. Of course, her choice is based on choosing the mate that will increase her chances of producing and rearing more offspring. Often the sexually dimorphic features of males are so elaborate or seemingly unwise (for example, loud vocalizations by frogs to attract mates, but may reveal their location to predators) as to hinder survival (Shaw, et al, 2005). However, even with this higher risk of not surviving, if a trait greatly enhances mating/fertilization of an individual, then that individual has the potential to increase its offspring (and thus the trait in question) in a population due to sexual selection (Shaw, et al, 2005).

What has now become somewhat of a classic example of a study of sexual selection is the work by John Endler in the 1970s on the guppy in freshwater mountain streams in Trinidad (Losos, 2017). The guppy (*Poecilia reticulata*) is a common aquarium fish that is familiar to most of us. This fish is native to the small streams in Venezuela and the nearby island of Trinidad (Endler, 1978; Endler, 1980; Reznick and Endler, 1982; Kardong, 2008; Losos, 2017). Guppies reproduce relatively rapid, so they are ideal animals to do both controlled laboratory experiments and field experiments, rather than to just observe natural populations in the field. Experimental studies allow additional testing to determine if natural selection is working on the guppies in their natural environment, and if so, the rate at which selection pressure results in evolutionary change (Losos, 2017). John Endler in the 1970s was one of the first to study guppies in the mountain streams of Trinidad. In these streams in Trinidad, waterfalls separate shallow pools and the pools harbor guppies. Endler (1978) noted in field investigations that there were differences in the secondary sexual characteristics of male guppies. Male guppies varied from stream to stream and within the same stream in different pools. Male guppies in upstream pools near the source of a stream were brightly colored, with many colorful spots, and also had enlarged tail fins that exhibited flashy colors (Endler, 1978; Kardong, 2008). However, farther downstream (and below waterfalls), where more predator fish, such as the pike cichlid (*Crenicichla alta*) were abundant, male guppies had few spots and were much drabber in color (Endler, 1978; Endler, 1980; Reznick and Endler, 1982; Kardong, 2008; Losos, 2017). Endler (1978) also noticed that there were differences in color and spots of male guppies, even in the same larger downstream pool (where abundant predators were present), with variations in the substrate (coarser gravel versus finer gravel). Where predators were present, guppies tended to blend in with their substrate; coarser substrate, larger and darker spots, as compared to finer and lighter spots in finer substrate (Endler, 1978; Endler, 1980; Reznick and Endler, 1982; Losos, 2017). Endler (1978) hypothesized that the secondary sexual characteristics expressed by the males, particularly in the upstream pools where not as many predators could get over the waterfalls, was due to sexual selection and that brighter colored male guppies were most often chosen by females and thus would pass on their traits to the next generation. He further hypothesized that natural selection was at work in a predator-prey situation. Brightly colored male guppies would be more visible targets for the predators. Thus, where predators were plentiful, drabber guppies were selected for (via natural selection), whereas where predators were not present or few in number, females selected the more colorful male guppies for mating (Endler, 1978; Endler, 1980).

In order to test this hypothesis, and others about substrate and life-history strategies, Endler and his associates did a series of both laboratory and field experiments (Endler, 1978; Endler, 1980; Reznick and Endler, 1982; Losos, 2017). Endler and his colleagues built several large artificial ponds in a laboratory greenhouse to test their hypotheses. They could control which ponds had predatory fish and how many. They could also control the substrate and other variables. In a parallel field experiment in Trinidad, Endler and associates took drab colored guppies from a predatory infested pool and placed them in an upstream pool that had no predators. Over several years, and based on observations of the laboratory ponds and the field pool, Endler and his associates collected data that was consistent with their hypotheses. That is, where predators are not present (or fewer and/or of lower aggressiveness towards guppies), the male guppies tend to become large, with many spots, and are very colorful. Where predators are present in significant numbers (in particular the pike cichlid), as time goes on male guppies evolve a very drab coloration with few spots and are small (Endler, 1978; Endler, 1980; Reznick and Endler, 1982; Kardong, 2008; Losos, 2017).

These studies by Endler and associates illustrate the nip-and-tug coevolutionary relationship between sexual selection in guppy populations and their common predators. Endler and colleagues show that natural selection may show differences in reproductive rate, not just survival (Endler, 1978; Endler, 1980; Reznick and Endler, 1982 Futuyma, 2009; Losos, 2017). These studies show that characteristics of organisms may be subject to conflicting selection pressures and trade-offs between advantageous characters (in this case, mating success of male guppies due to dimorphic features that attract female guppies) and the corresponding disadvantages to survival of possessing these characters (in this case, dimorphic coloration and spotting that attracts predators as well as female guppies) (Futuyma, 2009, p. 288). Endler and his colleagues found that if pike cichlid predators were present that eat larger males, a guppy life history pattern evolved such that male guppies tended to be small, mature early, and father large broods (i.e. with more effort going into reproduction) (Endler, 1978; Endler, 1980; Reznick and Endler, 1982; Kardong, 2008; Losos, 2017). However, in pools that had low predation or moderate predation with predators that preferred juvenile guppies, male guppies tended to grow larger, mature later, and father smaller broods (i.e. with less effort put into reproduction per say, but more effort into growth and developing more elaborate sexual dimorphic features) (Endler, 1978; Endler, 1980; Reznick and Endler, 1982; Kardong, 2008; Losos, 2017). In another study of guppies (Godin and Dugatkin, 1996), researchers doing experimental studies on male and female guppy behavior determined that females appear to choose more bold males (relative to evading predators) regardless of their coloration. The inference being that if a male is bolder in approaching predators, he may also be more fit and a good choice as a mating partner to insure more fit offspring (Godin and Dugatkin, 1996).

GENETIC DRIFT

The other major mechanism by which evolution can take place is genetic drift. However, unlike natural selection (including sexual selection), genetic drift is a random process and does not result in adaptation. Genetic drift is just the chance that a particular gene (allele) frequency will be passed on to the next generation. If a beneficial mutation arises in a population in only one individual, there is a high probability that the new form of the gene (allele), just by chance, will not become fixed in the population (particularly in a large population). The tendency of genetic drift is to reduce variation in a population. Of course, this reduces a population's ability to evolve in a changing environment with new selection pressures. If a population has suffered reduction in the number of individuals due to environmental factors (or human induced factors) such that the population size is greatly decreased, then a bottleneck event is said to have occurred, and because of genetic drift the genetic variation within the population will be greatly reduced. This has happened to the African cheetahs in the recent past and may result in their extinction if new selection pressures are imposed on them (Kardong, 2008). In small isolated populations (that do not have the variability of the larger population that it has become isolated from), genetic drift (particularly when combined with new mutations) may result in speciation. This is sometimes referred to as the founder effect.

So, what is the origination of the variations in populations that result in evolutionary change by natural selection or genetic drift? As pointed out by Futuyma (2009), we now know that inherited variation arises from mutations of genes that alter the DNA sequences. Mutations result in a change in the hereditary information that is passed on to offspring. Mutations that take place in sex cells are inheritable, whether they are chromosomal mutations, affecting a large segment of a chromosome, or point mutations that result in individual changes in particular genes (nucleotide substitutions, deletions, or insertions).

Mutations are random with respect to fitness and they may be beneficial, neutral, or harmful (Kardong, 2008). Because mutations are random, they do not arise because of "need". As pointed out by Futuyma (2009), "...an environmental change does not increase the likelihood that just the right mutation will occur."

Why are organisms on Earth so tremendously diverse, even though (as shown recently with gene sequencing techniques) much of the DNA in many organisms is quite similar? The recent advances in evolutionary development (evo-devo) of the embryo for sexually reproducing organisms may hold some of the answers. We have known for a long time that embryos in vertebrates (and within other groups also) are very similar at early stages of embryological development (Kardong, 2008). So why are these organisms so different as adults, even though they contain many very similar genes (and DNA sequences). This may be because, in addition to structural genes that code for particular amino acids that eventually build specific proteins, there are also major regulatory genes (like the Hox genes, a set of regulatory genes that partition the body of complex organisms) that control the timing and place within the embryo where major switches are located that turn on (or off) structural genes that code for amino acids (and thus protein formation). A particular region of the sequence of Hox genes in a mouse is similar to the same region in a human and serves a similar function (to turn on the structural genes at a particular place and growth time to develop a mouse foot, for example, rather than another structure or organ at that position). Mutations in regulatory genes may result in big changes in the developing embryo. Most of these would most likely be lethal or disadvantages, but sometimes they might provide a significant advantage to an organism so that natural selection (and/or genetic drift in small populations) could work on fixing that change into the genome (Kardong, 2008).

Recently, I watched a Nova documentary on PBS entitled *"What Darwin Never Knew."* This was an excellent video on evolution in general and covered many topics that we have been discussing in this book. However, it was particularly heavy on evolutionary development (evo devo) and the development of embryos in complex organisms. One part dealt with the idea that hominins (humans and human ancestors and cousins that evolved after the split with chimpanzees) may have evolved big brains because of some mutations in regulatory genes that reduced the size of the jaw muscles (as compared to chimps and gorillas), that left more room and less constraint on the skull bones so that a larger brain could develop. Of course, if this happened in the past, and the isolation of a region of the genome that controls this has apparently been found by researchers, then natural selection would certainly have favored this trait in our ancestors. (The video of the Nova documentary on this topic can be found at the following site: http://www.pbs.org/wgbh/nova/evolution/darwin-never-knew.html I highly recommend it, but it is about 2 hours long.)

NATURE VERSUS NURTURE (OR GENES VERSUS ENVIRNOMENT)

Although all organisms have a genetic potential that to a large extent determines what that organism will develop into, the environment often has a large imprint upon the specific development of an organism. Not only does the environment (including the embryological environment for sexually reproducing organisms) play a large role in determining the timing and activity of regulatory genes 1) in turning on and off structural genes and 2) in regulating the potential expression of structural genes, the environment may also impact the way that genes interact with each other to yield a particular phenotypic expression. Thus, organisms of a particular species with very similar genotypes have great phenotypic plasticity for some traits due to environmental influence (Englebrecht, 2003).

Recent advances in the study of epigenetics document how environmental influences can alter gene expression without changing the genetic sequence. Epigenetics is the study of chemical changes to DNA and its associated proteins, the histones, which can result in modifications of gene expression (Englebrecht, 2003; Carey, 2012). Evidently this chemical alteration (to the structure of the DNA and histones by methylation and acetylation) due to environmental influences does not change the DNA sequence, but the phenotypic expressions of the genes are altered and this alteration can be transmitted to the next generation (Englebrecht, 2003; Carey, 2012). Of course, the best example of epigenetics is the alteration of stem cells in eukaryotic organisms to various other types of cells within the body; thus, once a liver cell has been epigenetically altered it will produce other liver cells during reproduction (mitosis) (Futuyma, 2009). However, epigenetic changes are not an example of inheritance of acquired characteristics (as one might get the impression from what is stated above). One of the causes of epigenetic change is methylation of a portion of the DNA sequence, in which a methyl group (CH_3) is connected to a cytosine (C) in a CG pair (Futuyma, 2009; Carey, 2012). "This methylation process is repeated during DNA replication, and the methylated state, which often reduces or eliminates gene transcription, may be transmitted for few or many generations" (Futuyma, 2009). There is no evidence to date that epigenetic changes cause evolutionary change, however, there is considerable debate about whether or not it is likely to result in evolutionary change (Futuyma, 2009). Epigenetic changes to the DNA are really forms of mutational change that are induced by environmental factors (Englebrecht, 2003; Futuyma, 2009).

The Mexican salamander *Ambystoma* illustrates a good example of how the environment can influence the development of an individual organism. Usually, during normal environmental conditions for the salamander, it retains its larval gills into adulthood (i.e. it does not undergo metamorphosis to become a lung breathing adult salamander) and experiences the same aquatic lifestyle as the larva. However, if the environmental conditions deteriorate such that the water becomes stagnant, the salamander will undergo metamorphosis to become a true adult form with functional lungs, allowing it to find a new body of fresh, unstagnated water and survive (Prothero, 2004). This ability to postpone or hasten ontogenetic changes due to environmental influences gives this salamander great ecological and phenotypic flexibility. This ability and the resulting change in phenotypic characteristics are often referred to as ecological or phenotypic plasticity resulting in ecophenotypic variation. There are many examples of organisms in nature where changes in environmental parameters, such as nutrient availability, light, temperature, chemistry of the environment, etc., result in phenotypic plasticity producing ecophenotypic variation (Prothero, 2004). Think about social insects, such as ants. All ants in a colony have almost identical genotypes. However, because of the way they are nourished and given other chemical stimuli they develop into workers, soldiers, winged males, etc., thus they have great phenotypic variation due to environmental influences (Prothero, 2004).

Although there is great variation in some traits because of environmental influences, other traits do appear to be under strong genetic control. But even here, the environment can influence the ultimate phenotypic expression to some extent. For example, as Englebrecht (2003) points out, the only naturally occurring genetically identical humans are identical twins or triplets or quadruplets, etc. If we compare identical twins that grew up together versus identical twins that lived in different places during their lives, we can get some idea of the nature (genetic factors)/nurture (environmental factors) differences in traits. The data from identical twin studies has also been compared for fraternal twin and sibling data gathered in a similar manner (Englebrecht, 2003). One study that was mentioned by Englebrecht (2003) for identical twins seemed to suggest that intelligence may be strongly correlated with genetics rather than environmental influence (but perhaps some by environmental influence). In this study of identical twins, it was concluded that 85%

of twins that lived together had the same IQ, whereas only 74% of twins that were raised apart had the same IQ.

We certainly know that there are certain mutations that occur to genes or chromosomes that override the environmental influences and cause genetic disorders or diseases; diseases and disorders such as sickle cell anemia, hemophilia, Turner syndrome (females that only have one X chromosome and do not acquire secondary sex characteristics)(Robinson, 2005), cystic fibrosis, some forms of breast cancer, and alkaptonuria (urine turns black when exposed to air – Griffiths, et al, 2002), to name a few. As pointed out by DeSalle and Yudell (2005), the identification of perfect pitch appears to be controlled exclusively by a person's genetic make-up. So, it appears that many traits are fairly strongly controlled by our genes, but for many other traits there appears to be a complex interplay of interacting genes where environmental influences are important in terms of the expressions of traits associated with these genes.

The nature/nurture debate has continued because it is critical to understand the influence of both the environmental influences on organisms (particularly humans) and the interplay of genetic interactions with the environment in order to adequately understand how individuals develop (their ontogeny) and also how populations evolve. According to DeSalle and Yudell (2005), Harvard University evolutionary biologists Richard Lewontin "sees this interaction as a triple helix, suggesting that an organism is the product of a 'unique interaction between the genes it carries,' the 'external environments through which it passes during its life,' and the random 'molecular interactions within individual cells.'" For humans, understanding these relationships is critical for diagnostic and preventive medical practices relative to our genetic make-up and for the treatment of various genetic disorders and diseases.

The mapping of the human genome and the rapid improvement of DNA sequencing technology will hasten the research on gene interactions and how the environment impacts these interactions. This will allow for the development of better medical treatments and adjustments of individual environments to prevent some adverse environmental influences on particular genes and gene interactions. Of course, as we have already been discussing, there will be major ethical issues that will arise (and that are currently arising) as more people have their genomes sequenced.

CHAPTER 9

SPECIES AND SPECIATION

INTRODUCTION

How much variation in living organisms and fossils is a result of ontogeny, sexual dimorphism, ecophenotypic variation (different phenotypic expression due to environmental factors), population variation, deformation of fossils, etc. (Prothero, 2004)? For modern organisms, how does morphologic, physiologic, and behavioral characteristics relate to differences in DNA sequences between organisms? And how much variation can be the result of individuals belonging to a different species? These are very important questions that biologists and paleontologists must grapple with when identifying, classifying, and determining the evolutionary relationships of organisms. Biologists and paleontologists must be able to define what a species represents and must understand how speciation occurs. The species concept and the process of speciation is very important in systematics (including taxonomy) and working out phylogenetic relationships (i.e., the tree of life), in ecology and paleoecology, determining the diversity of life (both in the present and the past), in determining the severity of mass extinction events, and in wildlife conservation (for example, how do we define the biologic unit that should be protected as endangered species). Understanding speciation and what constitutes a species is essential if biologists and paleontologists are to understand evolutionary patterns and how this relates to natural selection and other mechanisms of evolution (Kardong, 2008).

The formation of new species is the basic response to mechanisms of evolutionary change. However, how we define species is another matter. The pre-Darwinian view of species is that God created the ideal type for each species. This is a holdover from the ancient Greek idea of a Typological Species Concept. Any deviation from the ideal type was viewed as an imperfection from God's blueprint (Prothero, 2004). Species were then thought to be fixed and unchanging (or if variations were present they were viewed as degradational variations from the ideal type). With Darwin's publication of *On the Origin of Species*, it became apparent that species were not static and did not fit a "blueprint" type. Darwin (1859) realized that because of the natural variation in populations, changes in species would occur with time because of the continued struggle for existence and the continued selection for the more fit. Thus, variation is the basis for evolution and natural selection works on population variation to generate new species (Prothero, 2004).

DEFINITION OF SPECIES

Although, over the last several decades, there have been many definitions of species and much debate about what species are and ideas on how speciation takes place; modern biologists and paleontologists basically use two differing concepts for the definition of what a species is and how speciation occurs. The current controversy is basically whether the biologic species concept or the phylogenetic species concept is the best explanation for the definition of what species are (Zink, 2005). It appears that geneticists and population biologist (for the most part) accept the biological species concept, in which reproductive isolation from closely related species is paramount. On the other hand, modern systematists and conservationists seem to prefer the phylogenetic species concept, in which reproductive isolation is not as important as separating closely related species based on diagnostically distinct characters (whether they are morphological, behavioral, physiological, or genetic/molecular).

Ernst Mayr (1963) defined the biological species concept as follows: "a species is an array of populations which are actually or potentially interbreeding, and which are reproductively isolated from other such arrays under natural conditions." Or, in other words, a species is a group of naturally interbreeding or potentially interbreeding populations with a common gene pool and reproductively isolated from other species. According to Zink (2005), an alternative to the biologic species concept is the phylogenetic species concept that may be defined as a group of individuals with a unique evolutionary history, which is established by the presence of one or more diagnostic traits (features). I prefer this concept because it fits best with the morphologic species concept used by paleontologists (see discussion below). Also, the biological species concept is very restrictive and does not work well for plants due to common hybridization and polyploidy. The biological species concept does not work at all for asexually reproducing organisms.

SPECIATION

For both the biologic species concept and the phylogenetic species concept, geographic isolation is the most important criteria in speciation (Zink, 2005). Speciation resulting from geographic isolation is referred to as allopatric speciation (Mayr, 1963; Prothero, 2004; Zink, 2005). This typically results when a barrier of some kind develops that separate members of a population into geographically isolated portions of a once continuous population. This would include the founder principle, in which a few individuals become isolated from the main population and take a different evolutionary path (Prothero, 2004). For example, oceanic islands that contain small isolated populations with small gene pools; these populations may have gotten there by accident. These small populations evolve rapidly compared to the larger gene pool on the mainland (Example: Galapagos Islands compared to mainland South America). Over time, these isolated variants become diagnostically different (or in some cases reproductively isolated) from the parent population and become new species. Most speciation is thought to result from allopatric speciation (Mayr, 1963; Prothero, 2004; Zink, 2005). This type of speciation is not always due to geographic barriers (body of water, mountains, desert, etc.). Climatic differences can also result in geographically isolated portions of populations on the periphery of the main population (peripheral isolates that do not share the entire gene pool of the larger population). For example, in cold environments a certain species may grow larger, thus increasing volume at a faster rate than surface area (smaller SA/V ratio) to retain more heat (for vertebrates this might mean having a larger body torso, with shorter, stubbier ears and limbs), whereas in hotter climates the species may evolve toward being smaller to have a larger SA/V ratio (for vertebrates this might mean having a smaller body torso, with longer, slimmer ears and limbs for dissipating heat more quickly). Eventually such climatic difference of a once continuous population may result in diagnostically different traits or reproductive isolation (Prothero, 2004). An example would be the arctic bunny compared to a jackrabbit of the southwest U.S.

Peripherally isolated populations result if a portion of the population becomes isolated to the fringe of the main population, then the gene pool is smaller and any unusual gene frequencies (say from original variability in the population or from new alleles introduced by mutations) have a higher probability of becoming dominant in the peripheral population (i.e. there is no longer, or very limited, gene flow with the main population). A novel beneficial mutation in a large gene pool may be hybridized out (or at least may not spread rapidly). But in a small peripherally isolated population (with a small gene pool), a novel beneficial mutation may soon become dominant in the population by natural selection. Eventually, the gene pool of the peripheral population may become so different that they are reproductively isolated from the main population, even if other isolating barriers (geographic, climatic, etc.) are removed.

However, sympatric speciation (same place speciation) does sometimes occur, but is not thought to be very common (Cracraft, 2005). This results from some sort of reproductive isolation of a portion of the population within the same geographic area. Zink (2005) and Kardong (2008) give the example of flies that parasitize apples and hawthorns. With the introduction of apples into North America a few hundred years ago, some flies started laying eggs on apples rather than hawthorn fruit; when the larvae matured, these flies tended to remain on the apples for all reproductive activities. Even though the flies that infest hawthorns and apples are very similar in appearance, they do have detectable genetic differences; thus, it appears that speciation has occurred (Zink, 2005). Plants also sometimes undergo sympatric speciation because of hybridization and polyploidy (see Zink, 2005 and Kardong, 2008).

PALEONTOLOGISTS AND THE CONCEPT OF SPECIES AND SPECIATION

Of course, paleontologists have a different problem than modern biologists. They must have a concept of species for ancient organisms that is based on the fossils that are found in the rock record. Often, the rock record is incomplete and in most cases the rock record is biased towards organisms that had hard parts. Another factor that further complicates concepts of speciation for paleontologists is the pattern of organisms observed in the rock record. Do we see gradual evolution and speciation (anagenesis) that has taken place in the past, or more abrupt, rapid speciation followed by periods of stasis (not much change in species with time)? This is certainly an important question for paleontologists to grapple with and affects their concept of what a good paleontological species is. If the rock record primarily shows gradualistic evolution, then where do we draw the line(s) of speciation within particular lineages (which have split, diverged from a common ancestor) that appear to be slowly changing, with perhaps the end members of the two lineages being significantly different in morphology? However, this would not be a problem, of course, if we do not find the intermediate forms. Is this due to the incompleteness of the fossil record? Or, as Eldredge and Gould (1972) claim, the pattern of speciation that actual takes place. They suggest that speciation occurs rapidly, perhaps in a few thousand years (an instant in geologic time), followed by long periods (several million years for most species) of no change (stasis) (Eldredge and Gould, 1972). This concept is referred to as punctuated equilibrium (singular, or punctuated equilibria, plural). It does appear that the fossil record (for the most part) supports this concept (Eldredge and Gould, 1972; Gould and Eldredge, 1977; Gould, 2002; Prothero, 2004). These authors further point out that most evolutionary change occurs during speciation events and that speciation events follow episodes of environmental disruption and/or allopatric speciation in small peripherally isolated populations (thus high selection pressure and significant genetic drift) (and for environmental disruption in particular following mass extinction events – also see Sheehan, 2005).

Because paleontologists cannot observe living organisms and whether ancient organisms were part of an array of interbreeding populations, the biological species concept does not work for them. Paleontologists use a species concept referred to as the morphological species concept (or morphospecies concept) (Note: they also cannot, usually, obtain DNA or other organic molecular samples from fossils). Thus, differences in morphology are the primary criteria used in naming and classifying fossils into species categories. This concept in reality appears to be very close to the phylogenetic species concept. In fact, according to Prothero (2004), the morphological species concept was first proposed by Eldredge and Cracraft in 1980 with the following definition: "...a species is a diagnosable cluster of individuals within which there is a pattern of ancestry and descent, and beyond which there is not". Of course, this definition implies morphologic similarity, but also evolutionary relationships. This definition sounds equivalent to what Zink (2005) is calling the phylogenetic species concept, except now for modern organisms we can also use DNA sequences as well as morphological, physiological, and behavioral traits to define diagnostic differences to establish species.

CLINAL VARIATION AND SPECIATION

Then there is the tricky business of "clinal species" and "ring species." I will give an example of clinal variation here and we will discuss an example of "ring species" later. A good example of clinal variation is seen in the leopard frog. Leopard frogs look similar across their range from Canada to Mexico (and Central America) and some biologists still refer to them as one species, *Rana pipiens*. However, several recent studies have shown that there are differences in their mating calls, a prezygotic reproductive isolating mechanism (prezygotic RIM) (in fact, there are at least four different geographically distinct mating calls [Kardong, 2008]), differences in the tolerance of tadpoles to different temperature ranges and survivability of hybrid tadpoles, and the incidence of defective embryos along the gradient from north to south, a postzygotic reproductive isolating mechanism (postzygotic RIM) (one study actually mated females from the north with males progressively farther south into Mexico and concluded that increased embryo and tadpole abnormalities occurred, from north to south) (see Kardong, 2008). Several of the biologists that studied these frogs had (during the latter part of the 20th century) divided the one species (at least for the clinal range from southern Canada to northern Mexico) into four separate species (Kardong, 2008). Of course, they were using the phylogenetic species concept. However, if one accepted the biologic species concept strictly, these would be called subspecies, and one would say that reproductive isolating mechanisms (RIMs) are not complete and that these frogs can potentially interbred across their entire range, therefore, they are a valid species, *Rana pipiens* with four subspecies. However, according to Kardong (2008) some biologists have separated the clinal variations in *Rana pipiens* into at least four species, those being: *Rana pipiens, Rana blairi, Rana utricularia*, and *Rana berlandieri*. More recently, with increasing numbers of field and laboratory studies (including DNA studies), the leopard frog has been divided into about thirty species (see Hillis and Wilcox, 2005).

To make things even more complicated for this frog genus (*Rana*), some taxonomists (and according to Pauly, Hillis, and Cannatella [2009], one in particular, Darrel R. Frost of the American Museum of Natural History) have suggested a change of the genus name to *Lithobates*, thus, for example, the Northern Leopard Frog would be called *Lithobates pipiens* rather than *Rana pipiens*. This name change would affect most of the true frogs of North America in the family Ranidae that have traditionally been included in the genus *Rana*, including the American Bullfrog and Northern Leopard Frog (Wikipedia, *Lithobates*, accessed 06/15/2014). According to Wikipedia, Lithobates, accessed 06/15/2014, "This proposed change has since been rejected by others, such as Stuart (2008)[7] and Pauly et al. (2009).[8] AmphibiaWeb[9], an online compendium of amphibian names, also does not recognize *Lithobates* as a distinct genus but Amphibian Species of the World 5.6, an Online Reference has accepted this genus as valid..." The following statement by Pauly, Hillis, and Cannatella (2009, p. 123) seems to be revealing relative to this problem:

> Hillis (2007) suggested that *Rana* be retained for all North American ranids and that *Lithobates* be recognized as one of several subclades of *Rana*. This proposal retains the existing species names but also provides more information about the phylogeny of these frogs. Hillis and Wilcox (2005)

originally defined *Lithobates* as a subclade equivalent to the *Rana palmipes* species group, and treated this taxon as a subgenus under the Linnean system. Moreover, this taxon concept of *Lithobates* is much more restricted than the genus called *Lithobates* by Frost et al. (2006a). The brief comments in Frost et al. (2008) give the reader no hint of these issues.

In addition, Pauly, Hillis, and Cannatella (2009, P. 124) state the following: "Use of *Lithobates* for the majority of North American *Rana* is not only unnecessary and destabilizing for taxonomy, it is also highly confusing." Hillis and Wilcox (2005) and Hillis (2007) offer further understanding of this taxonomic nomenclature problem.

There are several other members of the genus *Rana* that are experiencing clinal variation. A situation of clinal (geographic) variation that may represent another example of phylogenetic speciation may be the very common eastern North American frog species *Rana* (*Lithobates*) *clamitans* (or just *Rana clamitans*), commonly called the green frog (Figure 9-1). This species is currently divided into two subspecies: *Rana clamitans melanota*, the Northern Green Frog (see Figure 9-1), and *Rana clamitans clamitans*, the Bronze Frog.

Figure 9-1. Photograph of the Green Frog, *Rana clamitans*. Two subspecies are the Northern Green Frog, *Rana clamitans melanota*, and the Bronze Frog, *Rana clamitans clamitans*). This specimen is the Northern Green Frog, *Rana clamitans melanota*, because it was found in southeastern Ohio, well north of the range for the Bronze Frog. Bronze Frogs are only found in the southeastern portion of the United States, from southern North Carolina to the eastern third of Texas and southward, whereas Northern Green Frogs are found from northern North Carolina to southern Canada and westward to Minnesota. This specimen was caught by Wade Huck near Waterford, Ohio (southeastern Ohio). This frog is about 3.0 inches (7.6 cm) in length and is mostly medium brown on the back and legs with a green upper lip and a white underside. (Photo by E. L. Crisp, July 2014.)

Northern Green Frogs, *Rana clamitans melanota*, are very common frogs that inhabit ponds, wetlands, marshes, swamps, springs, small stream pools, larger slow-moving streams, deep road puddles, or roadside ditches in the eastern half of the U. S. from southern North Carolina northward (AmphibiaWeb, accessed 08/31/2014). They are medium-sized frogs, with adults ranging from about 2.0 inches (about 5.0 cm) to about 4.0 inches (about 10. cm) (Wikipedia, *Lithobates clamitans*, accessed 08/31/2014; Pauley, accessed 09/01/2014). The Northern Green Frog shown in Figure 9-1 is about 3.0 inches (about 7.6 cm) long, from tip of snout to vent (excluding the hind legs). Their backs are dark brown to tan to green and may have small irregular spots. Their snouts are often green (the specimen in Figure 9-1 has a green upper lip) and their undersides are whitish colored, with males having bright yellow throats (Wikipedia, *Lithobates clamitans*, accessed 08/31/2014). Females are slightly larger than males, and males have a tympanic membrane about twice the size of the eye, whereas in females the tympanum is about the size of the eye (Wikipedia, *Lithobates clamitans*, accessed 08/31/2014). Dorsolateral ridges (seam-like skin folds) extend down on each side of the back only about two-thirds the way to the groin, whereas in leopard frogs they extend all the way to the groin and the bullfrog completely lacks these ridges (except for a short ridge around the back of the eardrum) (Wikipedia, *Lithobates clamitans*, accessed 08/31/2014; Pauley, accessed 09/01/2014).

The geographical distribution of *Rana clamitans melanota* (the Northern Green Frog) grades relatively rapidly into *Rana clamitans clamitans* (the Bronze Frog) in moving southward in the eastern U. S., but does intergrade along the fall line in Georgia and Alabama (Warny, Meshaka, and Klippel, 2012). According to Warny, Meshaka, and Klippel (2012), in addition to color pattern differences (the bronze color), there are definite size differences between the two subspecies, with the Bronze Frog being smaller in both the adult and the metamorphosing size when compared to the Northern Green Frog. Of course, the breeding seasons also follow a north-south gradient, being shorter in the north. This of course affects the larval period for the frogs, with tadpoles of the more northern frog populations overwintering the following year (Warny, Meshaka, and Klippel, 2012). So, do the two frog subspecies have unique evolutionary histories and definite diagnostic properties that may be used to distinguish them? Thus, should they be elevated to the status of species, or not? Remember, some systematists are not very keen on subspecies and would rather just elevate subspecies to species categories.

In my opinion, the reason that all biologists do not jump on the "band wagon" for the phylogenetic species concept is the fear that every slight variation (whether in morphology, physiology, behavior, or DNA differences) will be used to name a new species. Certainly, we do have our "splitters" and "lumpers" in both biology and paleontology. So, I think we do need to be careful when naming new species (whether an extant species or one that is extinct). We should make sure that we definitely have a unique evolutionary history for the organism and that we can definitely recognize this organism in nature (or as a fossil) with diagnostic traits.

ADDITIONAL EXAMPLES OF POSSIBLE SPECIATION

Carl Zimmer (2010) in his recent book, *The Tangled Bank – An Introduction to Evolution*, gives a neat example of how the phylogenetic species concept is being used today by conservationist. As Zimmer points out, the giraffe has long been thought to be a single species endemic to Africa. Zimmer (2010) explains that the wildlife conservation geneticist , Rick Brenneman and his colleagues from the Henry Doorly Zoo in Omaha, Nebraska are trying to save the giraffe from extinction. Brenneman and his team have collected DNA samples from giraffes in six habitats in central, southern, and eastern Africa. They sequenced this DNA (and that of an okapi, the short necked relative of the giraffe, to use as an outgroup) and constructed a cladogram to show the relationships of the giraffes from the six habitats in Africa. The study indicated that the six populations of giraffes diverged from a common ancestor about a million years ago and they now can be divided into six groupings based on the DNA similarity (so the six groupings have diagnostic traits). These DNA sequences also agree with differences in coat blotch patterns of the six populations of giraffes. Thus, these diagnostic traits are consistent with there being six different species of modern giraffes, according to the phylogenetic species concept. Although giraffes will breed in captivity and yield viable offspring, they found limited hybridization (only three giraffes, less than 1% of the giraffes they darted and studied) in the three populations that are adjacent to each other in eastern Africa. The results of this study have important implications for saving giraffes from extinction. They will need to devise strategies for saving six different species with different habitats, rather than a single species spread across Africa (Zimmer, 2010).

Another example of speciation based on the phylogenetic species concept might be the Key Deer (*Odocoileus virginianus clavium*) of the Florida Keys (Figures 9-2 and 9-3). The Key Deer is endemic to the Florida Keys, originally represented in the Keys as White-tailed Deer (*Odocoileus virginianus*) (Figure 9-4) that migrated over a land bridge during the Wisconsin glaciation (Wikipedia, Key Deer, accessed 05/01/2014). The Key Deer is now the smallest of the 28 subspecies of the White-tailed Deer (in the past, some taxonomists have named as many as 40 subspecies of the White-tailed Deer based on morphologic variability over the extensive geographic range, from Alaska, across most of southern Canada, most of the U. S. and Mexico, across central America, to as far south as Peru [Wikipedia, White-tailed Deer, accessed 05/01/2014]). In fact, the Key Deer is the smallest deer in North America (Wikipedia, Key Deer, accessed 05/01/2014).

Figure 9-2. A Key Deer buck on No Name Key, Florida Keys. (Photo by E. L. Crisp, January 2014.)

Figure 9-3. The same Key Deer buck as in Figure 9-2. Susan Sowards serves as scale for this photo. (Photo by E. L. Crisp, January 2014.)

The Key Deer bucks range up to about 32 inches (about 81 cm) at the shoulder and weigh about 80 pounds (mass of about 36 Kg), whereas the does range up to about

28 inches (about 71 cm) and weigh about 65 pounds (mass of about 29.5 Kg) (U. S. Fish & Wildlife Service, Key Deer Fact Sheet, accessed 05/01/2014). These numbers compare to a height at the shoulder of up to about 47 inches (about 120 cm) and a weight of about 132 to 287 pounds (mass of about 60 to 130 Kg) for a typical North American White-tailed Deer buck (although significantly taller and heavier White-tailed Deer bucks have been reported) (Wikipedia, White-tailed Deer, accessed 05/01/2014).

Figure 9-4. A White-tailed Deer (doe) from West Virginia. (Photo by E. L. Crisp, March 2005.)

By the 1950s the Key Deer was almost extinct, down to about 27 individuals in 1957. It was placed on the endangered species list and in 1957 the National Key Deer Refuge (consisting of about 8500 acres in the lower Keys, with the headquarters on Big Pine Key) was established to protect and preserve Key Deer and other wildlife resources in the Florida Keys (U. S. Fish & Wildlife, National Key Deer Refuge; accessed 05/01/2014). Although Key Deer can easily swim from island to island in the Florida Keys, they commonly inhabit about 25 islands in the lower Keys, but the largest number of Key Deer is on Big Pine Key and No Name Key. The population has now rebounded to about 700 to 800 deer, with about 600 of the population on Big Pine Key and No Name Key (based on a three-year scientific study in 2000 [U. S. Fish & Wildlife, Key Deer Fact Sheet], accessed 05/01/2014). The Key Deer today are still threatened and have an uncertain future, primarily due to industrialization of the Florida Keys and lost habitat. But currently, 70% of the annual mortality rate is due to road kills (U. S. Fish & Wildlife Service, Key Deer Fact Sheet, accessed 05/01/2014).

ISLANDS, DWARFISM, GIGANTISM, AND SPECIATION

Insular Dwarfism

So, why are Key Deer very small on the islands that make up the Florida Keys? Is this a case of insular dwarfism? There are many cases of both extinct and extant mammals around the world that have migrated (usually by swimming) to islands and evolved into dwarfs, in particular, hippopotamuses, deer, and elephants. These are mammals that are noted for their ability to swim to new territory, such as offshore islands (Quammen, 1996, p. 153). As noted by Quammen, 1996 (p. 157):

> Take the Pleistocene elephant of Sicily, *Elephas falconeri*. It had descended from a mainland ancestor about the same size as a modern elephant, but on Sicily the animal was shrunk by three-quarters. Its skull was rounder than the skull of its ancestor, with proportionately more room for brain and less bone area for attachment of large muscles. *Elephas falconeri* was also stockier. Short-legged, tiny, bright, it stood as tall as a pony. In adapting to life on its island, the Sicilian elephant had changed into a dwarf.

Another example of insular dwarfism is the Pleistocene Pygmy Mammoth of the Channel Islands, offshore California. Agenbroad (2009) claims that the Pleistocene Pygmy Mammoth, *Mammuthus exilis*, was present on the Channel Islands for at least the time period of about 200,000 years ago to about 11,000 years ago. He also states that Columbian Mammoths (*Mammuthus columbi*) were also present on the islands during the same time span, but in lesser numbers (with approximately 90% *M. exilis* and 10% *M. columbi*). A survey by Agenbroad and associates of the fossil localities on the three Channel Islands of San Miguel, Santa Rosa, and Santa Cruz yielded 380 mammoth localities on the islands. During the last glacial age of the Pleistocene (at maximum ice conditions), sea level would have been approximately 130 meters (about 430 feet) lower than at present (Quammen, 1996), thus Agenbroad (2009) reasons that the three islands above would have combined to form the larger Pleistocene island of Santarosae during major advances of Pleistocene glaciation. But Santarosae would have still been several miles offshore, such that *M. columbi* would have had to swim to Santarosae from the mainland. Thus, as stated by Agenbroad (2009), "The conclusion is that *M. columbi* initially (and perhaps periodically) colonized the Pleistocene island, Santarosae. Shortly after colonization, the island selected for smaller forms, and quickly formed the island pygmy mammoths. Sondaar (1977) concluded that island forms reach a reduced size quickly and endure until extinction." Donald L. Johnson, a geomorphologist from the University of Illinois, was one of the first to present evidence that elephants are very good swimmers and based on detailed studies of the Channel Islands he determined that they were not colonized by land bridges from the mainland, but were colonized by Columbian Mammoths (*Mammuthus columbi*) that swam to the islands and relatively rapidly evolved into Pygmy Mammoths (*Mammuthus exilis*) (Quammen, 1996). In a recent (2013) memorial presentation about the almost 50 years of work on the Channel Islands by Donald L. Johnson, Daniel R. Muhs, of the U. S. Geological Survey, makes the following statement relative to the evolution of the Pygmy Mammoths on the Channel Islands: "….and (6) perhaps his greatest contribution: the origin of Channel Islands pygmy mammoths of Pleistocene age, due not to imagined ancient land bridges (for which there was never any structural or geomorphic evidence), but rather due to the superb swimming abilities of proboscideans, combined with lowered sea level, favorable paleowinds, and an attractive paleovegetation on the Channel Islands."

There are many other examples of swimming hippopotamuses, deer, and elephantids that have swum to islands (see Quammen, 1996). This pattern of biogeographic distribution of swimming hippos, deer, and elephantids also "coincides with a pattern of dwarfism" (Quammen, 1996). This is because once these mammals are geographically isolated on islands, the struggle for food and the lack of predators resulted in the evolution of smallness.

Insular Gigantism

Do all organisms that migrate to islands tend to become smaller? Well, no. Some actually become larger. Mammals tend towards becoming smaller on islands, whereas reptiles tend to become larger (often gigantic) (Quammen, 1996). However, some mammals, such as rodents, tend towards gigantism on islands and some birds become flightless. As for gigantism of reptiles, the giant tortoises of the Galapagos Islands are prime examples. As also are the endemic iguanids of the Galapagos, both the marine iguana *Amblyrhynchus cristatus* and land iguanas belonging to the genus *Conolophus* (Quammen, 1996). Many lizards also tend to be larger on islands than the adjacent mainlands, but they pale in comparison to the large monitor lizards on the Indonesian Islands of Komodo, Rinca, Gilimotang, and the western end of Flores referred to as Komodo dragons, belonging to the species *Varanus komodoensis* (Quammen, 1996). Komodo dragons are fierce carnivores that according to Quammen (1996) maintain their gigantic size "with great bursts of overland ferocity, punctuated by meat-eating gluttony." A full size Komodo dragon may reach 10 feet in length, and if it has just eaten may weigh in at 500 pounds, whereas a more typical nine-footer may weigh only 120 pounds (Quammen, 1996). The Komodo dragon ambushes its prey by hiding in the brush along trails followed by deer, wild boar, water buffalo, rats, dogs, goats, horses, other komodos, or sometimes humans (Quammen, 1996).

So, when and how did *V. komodoensis* become a giant? No one seems to know, and the late Pleistocene and Holocene fossil record of the islands (primarily Flores and Komodo) that it inhabits is very incomplete (Quammen, 1996). But according to Quammen (1996), Jared Diamond (a relatively well known evolutionary ecologist and popular science writer) proposed in the journal *Nature* in 1987 that Komodo dragons evolved to feed on pygmy elephants during the late Pleistocene. According to Diamond (1987), perhaps from about 50,000 years ago to the time that people arrived on the islands (by at least 4000 B. C.), ancestral komodos (monitor lizards that swam to the islands from Australia) evolved gigantism by preying on dwarf elephantids, especially juveniles. Diamond (1987) points out that during the late Pleistocene, the island of Flores supported two now extinct small elephantids (based on fossil finds), *Stegodon trigonocephalus*, and the pygmy *Stegodon sompoensis*. *S. sompoensis* was only about 5 feet high at the shoulders and weighed about as much as a buffalo (Quammen, 1996). Diamond (1987) points out that there was not much else on the Pleistocene island that the Komodo dragons could have eaten, since ungulates and humans had not arrived yet.

RING SPECIES

Earlier I discussed the clinal variation within species and one of the examples that I presented was the north-south variation in the leopard frog (*Rana pipiens*), from Canada and the northeast U. S. in the north to Mexico and Central America in the south. Sometimes there is also clinal variation due to changes in altitude, such as changes in the yarrow genus *Achillea* (a sunflower family member) from sea level to high in the Sierra Nevada Mountains (see Kardong, 2008. p. 169). However, sometimes clinal variation within a species does not take place linearly, but rather may be the result of a species migrating around a geographic barrier to then meet again on the other side of the barrier (Kardong, 2008). Often, when individuals of the species meet again after encircling the barrier, they will not interbreed because they have become reproductively isolated; even though, adjacent local populations of the species in the path around the barrier may interbreed (Kardong, 2008). This species level genetic divergence of populations that exist at the endpoints of the "ring" where the populations become sympatric, results in two populations that have become genetically isolated (or near so, with no or limited hybridization) from each other (Stebbins, 1957; Wake, 1997; Kutchta, et al, 2009a). If environmental factors change such that the intermediate populations become extinct, two species (the older species and the newer species) remain in sympatry where before there was only the older species (Stebbins, 1957). The new species is referred to as a ring species.

Although there are numerous examples of ring speciation, one of the best and well known examples, even a classic example, is the incipient ring speciation that has taken place in the Ensatina salamander (the common name is named after the genus name, *Ensatina*) that occupies the evergreen-deciduous (coniferous forests and oak woodland [Wake, 1997]) forest habitats of California's coastal ranges and the western flank of the Sierra Nevada Mountains (Stebbins, 1949; Stebbins, 1957; Wake, 1997; Kuchta, et al, 2009a; Kuchta, et al, 2009b; Pereira and Wake, 2009). These moist mountain forests surround the hot, dry Central Valley, primarily the Sacramento and San Joaquin valleys, of California where the salamanders cannot survive. The species *Ensatina eschscholtzii* is a lungless (plethodontid) salamander that originated in northwestern California and southwestern Oregon about 21.5 million years ago (Wake, 1997; Bergstrom and Dugatkin, 2012), but currently has a range from British Columbia in Canada to northern Baja California in Mexico (Stebbins, 1949; Wake, 1997; Bergstrom and Dugatkin, 2012). As the species migrated from the southwestern Oregon-northern California region, populations were established along the two mountainous corridors that surround the Central Valley, the California coastal ranges and the western slope of the Sierra Nevada Mountains. As the populations migrated southward, phenotypic and genetic divergence took place, giving rise to a high degree of morphologic variability in skin coloration (blotchiness, hue, and number of colored strips) (Bergstrom and Dugatkin, 2012) and other morphologic, ecologic, and behavioral traits (Stebbins, 1949; Stebbins, 1957). Intermediate populations adapted to local conditions and diverged from the ancestral populations as the species continued to migrate southward, but there was continuous genetic continuity between populations until they again came sympatrically together in the coastal mountains of southern California with limited asymmetric hybridization or no hybridization. Based primarily on ecophenotypic variations, Robert Stebbins (1949) divided the species *Ensatina eschscholtzii* into seven subspecies (Figure 9.5). After the studies by Stebbins (see Stebbins, 1949 and Stebbins, 1957), David Wake and his colleagues have extensively researched this classic ring species hypothesis originally proposed by Stebbins (1949), including protein molecular, mitochondrial DNA, and nuclear DNA studies that have provided large amounts of evidence in support of Stebbins' original *Ensatina eschscholtzii* ring species hypothesis (see Wake, 1997; Kuchta, et al, 2009a; Kuchta, et al, 2009b; Pereira and Wake, 2009; Pereira, et al, 2011; Bergstrom and Dugatkin, 2012). The following direct quote from Kuchta (2009a) will perhaps make the above clearer:

Together these subspecies are distributed in a ring around the Central Valley of California, which is hot and arid and currently presents an environment that is inhospitable to terrestrial salamanders (Fig. 1a). However, in the mountains of southern California, the unblotched subspecies *eschscholtzii* and the blotched subspecies *klauberi* are locally sympatric with either limited or no hybridization, indicating they have reached the species level of divergence (Fig. 1a; Stebbins, 1949, 1957; Brown, 1974; Wake et al., 1986). Stebbins (1949) developed an explicit biogeographical model to account for this taxonomic oddity of sympatric subspecies. He postulated that the *Ensatina* complex originated in present-day northern California and southern Oregon, perhaps from a *picta*-like ancestor. This ancestral stock then expanded its distribution as two arms southward down the Coast Ranges (unblotched subspecies) and the inland ranges (blotched subspecies), the arms adapting and diverging as they spread, until they re-established contact in southern California as reproductively isolated entities (Fig. 1a). Broad zones of phenotypic intergradation between adjacent subspecies were interpreted as representative of ongoing genetic connectivity (Dobzhansky, 1958), and the two sympatric subspecies in southern California were thereby viewed as linked together by a continuous sequence of interbreeding populations, thus forming a ring species.

Much molecular systematic work has been done on the *Ensatina* complex since Stebbins (1949). The results are complex in detail, but support the major tenets of the ring species hypothesis in finding that secondary contacts between the coastal and inland arms are characterized by species-level divergence, while secondary contacts within the arms exhibit patterns of intergradation and genetic merger (Wake and Yanev, 1986; Wake et al., 1986, 1989; Moritz et al., 1992; Jackman and Wake, 1994; Wake, 1997; Alexandrino et al., 2005).

Figure 1a in the quote above is practically the same as Figure 9.5 herein. Taxonomic issues within the ring remain as further targets of research (Wake, 1997). Some researchers consider *Ensatina eschscholtzii klauberi* to be a separate species (*Ensatina klauberi*) (Frost and Hillis, 1990) from the other ring members (Wake, 1997), as well it should be (at least an incipient species) if it is reproductively isolated (for the most part) from *E. e. eschscholtzii* and if species level divergence around the ring has actually taken place at the terminus of the ring.

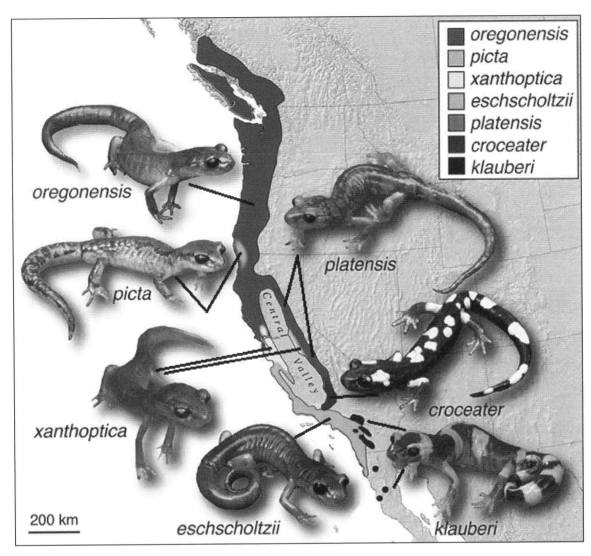

Figure 9.5. A diagram showing the geographic distribution of the seven extant subspecies of the *Ensatina eschscholtzii* ring species complex (as originally named and delineated by Stebbins, 1949) over the range of the salamander, from British Columbia, Canada to the northern Baja Peninsula, Mexico. There is gene flow around the ring and some viable hybridization of adjacent subspecies. However, where the subspecies *Ensatina eschscholtzii eschscholtzii* and *E. e. klauberi* join in sympatry at the end of the ring in the coastal mountains of southern California there is limited or no hybridization. A colored version of this figure at several resolutions may be obtained at the following website as accessed on 08/05/2015: https://commons.wikimedia.org/wiki/File:Ensatina_eschscholtzii_ring_species.jpg#mw-jump-to-license. (This figure has been slightly modified from Devitt, et al, 2011 by conversion of a colored version to shades of gray and the addition of lines to help show the geographic range of each subspecies. The source of this figure is the following: (2011). "Asymmetric reproductive isolation between terminal forms of the salamander ring species *Ensatina eschscholtzii* revealed by fine-scale genetic analysis of a hybrid zone". BMC Evolutionary Biology 11 (1): 245. DOI:10.1186/1471-2148-11-245. The authors of this figure are the following: Thomas J. Devitt, Stuart J.E. Baird and Craig Moritz.) (© 2011 Pereira et al; licensee BioMed Central Ltd. This figure is modified from an Open Access article distributed under the terms of the Creative Commons Attribution License [http://creativecommons.org/licenses/by/2.0], which permits unrestricted use, distribution, and reproduction in any medium, provided the original work is properly cited.)

CHAPTER 10

COEVOLUTION

INTRODUCTION

Coevolution is the condition that exists when two or more species evolve together in certain directions resulting from their ecological interactions (Kardong, 2008). Most biologists and ecologists use the term symbiosis to indicate purposeful direct interactions between two or more species living in contact with each other (Lang and Benbow, 2013). As such, Lang and Benbow (2013) state the following: "The term 'symbiosis' includes a broad range of species interactions but typically refers to three major types: mutualism, commensalism and parasitism." However, keep in mind that when some biologists and ecologists use the term symbiosis they are referring to mutualism only. According to Kardong (2008), and our usage here, coevolution results in a symbiotic relationship among two or more species that may take the following forms:

- **Competition** – two species competing for resources that often leads to an evolutionary arms race. Both species are harmed in the relationship (-/-).

- **Mutualism** – two or more species living together in a beneficial relationship. All species involved are benefited in the relationship (+/+).

- **Commensalism** - two species living together in a relationship where one benefits, but the other one is not harmed even though it receives little or no benefit from the symbiosis (+/0).

- **Predation**, **Parasitism**, and **Herbivory** - one species benefits and the other is harmed in a coevolutionary relationship (+/-).

Symbiosis, of course, often results in reciprocal selective pressures. This is often referred to as antagonistic coevolution in which each species undergoes a decrease in fitness due to evolutionary changes in the other species (Bergstrom and Dugatkin. 2012). One classic example of antagonistic coevolution is the predator-prey relationship, such as the wolf and the rabbit. Rabbits evolve that are faster – thus, to survive wolves evolve to be faster, etc. Thus, an evolutionary arms race results due to this antagonistic coevolutionary relationship (Kardong, 2008; Bergstrom and Dugatkin. 2012). As this example illustrates, many species are continually in conflict with each other and continually evolving to become better at surviving and being reproductively successful. This is particularly the case in competition for resources, predator-prey relationships, parasite-host relationships, and plant-herbivore relationships. Animals even exploit their mates to the selfish benefit of themselves in order to be more reproductively successful (Shaw, et al, 2005) (example: the spider female that eats her small male mate after fertilization). Organisms must continually change their genetic make-up to adapt to a changing environment if they are going to survive and reproduce. The competition for resources is fierce. Although organisms certainly must constantly adapt to a changing physical environment, uncommonly is the physical environment the reason for death or the cause of non-reproduction. Most often it is other organisms (parasites, predators, and competitors) that prevent survival and/or reproductive success (Ridley, 1993).

COMPETITION

Before we discuss competition, let's review some general ecological terms that will help in our discussion of competition among organisms. All organisms live within a biological (or an ecological) community (such as the benthic [bottom dwelling] community within a temperate lake), which consists of populations of two or more species that interact with each other either directly or indirectly in an ecosystem located in a specific geographic region (Johnson and Raven, 1998; Lang and Benbow, 2013; Urry et al., 2014). An ecosystem consists of one or more communities of organisms in an area and the physical environment that surrounds them. Ecosystem ecology deals with energy flow and chemical cycling between the physical environment and the communities of organisms within an ecosystem (Johnson and Raven, 1998; Lang and Benbow, 2013; Urry et al., 2014). A coral reef would be an example of a tropical marine ecosystem, whereas a tropical wet forest (rain forest) would be an example of a terrestrial ecosystem.

The role of a species in its ecosystem, or how a particular species functions in its ecosystem, is referred to as the niche of the species. Niche should not be confused with the habitat of an organism. To illustrate this with an analogy, the habitat of a species is where it lives in the ecosystem (for example, in a temperate meadow), whereas the niche is what an organism does to make a living in the ecosystem (for example, it eats certain insects that live on the grasses in the meadow). According to Urry et al., 2014 (p. 845), "American ecologist Eugene Odum used the following analogy to explain the niche concept: If an organism's habitat is its 'address,' the niche is the organism's 'profession.'"

It is a well known ecological principle that, because of the competition between species for limited resources, no two species can continuously occupy the same niche in an ecosystem of a particular area. If two species are locally competing for the exact same niche in a given ecosystem, one of the species (the weaker competitor) will be eliminated or the weaker competitor (over time) will have to modify its niche to decrease competition; if the weaker competitor is eliminated locally, ecologists say that the species has been competitively excluded (Johnson and Raven, 1998; Urry et al., 2014).

Thus, coevolutionary relationships between different species often result from competition for resources in the same or similar niches or in overlapping niches. This is referred to as interspecific competition (Johnson and Raven, 1998; Urry et al., 2014). However, the most extreme competition is often the competition between individuals in populations of the same species. Members of the same species occupy the same or very similar niches, such that there is fierce competition for limited resources. Competition between members of the same species is called intraspecific competition (Johnson and Raven, 1998; Urry et al., 2014). Resources that organisms compete for must be limited (otherwise there would be no competition) and the limiting resources may be food, territory (or space), water, light, or even mates. In nature, resources are almost always limited and competition for resources is present, if not there would be geometric or exponential growth of populations within species (Rockwood, 2006). Of course, new species that have evolved or migrated into an area (or brought there by humans, e. g., invasive species) often grow exponentially for some time until the available resources limit growth (However, the world population of humans has grown exponentially since about 1650 and continues to grow [Rockwood, 2006]). But many populations approximate logistic growth (see Rockwood, 2006 for a discussion of the logistic model for population growth); the logistic growth model predicts that growth of the population will cease when the carrying capacity of the population is reached for a particular environment. Rockwood (2006, p. 35) states the following relative to the carrying capacity according to the logistic growth model: "For a given species, in a specific environment, carrying capacity is defined as the number of individuals that can be maintained indefinitely." When organisms compete, either directly or indirectly for limited resources, there is a change in the fitness of the organisms that will generally have a larger negative effect on the weaker of the competitors (Rockwood, 2006; Lang and Benbow, 2013).

Intraspecific competition within populations of a species results in decreased availability of one or more resources in the environment. Thus, the competitive depletion of resources in the environment decreases growth, survivorship, or reproductive success (Rockwood, 2006). So, the individuals in a population that are better competitors are the ones that grow, survive, and are reproductively successful in their environment; these individuals are the most fit and have been (or are being) naturally selected.

There are two basic modes of competition: interference competition and exploitation competition (also referred to as depletion competition) (Rockwood, 2006; Lang and Benbow, 2013). Interference competition is direct competition where an individual directly changes the ability of another individual to obtain resources. Competition by males for females in many sexually dimorphic mammals is an example of interference competition. For example, male elk (*Cervus elaphus*) engage in direct combat to win the mating rites with female elk (Rockwood, 2006).

The winning male has increased his fitness by directly interfering with the success of the losing male. Or when the dominant male gorilla (*Gorilla gorilla*) aggressively prevents other males from mating with a female, he is directly altering their reproductive abilities; he is directly interfering with their sexual and reproductive resource (Lang and Benbow, 2013). When certain ants recruit soldiers to protect a supply of food on the forest floor by forming a ring around it to prevent other ant colony soldiers and workers from obtaining the food, they are directly interfering with the ability of the other ants to obtain a resource and are practicing interference competition (Rockwood, 2006). Exploitation (depletion) competition, on the other hand, just involves consuming or removing (depleting) the resource. Thus, individuals are competing indirectly to consume common resources, such as prey, food, water, territory, etc. (Lang and Benbow, 2013). The result of both direct and indirect competitive interaction is a change to lower fitness for the losers in intraspecific competition and some degree of competitive exclusion for the losers in interspecific competition.

In interspecific competition, a species' niche may be influenced. Remember, we said that no two species can occupy the exact same niche, or two species cannot coexist in the same niche permanently in a particular community if they have identical niches. However, when two similar species competitively interfere with each other, their niches in a community may be significantly altered over time such that they may coexist in that community (Urry et al, 2014). All species have a fundamental niche, which is the niche that an organism may potentially occupy in a community in the absence of competition. However, because of competition, the species may have to alter its fundamental niche to adequately coexist with another species in an ecosystem. This altered niche is called the realized niche, the portion of the fundamental niche that the species can actual occupy in a particular environmental setting (Johnson and Raven, 1998; Rockwood, 2006; Kardong, 2008; Urry et al., 2014).

This alteration of niches by organisms when they exist together in a community (i.e., they are sympatric) may also, in itself, indicate past competition between closely related species; this alteration of resource use is referred to as resource partitioning (Campbell, Reece, and Mitchell, 1999). Another line of indirect evidence for past competition for resources of closely related species is based on studies of populations of these species that are sometimes allopatric and sometimes sympatric. Often allopatric populations of closely related species use similar resources and occupy very similar niches, but of course they are separated from each other in allopatry. However, in sympatric populations of these same species there are differences in their exact niches and resource use that results in differences in body structures (like beak depth for seed eating birds). More divergence of characters in sympatric populations than in allopatric populations of the same species is referred to as character displacement (Campbell, Reece, and Mitchell, 1999; Rockwood, 2006; Urry et al., 2014). This divergence in characters reduces the competition for resources and allows the two species to coexist (Rockwood, 2006). A classic example of character displacement and indirect evidence of past competition for resources is the difference in beak depth of two of Darwin's finches on Santa Maria and San Cristóbal Islands in the Galápagos Archipelago. On these two islands (and others), *Geospiza fuliginosa* (the small ground finch) and *Geospiza fortis* (the medium ground finch) live in sympatry and compete with each other for seeds (their main diet). *G. fuliginosa* has a shallow, small beak and *G. fortis* has a deeper, larger beak; these adaptations favor eating seeds of different sizes, small ones for *G. fuliginosa* and larger ones for *G. fortis*. However, allopatric populations of these two species that live on Los Hermanos and Daphne Major, respectively, (and other) islands are not competing against each other and have similar sized beaks (both in depth and length), an adaptation for eating seeds of similar size (see Urry et al., 2014, p. 847, Figure 41.4 and Futuyma, 2009, p. 517, Figure 19.2; also see Grant and Grant, 2006; Grant and Grant, 2008; Bierema and Rudge, 2014; and Grant and Grant, 2014 for other examples and information on character displacement in Darwin's finches).

Although the above circumstantial evidence for past competition for resources and niche realization is strong, direct controlled field experimentation gives stronger evidence that one species can affect the niche and resource utilization (thus density and distribution) of another species when the two species are in resource competition (Campbell, Reece, and Mitchell, 1999). The classic field experimental study of interspecific competition for niche resources of two similar species was done by Joseph Connell (1961) on the barnacle species *Chthamalus stellatus* and *Semibalanus balanoides* (then called *Balanus balanoides*) (Connell, 1961; Johnson and Raven, 1998; Campbell, Reece, and Mitchell, 1999; Rockwood, 2006; Kardong, 2008; Urry et al., 2014). These two species compete for space and resources in the rocky intertidal zone along the coast of Scotland. *Semibalanus* is most concentrated in the lower intertidal to mean tide level or slightly above, whereas *Chthamalus* is found in the high tide region. The larvae of the two species settle and begin to grow over most of the rocky intertidal zone. However, the adults of these two species have distinctly separate distributions. Connell (1961) questioned whether the distribution of the two barnacles was due to different fundamental niches or was the difference due to interspecific competition. He performed a number of experiments on the two barnacle species that involved moving the barnacles to higher and lower positions in the rocky intertidal zone. He further removed one species or the other where they were growing together or placed them together when they were growing separately. Connell (1961) found that when he completely removed *Semibalanus* from the lower intertidal zone, *Chthamalus* was able to grow in the entire intertidal zone from high tide to low tide. However, when *Semibalanus* was present it grew faster than *Chthamalus* and would actually force any specimens of *Chthamalus* off the rocky surfaces by undercutting them, or it would smother and crush them. But when specimens of *Semibalanus* were transplanted to the high tide region, they were not able to survive the long exposures of several hours to air during low tides and thus desiccated. Connell (1961) concluded that the fundamental niche and the realized niche of *Semibalanus* was about the same, but the realized niche of *Chthamalus* was only a fraction of its fundamental niche due to competition with *Semibalanus* (see Campbell, Reece, and Mitchell, 1999; p. 1117, Figure 53.13).

THE RED QUEEN HYPOTHESIS

According to Prothero (2004), Leigh Van Valen (1973) referred to the concept (he considered it to be a new evolutionary law) of the need for continual change of the genome of an organism as the "Red Queen Hypothesis". The Red Queen is a reference to the Red Queen chess piece in *Alice Through the Looking Glass* (by Lewis Carroll, 1872), who told Alice that here (in this land) "...it takes all the running you can do, to keep in the same place" (Ridley, 1993; Prothero, 2004) (Figure 10-1). So, if species are going to avoid extinction, they must constantly improve their survival and reproductive fitness (Prothero, 2004). Thus, an evolutionary arms race can coevolve between a predator and its prey, parasites and their host, or competitors for the same resources (Van Valen, 1973; Bergstrom and Dugatkin, 2012). For example, a newt develops a toxin within its flesh so that predators will avoid it. However, natural selection will favor the variants of the predator that are more resistant to the toxin and that can eat the newt and withstand the toxin. So, the newt becomes more toxic to the predator. The predator evolves more resistance to the toxin, and so on (Univ. Calif. Museum Paleo., Biowarfare, accessed 05/09/2014). An example of such a predator-prey relationship undergoing an arms race is the one that exists between the red-sided (common) garter snake (*Thamnophis sirtalis*) and the rough-skinned newt (*Taricha granulosa*) (Brodie and Brodie, 1999; Brodie, et al, 2004; Williams, et al, 2010). The skin of the rough-skinned newt contains very high and variable concentrations of the neurotoxin tetrodotoxin (TTX) (Brodie, 1968). One rough-skinned newt contains enough toxin to kill 25,000 white mice (Brodie and Brodie, 1999). The common garter snake has evolved variable resistance to the toxin of the rough-skinned newt, but the resistance to toxin is not without consequences. Once the garter snake has attacked and eaten a newt, the TTX will result in some degree of loss of muscle control and mobility. After swallowing a newt, some surviving garter snakes have lost mobility for up to seven hours (Brodie and Brodie, 1999). This loss of mobility may result in the inability to thermoregulate body temperature (thus may result in overheating and death) or inability to escape predators (Brodie and Brodie, 1999).

Figure 10-1. "They were running hand in hand, and the Queen went so fast that it was all she could do to keep up with her...The most curious part of the thing was, that the trees and the other things around them never seemed to change their places at all." (From Lewis Carroll's 1872 *Alice Through the Looking Glass*). The National University of Singapore scanned this image and added text under the supervision of George P. Landow. (This image is free to use without prior permission for any scholarly or educational purpose as long as (1) the site is credited and (2) linked to the image website URL in a web document or citation of the Victorian Web in a print one). (From the Victorian Web at http://www.victorianweb.org/art/illustration/tenniel/lookingglass/2.4.html , accessed 05/09/2014).

There are many, many examples in nature of this evolutionary arms race. Here is another example that is important to humans. According to the Red Queen Hypothesis, sexual reproduction persists, in some cases, because it enables many species to rapidly evolve new genetic defenses against parasites that attempt to live off them (Ridley, 1993). For example, if a certain bacterium species infects humans, some people are more resistant to the bacteria than others. The bacterium species may kill large numbers of people, but those exposed that were not killed have a higher degree of immunity to the effects of the bacteria, however, they still may get sick. But their defense mechanisms (such as antibodies, white blood cells, etc.) were better able to kill the bacteria that invaded their bodies. However, the variants of the bacterium species that were not killed by the body defenses of the more resistant humans multiplied and became even more deadly. The humans that were not killed by the deadlier form of the bacterium species produced offspring that inherited their higher degree of resistance to these bacteria. Any mutations in the bacteria to make them better able to cope with the defenses of the humans would be selected for. The same is true for the humans; any mutations or different variants of defense mechanisms that arose through sexual reproduction (and reshuffling of the genes) that favored resistance to the bacteria would be favored and passed on to offspring. Thus, the arms race would go, but in the end neither the bacterium species nor the human hosts would have gained much ground (a zero sum game) (Ridley, 1993). However, Van Valen (1973), with reference to fossil patterns within different taxa based on survival rates versus the geologic age of the taxa, also came to the conclusion that just because a particular taxon had keep pace in a coevolutionary relationship (with another or several other taxa), that did not increase the probability that a particular taxon would not go extinct (Futuyma, 2009; Ridley, 1993). Van Valen (1973) (based on his Red Queen Hypothesis) suggested that because a particular species has competitors, predators, and parasites that also continually evolve, there is always a chance that it will not be able to keep up and will fail to do so and become extinct (Futuyma, 2009, p. 168).

MUTUALISM

Of course, sometimes it is more advantageous for a cooperative relationship between members of different species, even though typically selfishness is involved here, but because the cooperation is better for both (or all) the species, they benefit from the relationship (Futuyma, 2009). As we have already stated, a cooperative beneficial symbiotic relationship between two or more different species is referred to as a mutualism. A prime example of a mutualism that most of us have probably heard of is the intimate, obligatory symbiosis between certain types of fungi and algae and/or cyanobacteria (often both are present) to form the composite or "compound" organisms that we call lichens. Another important example is the endosymbiotic (inside + symbiosis) relationship between certain types of unicellular algae referred to as zooxanthellae (dinoflagellates, primarily of the genus *Symbiodinium*) and reef-forming corals (Baker, 2003). Without this coral-algal endosymbiosis, coral reefs would likely not exist (Rowan, 1998); reef-forming corals die without the algae. These algal endosymbionts photosynthesize to form nutrients for the corals and the corals provide a home within their flesh for the algae. According to Baker (2003, p. 661): "These symbionts are critical components of coral reef ecosystems whose loss during stress-related 'bleaching' events can lead to mass mortality of coral hosts and associated collapse of reef ecosystems." Baker further states (p. 661) the following relative to the importance of studying the relationship between modern coral reef ecosystems and their endosymbionts: "Monitoring symbiont communities worldwide is essential to understanding the long-term response of reefs to global climate change because it will help resolve current controversy over the timescales over which symbiont change might occur. Symbiont diversity should be explicitly incorporated into the design of coral reef Marine Protected Areas (MPAs) where resistance or resilience to bleaching is a consideration."

Another example of mutualistic endosymbiosis we have already discussed in Chapter 7, the origin of the eukaryotic cell about 2.0 billion years ago. In this process, according to the Endosymbiotic Theory for the formation of eukaryotic cells, host prokaryote cells engulfed aerobic bacteria (such as purple bacteria). This ultimately became an obligatory relationship and the bacteria parasites evolved into mitochondria to allow oxygenic respiration. A similar process of mutualistic endosymbiosis also resulted in the evolution of photosynthesizing eukaryotes when prokaryote host cells engulfed cyanobacteria that eventually evolved into chloroplasts. Indeed, according to Futuyma (2009, p. 515) relative to the evolution of mutualisms in forming adaptations and biochemical complexity: "The best-known examples are the evolution of mitochondria from purple bacteria and chloroplasts from cyanobacteria..."

In mutualistic relationships, the amount of cooperation may vary geographically for a particular coevolutionary relationship, such as the Greya moth (*Greya politella* and *Greya obscura*) and woodland star wildflower (several species of *Lithophragma*) relationship studied by John Thompson of the University of California at Santa Cruz (Thompson, 2013). Thompson (2013) found that the coevolutionary relationship varied from a mutually beneficial relationship (mutualism), where both benefited from the Greya moth laying her eggs in the woodland star flowers and pollinating the plant, and the moth adult getting nectar and moth larvae getting seeds to eat; to commensal (the flower was pollinated by other pollinators, so a rather neutral relationship evolved between the plant and the moth, but of course, the moth was still dependent on the wildflower for food; to antagonistic with the plant aborting the seed capsules containing the moths because they had less evasive pollinators (Thompson, 2013). Thompson's studies show that coevolution is a major force in organizing biological communities and he suggests, "Every major ecosystem in the world is fundamentally based on these kinds of tightly coevolved, often mutual relationships between species." (Thompson, 2013)

Darwin's Hawk Moth

As indicated above, in addition to bees (Figure 10-2), birds (particularly hummingbirds, Figure 10-3), and butterflies (Figure 10-4), moths often serve as pollinators of flowering plants (Thompson, 2013). One unusual example of a coevolutionary mutualistic relationship between a particular Madagascan orchid species and a Madagascan moth species was predicted by Charles Darwin in 1862 in his book *On the Various Contrivances by Which Orchids are Fertilized by Insects.* One orchid that Darwin had observed was the star orchid *Angraecum sesquipedale* from Madagascar (which he had seen in a London greenhouse). This orchid species contained a nectar tube (nectary or nectar spur) that was about 30 cm (about 12 inches) long. After observing this orchid, Darwin predicted that a moth would be found in Madagascar with a proboscis of at least 10 to 11 inches long, so that it would be able to get the nectar at the base of the nectar tube (on these long nectar orchids, the nectar is in the bottom inch or so of the nectar tube). According to Kritsky (2001), Darwin wrote the following in his book of 1862: "in Madagascar there must be moths with proboscises capable of extension to a length of between ten and eleven inches!" Darwin's peers (particularly entomologists) thought that he was nuts and that this was a ridiculous prediction. However, by 1903 a large Madagascan hawk moth (also called sphinx moth) was discovered that had a wingspan of about 6 inches and a proboscis of about 12 inches. The moth was appropriately named *Xanthopan morgani praedicta* (Kritsky, 2001). It was not until very recently that this moth was actually observed fertilizing the orchid *Angraecum sesquipedale* in Madagascar.

Figure 10-2. A bumble bee (*Bombus sp.)* feeding on and pollinating the flower of a pink swamp rose mallow (*Hibiscus moscheutos*). (Photo taken within the Kroger Wetlands in Marietta, Ohio by E. L. Crisp, September 2012.)

Figure 10-3. Bronze-tailed plumeleteer (*Chalybura urochrysia*) resting on a *Heliconia sp.* Hummingbirds are important in the pollination of heliconias. (Photo taken in Costa Rica by E. L. Crisp, August 2007.)

Figure 10-4. A female eastern tiger swallowtail (*Papilio glaucus*) butterfly feeding on and pollinating a bull thistle (*Cirsium vulgare*) flower. (Photo taken in Parkersburg, West Virginia by E. L. Crisp, 2004.)

Ants and Aphids

Certainly, a common form of mutualism that most of us have heard about is the relationship between ants and aphids. Aphids are small insects that suck sugary fluids from plants with their piercing mouthparts. They remove sugar and other nutrients from the plant fluids. However, much of the altered sugary fluid passes through the digestive system of the aphids and runs out their anus, often due to stroking of their abdomens by their mutualistic ants (Novgorodova, 2002). This fluid, referred to as honeydew, is "milked" from the aphids by certain types of ants and used as a food for the ants. In effect, in the ant-aphid mutualistic symbiosis, the ants have domesticated the aphids and use them like herds of dairy cows, moving them from one plant to another (Cohen, et al, 1998; Kardong, 2008). Of course, in this mutualistic relationship, the ants, in general, protect and care for the aphids.

When ants and aphids form a mutualistic relationship (only 60% of aphid species form a mutualistic relationship with ants [Shingleton, et al, 2005]), the ants protect the aphids from their predators, transport and brood their eggs, reduce their contamination of waste products by directly consuming the honeydew, remove dead individuals and exuviae, and provide transport to feeding sites (Offenberg, 2001; Shingleton, et al, 2005; Kardong, 2008). The aphids supply carbohydrates (in the form of honeydew) to the ants (Shingleton, et al, 2005; Kardong, 2008) and in some cases (if the ants are getting too much sugar and not enough protein) the ants will eat some of the aphids for protein (Offenberg, 2001). Of course, this last behavior is not mutualistic, but exploitative predation by the ants (Offenberg, 2001). However, this behavior appears to be rare in most ant-aphid mutualistic symbiotic relationships.

Leaf-cutter Ant/Fungus Mutualistic Symbiotic Relationship

The leaf-cutter ant/fungus relationship was thought until recently to be between four types of organisms: the ants, the fungus that they cultivate for food, certain molds (microfungi) that attack the fungus, and bacteria that live on the bodies of the ants that produce antibiotics that kill the molds on the fungus. However, several recent studies have identified another symbiont that is present in this relationship. This recently recognized symbiont is a black yeast that attacks the bacteria that live on the cuticle of the ants (Caldera, et al, 2009).

This is a very complex symbiotic coevolutionary relationship that has evolved over millions of years. The first ants to evolve from a hunter-gatherer existence to agriculture, culturing of fungi, were the attine ants (tribe Attini) of Central and South America (Holldobler and Wilson, 2009; Holldobler and Wilson, 2011). This took place about 50 to 60 million years ago (Holldobler and Wilson, 2009; Holldobler and Wilson, 2011). Most of the lower or more primitive members of the attine ants gather and process dead rotting leaves and dead organic material to fertilize their fungal gardens, and are not as specific relative to the fungus that they grow. However, the more derived attine ants, the leaf-cutter ants, evolved leaf-cutter agriculture about 10 million years ago (see Holldobler and Wilson, 2011, Plate 9, p. 22-23), the technique of cutting and harvesting live plant material (primarily cut pieces of leaves) for their gardens (Figure 10-5) (Holldobler and Wilson, 2011). This opened up a new nutritional niche for these ants of the genera *Acromyrmex* and *Atta* (Holldobler and Wilson, 2009; Holldobler and Wilson, 2011). These leaf-cutter ants do not eat the leaves that they harvest, instead they transfer the pieces of leaves along their superhighways (Figure 10-6) (kept clean of debris by worker ants) through the forest to their underground caverns (as much as three meters deep) where small worker ants chew the leaf material into tiny fragments (about 1mm) and then insert the tiny pieces into their garden plots as fertilizer for their fungus (Futuyma, 2009). The fungus grows rapidly and within a day has overgrown the leaf fertilizer and the ants harvest the crop as food for both the adult and larval ants (Futuyma, 2009; Kadong, 2008). The fungi cultivated by leaf-cutters produce specialized swellings at the hyphal tips, called gongylidia, which are consumed by the ants (Caldera, et al, 2009). Attine ants have also evolved over time for more complex agriculture, with the more basal lineages of attine ants, having fewer casts and smaller colonies. "In addition, the social structure of ant colonies follows a pattern of increasing complexity, with monomorphic worker size in basal genera and strong worker size polymorphism in the more phylogenetically derived genera, especially *Acromyrmex* and *Atta*." (Caldera, et al, 2009)

Figure 10-5. A leaf-cutter ant carrying a piece of fresh leaf. (Photo by E. L. Crisp, July 2008, in the Rincon de la Vieja area of northwestern Costa Rica.)

Figure 10-6. A leaf-cutter ant carrying a piece of fresh leaf along an ant superhighway. Sometimes hundreds of ants may be seen traveling along these superhighways through the rainforest (Photo by S. L. Sowards, July 2008, in a Caribbean lowlands tropical wet forest near Chilamate, Costa Rica.)

Most of the cultivated fungi of leaf-cutter ants (*Acromyrmex* and *Atta*) belong to

two related genera, *Leucoagaricus* and *Leucocoprinus* (Holldobler and Wilson, 2009;

Holldobler and Wilson, 2011). This is an obligate, coevolved, mutualistic symbiotic relationship; neither the ants nor the fungus grown by the ants can survive without the other (Holldobler and Wilson, 2009; Holldobler and Wilson, 2011). However, there are other coevolved organisms that are present in the leaf-cutter ant–fungus mutualism. The microfungus genus *Escovopsis* is a weed in the fungal garden that on contact with the fungus grown by the ants secrets chemicals that degrade the cells to allow *Escovopsis* to then absorb the nutrients (Caldera, et al, 2009; Holldobler and Wilson, 2009; Holldobler and Wilson, 2011). Thus, a parasitic symbiotic relationship exists between the ant's fungus and *Escovopsis*. However, the ants do tend their gardens and remove spores of *Escovopsis* as well as remove infected portions of the fungal garden (Caldera, et al, 2009; Holldobler and Wilson, 2009; Holldobler and Wilson, 2011), but this alone will not prevent the microfungus from overtaking and killing the fungal garden. In addition, the ants harbor actinomycete bacteria, *Pseudonocardia* and *Streptomyces*, on their cuticle, which secrete antibiotics that inhibit the growth of *Escovopsis* on the ant's fungal garden (Caldera, et al, 2009; Futuyma, 2009; Holldobler and Wilson, 2009; Holldobler and Wilson, 2011). This is a coevolved mutualistic relationship between the bacteria and the ants; the ants benefit by the bacteria killing the weeds (primarily *Escovopsis*) in their garden and the bacteria benefit by secretions from the pores within the cuticle of the ants that nourish the bacteria. But all is not bliss in this complex symbiotic web; recently advances in microbiological and molecular techniques have identified another microbial symbiont in the ant-fungus system. A black yeast (Ascomycota; *Phialophora*) has been found growing on the same locations of the cuticle of the ants that harbors the antibiotic secreting bacteria. The black yeast parasitizes the bacteria by robbing nutrients from the bacteria, which reduces the ability of the *Pseudonocardia* and *Streptomyces* to suppress the microfungus *Escovopsis* (Caldera, et al, 2009).

Recently, even another component to this complex symbiotic network has been discovered by scientists associated with Cameron Currie in his lab at the University of Wisconsin (see Currie Lab, University of Wisconsin-Madison at https://currielab.wisc.edu/index.php). These researchers have isolated nitrogen-fixing bacteria from the ant fungus gardens of eighty leaf-cutter colonies (Holldobler and Wilson, 2011). These nitrogen-fixing bacteria convert atmospheric nitrogen (N_2) into ammonia (NH_3), which is critical for the ant's fungus gardens because ammonia is the primary building block for amino acids and proteins. According to Holldobler and Wilson, (2011, p. 106): "Currie and his co-workers demonstrated that atmospheric nitrogen-fixing bacteria (most likely of the genus *Klebsiella*) live in the ants' fungus gardens and facilitate the cultivation of the leafcutters' symbiotic fungus.[137]" (Note: The citation 137 is Pinto-Tomás et al., 2009 [see References Cited].)

Phylogenetic analysis using DNA genome sequencing of the organisms involved in this complex relationship are beginning to be studied and there have been a few publications to date attempting to show the evolutionary history of the symbionts involved in the ant-fungus system (see Currie Lab, University of Wisconsin-Madison at https://currielab.wisc.edu/index.php). There has also been some genetic research on the species of fungi, molds, ants, and bacteria that have this type of symbiotic relationship and phylogenetic trees have been constructed of the ants and their symbionts (Hinkle, et al, 1994; Poulsen and Currie, 2010; Bergstrom and Dugatkin, 2012). There is some data to suggest a long-term coevolution of derived leaf cutting attine ants and their fungal symbionts (see Figure 2 from Hinkle, et al, 1994 – note the parallel branching pattern of the ant phylogeny and the fungal phylogeny). Other work is continuing to show the coevolutionary relationship of other symbionts in the ant-fungus relationship, particularly by Cameron Currie and his associates at the University of Wisconsin's Currie Lab (see this web site: http://currielab.wisc.edu/). These studies are still in their infancy, but offer much promise, not only just to see how this relationship evolved, but for practical purposes involving the discovery of new pathogens and the development of new antibiotics in the medical field.

Ants and Acacias in Costa Rica

In Costa Rica, bull's-horn acacias (*Acacia collinsii*) (Figure 10-7), small trees that are usually less than 3 meters (about 9.8 feet) tall and have brown to tan to copper thorns, have a mutualistic relationship with ants. The mature thorns are about 4 centimeters (about 1.6 inches) long. These acacias range from Mexico to Columbia and in Costa Rica are located on the Pacific slope in dry forests from sea level to 1000 meters (about 3,280 feet). They are common in the Guanacaste region of Costa Rica (Zuchowski, 2007).

The ant-acacia tree relationship is a classic example of evolved mutualistic symbiosis between the bull's-horn acacia (*Acacia collinsii*) and ants of the genus *Pseudomyrmex* (Janzen, 1966; Zuchowski, 2007; Kardong, 2008). The ants clear (defoliate) the area around the acacia of competing plants and vines. They also viciously attack herbivores (like browsing mammals) and phytophagous insects that may want to feed on the acacia trees (Janzen, 1966; Thompson, 2013; Kardong, 2008). In return, the ants, in addition to their hollow conical thorn nests (see Figure 10-7), receive sugary secretions from nectaries on leaf stalks and protein and lipids from Beltian bodies (Figure 10-8) at the tips of the leaflets (Janzen, 1966; Zuchowski, 2007; Thompson, 2013). The Beltian bodies are picked and feed to the ant larvae (Kardong, 2008). This complex interdependence of the acacias and ants most likely evolved over a relatively long period of time, in small steps that were selectively advantageous to both the ants and the acacias (Janzen, 1966; Kardong, 2008).

Figure 10-7. The bull's-horn acacia tree with conical hollow thorns that *Pseudomyrmex* ants drill into to make their nests. (Photo by E. L. Crisp, July 2008 in a coastal mangrove forest of western Costa Rica.)

Figure 10-8. *Pseudomyrmex* ants gathering fat and protein rich Beltian bodies from bull's horn acacias to feed to their larvae. (Photo by E. L. Crisp, July 2008 in a coastal mangrove forest of western Costa Rica.)

COMMENSALISM

Commensalism, as we stated before, is a symbiotic relationship between two species where one benefits and the other does not benefit, but is not harmed in the relationship. This is a much rarer form of symbiosis as compared to mutualism. In some cases, it may be hard to determine if the other species is really not harmed, which would then be a case of parasitism.

One example of commensalism may be the symbiotic relationship between tropical trees and tropical epiphytes (mostly as the 20,000 to 25,000 tropical vascular epiphytes [Zotz and Hietz, 2001]) (Figure 10-9), such as many orchids, bromeliads, and philodendrons. To illustrate the relationship between tropical trees and tropical vascular epiphytes (ferns, gymnosperms, and flowering plants), as opposed to primarily nonvascular epiphytes (such as lichens, mosses, and liverworts) in temperate forests (Hicks, 1995), we will only concern ourselves here with one tropical vascular epiphyte family, the bromeliads (Bromeliaceae), which includes the common bromeliads (Figure 10-10) and also pineapples (Zuchowski, 2007).

Figure 10-9. A photograph showing very abundant epiphytes growing on the limbs of the tree in the background. Notice the large bromeliad growing on a limb in the right central portion of the photograph. (Photo by E. L. Crisp, July 2008, shot in a Caribbean lowlands tropical wet forest near Chilamate, northeastern Costa Rica.)

Figure 10-10. Photograph of a bromeliad in bloom growing in the fork of a tree. (Photo by E. L. Crisp, July 2008, shot in a Caribbean lowlands tropical wet forest near Chilamate, northeastern Costa Rica.)

During the early and mid-1800s, vascular epiphytes (including bromeliads) were thought to be parasitic, but by the late 1800s research on these epiphytes had shown that they lack parasitic roots and do not tap into the fluid transport system of their host trees (Hicks, 1995). Bromeliads, as with all vascular epiphytes, are not parasitic on their support trees, but appear to be commensalistic (see Sáyago et al., 2013). At least, epiphytic bromeliads are not robbing nutrients from the host support tree. However, the bromeliads are receiving support from their host trees, thus a commensalistic symbiotic relationship exists.

Bromeliads also appear to be in commensalistic relationships with poison-dart frogs, as well as with various insect larvae, such as mosquito larvae (Frank, 1990). As an example of commensalism, let's consider the relationship between bromeliads and poison-dart frogs of Central and South American tropical forests. As we will discuss in more detail later in this chapter, these frogs sequester various alkaloid skin toxins (Meyers and Daley, 1983; Meyers, 1987) and have warning coloration to advertise their unpalatability to predators. Poison-dart frogs commonly spend their larval stage in water-filled leaf fold axils of bromeliads (and more rarely in water-filled brachial cups of *Heliconia* plants, or other small bodies of water) (Meyers and Daly, 1983; Meyers, Daly, and Martinez; 1984; Summers et al., 1997; Pröhl and Hödl, 1999; Leenders, 2001; Guyer and Donnelly, 2005; Brown, Morales, and Summers,

2010; Meuche et al., 2013; Ringler et al., 2013). The bromeliad host plants that supply the pools of water for the development and parental care of poison-dart frogs do not appear to benefit from this relationship (as most certainly the poison-dart frogs do), but are not harmed in this relationship. Thus, this appears to be a commensalistic relationship.

Poison-dart frogs (in the family Dendrobatidae) of the Neotropics of Central and South America have evolved unique parental behaviors for the rearing of their young (Meyers and Daly, 1983; Summers, 1991; Summers et al., 1997; Ringler et al., 2013). These frogs are very small (see discussion later in this chapter), diurnal frogs that live in diverse habitats; bordering streams or away from the water on the ground in low lying or mountainous wet tropical forests. Some species even live mostly in trees, whereas a few species live under moist shade of low vegetation in open country (Meyers and Daly, 1983). However, most live in low tropical wet forests on land and breed and lay their eggs under leaves in moist leaf litter (Meyers and Daly, 1983). They watch their eggs until they hatch, then either the mother or the father (or in rare cases either) act as a "nurse" frog and carry the tadpoles on their backs to a small pool of water, typically in a bromeliad (sometimes high up in the canopy of the tropical forest) (Meyers and Daly, 1983; Summers et al., 1997; Ringler et al., 2013).

To illustrate an example of the distinctive parental care of poison-dart frogs and their commensalistic relationship with bromeliads or other tropical plants (for many of the species of poison-dart frogs), let's look at the blue-jeans poison-dart frog (also called the strawberry poison-dart frog), *Oophaga* (=*Dendrobates*) *pumilio*. The male of all dendrobatids is very territorial and somewhat smaller than the female. He defends his territory by wrestling with other males that invade his premises and during the day he is constantly chirping, peeping, or trilling to gain the attention of the female (Meyers and Dailey, 1983). Once a female blue-jeans poison-dart frog has decided to mate with a particular male, she will approach him and they will position themselves vent to vent. The 3 to 4 eggs laid by the female (different species of dendrobatids lay different numbers of eggs) under a leaf in the moist leaf litter on the forest floor, usually after a tropical rain, will be immediately fertilized by the male (Leenders, 2001). The male remains with the clutch of eggs; occasionally he urinates on the eggs to keep them moist. After fertilization, the eggs hatch in 5-7 days into tadpoles. The female blue-jeans poison-dart frog returns to the clutch and waits among her offspring until one of the tadpoles slithers onto her back (she [in some dendrobatids the male acts as the nurse frog {Summers et al., 1997; Ringler et al., 2013}] secrets a mucus-like glue that helps retain the tadpole on her back) (Meyers and Dailey, 1984; Leenders, 2001). The female then climbs up a tree and puts the tadpole in a small pool or cup of water made at the juncture between a leaf stem and the supporting stem of a plant, most commonly a bromeliad. The female then returns to her clutch and one by one carries all the tadpoles to a water-filled pool (some species of dendrobatids carry many more than one tadpole at a time [Meyers and Dailey, 1983]), with each tadpole getting its own little pool of water (dendrobatid tadpoles are cannibalistic), not always on the same plant. The tadpoles metamorphose into young frogs after about 43-52 days (Leenders, 2001). Rainwater in small pools of water in bromeliads or other plants have low concentrations of nutrients, thus unfertilized eggs provided by the mother feed the tadpoles (Meyer and Dailey, 1983; Leenders, 2001). During their development into young adults, the female blue-jeans poison-dart frog visits each of her tadpoles about every 4 days to deposit an unfertilized egg for them to eat.

PREDATOR-PREY COEVOLUTION

Camouflage

Camouflage involves protective (or cryptic) coloration and/or shape of an organism that allows it to be concealed as a would-be meal, such that a predator will not be able to see it. Earlier in this book we discussed the peppered moth, which due to cryptic coloration (whether in the peppered form or melanic form) could sleep on limbs and trunks of trees during the daytime without being seen by bird predators. The lichen covered trees in unpolluted areas hide the peppered form of the moth, whereas, in heavily polluted industrial areas of the late 19th and early 20th centuries in Great Britain and the U. S., the melanic form was practically invisible to bird predators. There are many more examples of camouflage to hide organisms from predators. The arctic hare is snow white to blend in with the background of snow on the terrain. However, the polar bear is also snow white to hide himself from his prey until he is close enough to attack.

Leaf-mimicking katydids (*Orophous consperusus*) (Figure 10-11) are common and familiar in Costa Rica and other tropical regions and are another example of camouflage or cryptic coloration and shape in order to conceal themselves from enemies, such as insectivorous birds (Kardong, 2008). They sometimes perfect their disguise as leaves with simulated leaf veins, mold spots, and shriveled edges (Fogden and Fogden, 1997).

Figure 10-11. Photo of a leaf-mimicking katydid (*Orophous consperusus*). This specimen is leaf green in color, has simulated leaf veins, and is about 4.0 cm (about 1.6 inches) in length. (Photo taken by E. L. Crisp in a tropical wet forest within the Caribbean lowlands of northeastern Costa Rica, July 2008.)

Aposematic (Warning) Coloration

Blue-jeans poison-dart frogs (also called strawberry poison-dart frogs), *Oophaga* (=*Dendrobates*) *pumilio* (Figure 10-12 and 10-13), and green and black poison-dart frogs, *Dendrobates auratus* (Figure 10-14) are common in the tropical wet forests in the Caribbean lowlands of northeastern Costa Rica (Lenders, 2001; Guyer and Donnelly, 2005). Other poison-dart frogs are common in wet tropical forests of other lower Central American and South American countries (Meyers and Daly, 1983; Grant et al., 2006). Poison-dart frogs, such as *Oophaga* (=*Dendrobates*) *pumilio*, have bright coloration to signal a warning to predators that they are toxic and should be avoided by the predator (Saporito et al., 2007). Poison-dart frogs are also called arrow-poison frogs (or poison-arrow frogs) (Newman, 2002; Nichol, 1990), dart-poison frogs, or poison frogs (Guyer and Donnelly, 2005). These frogs belong to the family Dendrobatidae that contains about 247 species distributed from the tropical wet forests of Nicaragua to Bolivia and the Atlantic tropical wet forests of Brazil, also along the Pacific coast of South America to Martinique in the French Antilles (Guyer and Donnelly, 2005; Grant et al., 2006). Dendrobatids are all small (see Figures 10-12, 10-13, and 10-14), diurnal frogs. About a third of the species of the family Dendrobatidae harbor alkaloid skin toxins and form a monophyletic clade of brightly colored poison-dart frogs (Grant et al, 2006). The remaining two-thirds, or so, of dendrobatid frogs form a group of brown, nontoxic frogs (Grant et al., 2006).

The frogs are called poison-dart frogs, or dart-poison frogs, or poison-arrow frogs, etc., because some native cultures use these frogs as a source of poison for blowgun darts or arrows that are used in hunting (Guyer and Donnelly, 2005; Kardong, 2008). For example, the Emberá indigenous people of the Chocó region of western Colombia have traditionally used the skin toxins of three species of dendrobatids to apply poison to the tips of their blow-gun darts used in hunting (Meyers and Daly, 1983; Grant et al, 2006). According to Nichol (1990), the skin toxins contained in these frogs are sometimes extracted in an extremely cruel fashion. The poor frog is spitted alive on a long wooden skewer and slowly roasted over a fire. The chemical poison oozes from pores in the skin and is scraped off and added to other ingredients to be cooked into a paste, which is used to dip the tips of arrows or blowgun darts. Indigenous natives hunt for animals that supply meat to them (such as monkeys or sloths) with the poison darts or arrows (Leenders, 2001). The poison on the tips of the darts or arrows is a strong neurotoxin (obtained by the frogs from a diet of mostly ants that contain these alkaloid toxins) that typically causes paralysis, or sometimes cardiac arrest, quickly (Leenders, 2001). Skin toxins of the species *Phyllobates terribilis* are so poisonous that the toxin from one frog can kill ten thousand mice or ten adult humans (Leenders, 2001). In fact, this is the most potent animal toxin known to biologists (Leenders, 2001). Columbian natives (the southern Choco Indians) only have to rub the tips of their darts across the back of *Phyllobates terribilis* to make the darts lethal to their hunted prey (Meyer and Daily, 1983). The cooked meat of hunted animals killed with these toxins on darts or arrows is not poisonous because the heating denatures the poisons (Kardong, 2008).

The bright coloration of poison-dart frogs is an example of aposematic (warning) coloration (Saporito et al., 2007; Kardong, 2008; Maan and Cummings, 2009). The bright coloration is associated with the skin toxins containing a variety of alkaloids. This bright coloration also advertises to a would-be predator that the frog is poisonous. A conspicuous trait, like bright coloration, would be naturally selected for (as diversifying selection) (Saporito et al., 2007; Maan and Cummings, 2009; Brown et al., 2010), because even though a young predator may try to eat a poison dart frog, it would quickly release it because of the distasteful and nauseating effect of the skin toxins. In the future, the predator would tend to avoid these bright colored frogs (Miller, 2007; Kardong, 2008). In addition, sexual selection may be a factor in some cases (Maan and Cummings, 2009; Richards-Zawacki et al., 2012). In a study by Maan and Cummings (2009), it appears that females may prefer more brightly colored male poison-dart frogs, at least in the case of the strawberry poison-dart frog, *Oophaga* (=*Dendrobates*) *pumilio*, in Panama's Bocas del Toro archipelago (Summers et al., 1997; Maan and Cummings, 2009; Richards-Zawacki and Cummings, 2010). This has generated more color variation and sexual dimorphism (due to sexual selection for brighter males) in this archipelago as compared to the mainland (Maan and Cummings, 2009; Richards-Zawacki and Cummings, 2010).

Aposematic Mimicry

Aposematic coloration is seen in many other organisms also. The monarch butterfly displays a bright orange colored body to advertise to birds (such as the blue jay) that it is toxic and unpalatable. The viceroy butterfly also possesses bright orange aposematic coloration and is also toxic and unpalatable to vertebrates (primarily birds). This type of mimicry in which two (or more) different toxic species mimic the aposematic coloration of each other is referred to as Müllerian mimicry. The other major type of aposematic mimicry is Batesian mimicry. In Batesian mimicry, the aposematic model is the species that is toxic or distasteful if eaten and has aposematic coloration to warn likely predators. The other species that imitates the appearance of the model, but is not toxic or distasteful, is the aposematic mimic. Batesian mimicry is named after Henry Walter Bates, a nineteenth century naturalist that first recognized this type of mimicry in 1862, in fact, he first recognized aposematic mimicry of any type (Merrill et al., 2015; also see Bates, 1862), whereas Müllerian mimicry was first named in 1878 by another 19th century zoologist and naturalist, Fritz Müller (Sherratt, 2008; Merrill et al., 2015; also see Müller, 1879).

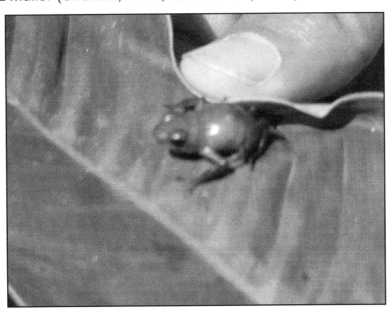

Figure 10-12. Photograph of a blue-jeans poison-dart frog (also called strawberry poison- dart frog), *Oophaga (= Dendrobates) pumilio,* **to emphasize the minute size. All poison-dart frogs are very small, but** *Oophaga (= Dendrobates) pumilio* **has a largest length (from tip of snout to tip of vent) of only about 24 mm (0.94 inches) (Grant, et al., 2006). (Photo by E. L. Crisp, August 2007, shot in a Caribbean lowlands tropical wet forest near Chilamate, northeastern Costa Rica.**

Figure 10-13. Photograph of a blue-jeans poison-dart frog (also called strawberry poison-dart frog), *Oophaga (= Dendrobates) pumilio*. The body of this frog is primarily a bright red, the back legs from the hips down are sky blue to dark blue, and the front legs from the elbow down are dark blue. Although the aposematic coloration of this frog may vary somewhat from specimen to specimen, this is a typical color for this species. (Photo by E. L. Crisp, July 2008, shot in a Caribbean lowlands tropical wet forest near Chilamate, northeastern Costa Rica.)

Figure 10-14. Photograph of a green and black poison-dart frog, *Dendrobates auratus*. This frog typically has a variable pattern of green, yellow-green, bluish-green, or blue spots on a black background (Leenders, 2001) (this specimen is only green and black). This species is one of the larger poison-dart frogs (the largest in Costa Rica), with adults measuring 30 to 39 mm (1.1 to 1.5 inches), but ranging up to 42 mm (1.7 inches) (Leenders, 2001; Guyer and Donnelly, 2005). (Photo by E. L. Crisp, July 2008, shot in a Caribbean lowlands tropical wet forest near Chilamate, northeastern Costa Rica.)

Traditionally, the classic example of Batesian mimicry was the imitation of the wing color and pattern of the monarch butterfly (*Danaus plexippus*) by the viceroy butterfly (*Limenitis archippus*) (Ritland and Brower, 1991) (see Figure 10-15). However, experimentation by Ritland and Brower (1991) has shown that the viceroy is as unpalatable as the monarch to bird predators, such as the blue jay; thus, the two butterfly species are examples of Müllerian comimics and not examples of Batesian mimicry.

Of course, Müllerian comimicry is a remarkable form of convergent evolution that can occur between organisms that are rather closely related, for example in the same genus, such as butterflies of the genus *Heliconius* (see Figure 16) from the Neotropics of Central and South America (Merrill et al., 2015), or more distantly related organisms (Zimmer and Emlen, 2013). This type of mimicry takes place when two or more (often several) toxic or unpalatable species mimic the appearance of each other, thus helping to accomplish the learned avoidance of them by predators (Zimmer and Emlen, 2013); indeed, they share mortality costs associated with teaching young, inexperienced predators to avoid them (Sherratt, 2008).

Figure 10-15. Photograph of the monarch butterfly, *Danaus plexippus* (left, no. 43), and the viceroy butterfly, *Limenitis archippus* (right, no. 44). Both butterflies have an aposematic orange color. These butterflies are an example of Müllerian mimicry (Photo taken by E. L. Crisp at the Carnegie Museum of Natural History in Pittsburgh, PA.)

Figure 10-16. Photograph of a butterfly of the genus *Heliconius* (*Heliconius hecale*) on a trail in the dry forest near Rincon de la Vieja National Park, Costa Rica. Butterflies of the genus *Heliconius* are well known for their convergent coevolution to establish Müllerian mimicry. This specimen is black with white spots and a large area of bright orange/red in the central portion of the lower wings (Photo by E. L. Crisp, July 2008.)

In Batesian mimicry, in contrast to Müllerian mimicry, a palatable species mimics a toxic or unpalatable species, therefore receiving protection from predators. This is a rather dishonest form of mimicry with the mimic parasitizing the model; i. e., the mimic benefits, but does not share in the cost of maintaining the avoidance of predators. This type of mimicry degrades the effectiveness of the model signal, as a result Batesian mimicry works best when the model is much more abundant than the mimic (or i. e., the mimic is rare relative to the model) (Kuchta et al., 2008). Indeed, if the mimic becomes too abundant and is eaten more often by the predator without adverse effects, the predator will attack the model more often. Because of this, there will be selection pressure for the model to divergently evolve differences to the mimic, and selection pressure for the mimic to evolve to catch up (in appearance, or other aposematic characters) to the model (Kuchta et al., 2008; Zimmer and Emlen, 2013). According to Zimmer and Emlen (2013, p. 491) "...selection will favor unpalatable individuals that diverge from the original color pattern. Selection will then favor palatable individuals that converge on the new pattern. And then the unpalatable species will diverge again. The result is a coevolutionary chase from one color pattern to another to another." Zimmer and Emlen (2013, p. 491-492) further state the following that is the case when mimics are rare: "When mimics are rare, on the other hand, naïve predators hunting for prey will be most likely to encounter the unpalatable model species. They will learn to associate the color pattern with the model species, and so its existing pattern will be strongly selected. As a result, stabilizing selection will leave both the model species and the mimic unchanged..."

Although it is now known that the monarch/viceroy relationship is not an example of Batesian mimicry, as originally thought, there are many other examples of Batesian mimicry. Earlier we discussed the polymorphic salamander ring species *Ensatina eschscholtzii* that forms a ring of seven subspecies in the mountains surrounding the Central Valley of California. One of these subspecies, *Ensatina eschscholtzii xanthoptica* (the yellow-eyed salamander), is hypothesized to mimic, via Batesian mimicry, the aposematic and highly toxic newt genus *Taricha* (remember, we also discussed this toxic newt previously), consisting of two species, *T. granulosa* (the rough-skinned newt) and *T. torosa* (the California newt), that overlap in sympatry (have overlapping geographic ranges) (Kuchta, 2005; Kuchta et al., 2008). All three species possess aposematic (warning) coloration of rather striking orange undersides and bold yellow patches in the iris of the eyes (see Kuchta et al., 2008, p.985, Fig. 1). Of course, only in the two newt species of *Taricha* (the models) is the warning coloration backed-up by high levels of skin toxins that result in unpalatability to bird predators. The yellow-eyed salamander (*E. e. xanthoptica*) is the mimic in this case and the aposematic warning coloration serves to deter attacks by bird predators. Kuchta et al., (2008) used caged Western Scrub-Jays as the predator in experiments where specimens of the known edible salamander of the genus *Batrachoseps* (as a control to show that the birds will eat salamanders), the newt species (of the genus *Taricha*), the yellow-eyed salamander (*E. e. xanthoptica*), and a nonaposematic brown colored *Ensatina* salamander (*E. e. oregonensis*), as another control, were fed to the birds. The results of these experiments provided strong evidence that the relationship between the two newt species and the yellow-eyed salamander are Batesian mimicry. Kuchta et al., 2008 (p. 987) state the following relative to their results:

> Using predation experiments with captive Western Scrub-Jays, we tested the hypothesis that one of the subspecies of the polymorphic ring species *E. e. xanthoptica* is a mimic of highly toxic newts in the genus *Taricha*... These results support the hypothesis that the coloration of *E. e. xanthoptica* effectively mimics the aposematic coloration of *Taricha*. In our experiment, however, all the Western Scrub-Jays were first presented with a model species, *T. torosa*, to provide a standardized prior exposure... Indeed, the results of our feeding trials suggest that the mimetic relationship is Batesian rather than Müllerian.

As we stated previously, the monarch/viceroy butterfly mimic pair are now considered a case of Müllerian mimicry and share the cost of teaching predator avoidance. However, these comimics obtain their toxic unpalatability in different manners. During their larval stages (caterpillars), monarch butterflies (*Danaus plexippus*, family Danaidae) sequester cardiac glycosides, also called cardenolides (toxic steroids), from milkweeds of the family Asclepiadaceae (Brower and Glazier, 1975; Kardong, 2008). There are several different species of milkweeds that range from the tropics to high mid-latitudes, primarily of the genus *Asclepias* (see Figure 10-17, the common milkweed, *Asclepias syriaca*). Adult monarch butterflies (see Figure 10-18) retain the cardiac glycosides sequestered in different concentrations in different parts of their bodies (Brower and Glazier, 1975). Milkweed plants initially evolved high contents of cardiac glycosides as an adaptive advantage that they received from these poisons, which discouraged phytophagous insects (plant-eating insects) and other herbivores from consuming them (Kardong, 2008). Next the monarch butterfly larvae evolved resistance (or tolerance) to the milkweed toxins (the ability to safely sequester the toxins into storage in the body), thus providing a source of food for the caterpillar that other species tend to avoid (Brower and Glazier, 1975; Kardong, 2008). These toxins are then passed on from the larval stages, through the chrysalis stage to the adult monarch. Now the adult monarch has toxins that will cause a nauseating reaction in its predators, and thus will be avoided by them (Kardong, 2008). As stated by Kardong (2008, p.188): "At this point any individual with a distinctive color, easily recognizable by visual predators, would be favored because it avoids being mistaken for a tasty butterfly. This would favor the *evolution of warning coloration* by further reducing the chances of a mistaken attack. An experienced predator, ...would tend to avoid any other species of butterfly that can be mistaken for the monarch." (Note: The italics are by the original author.) Several other aposematic insects, in addition to the monarch butterfly, share a coevolutionary relationship with milkweed plants, for example the red milkweed beetle (*Tetraopes tetraopthalmus*) (Figure 10-19), and contain cardiac glycosides (Holdrege, 2010).

Cardiac glycosides may result in cardiac arrest if consumed by a vertebrate in sufficient quantity, but also initiate severe vomiting when consumed by vertebrates (Brower et al., 1968; Brower and Glazier, 1975; Kardong, 2008). Because of this, monarch butterflies are toxic and unpalatable to bird predators (for example, blue jays [*Cyanocitta cristata*]), which quickly learn to reject the butterflies after one or more vomiting episodes due to ingestion of a monarch (Brower et al., 1968; Brower and Glazier, 1975; Kardong, 2008).

Figure 10-17. Photo of the common milkweed, *Asclepias syriaca*. Monarch butterflies have an intimate coevolutionary relationship with milkweeds (of several species of the genus *Asclepias*). They will only lay their eggs on milkweeds, and the larvae (caterpillars) will only feed on and mature through a chrysalis (pupa) stage to an adult monarch on these plants. About five fruit pods that contain the seeds are shown on this plant. The fruit pod in common milkweed typically grows to about 10 cm (3.9 inches). (Photo taken by E. L. Crisp on September 15, 2012 at Williamstown Wetlands, Williamstown, West Virginia.)

Figure 10-18. Photograph of a monarch butterfly seeking nectar on purple loosestrife (*Lythrum salicaria*). Purple loosestrife is a very troublesome invasive plant species in North America, where it disrupts flow in waterways and greatly decreases biological diversity in wetlands by crowding out native species, in particular cattails. (Photo taken by E. L. Crisp on August 26, 2012 in Williamstown Wetlands, Williamstown, West Virginia.)

Figure 10-19. Photograph of male and female red milkweed beetles (*Tetraopes tetraophthalmus*), shown breeding on a common milkweed leaf. The red milkweed beetle spends its entire lifetime on its host, the common milkweed (*Asclepias syriaca*) (Rasmann and Agrawal, 2011). This beetle has aposematic (warning) coloration of bright red with black spots (Photo by E. L. Crisp taken on September 15, 2012 at the Kroger Wetlands, Marietta, Ohio.)

Viceroy butterflies (*Limenitis archippus*), as we have previously stated, are also unpalatable to their predators due to sequestered toxins in their bodies. But viceroys have acquired different toxins than monarchs and have evolved a coevolutionary relationship primarily with members of the willow family, Salicaceae (Prudic, et al., 2007); including willows of the genus *Salix*, poplars of the genus *Populus*, and cottonwoods, also of the genus *Populus*. The larval (caterpillar) stage of the viceroy sequesters deterrent phenolic compounds (Prudic, et al., 2007) (including the phenolic compound, salicylic acid [Rivas-San Vicente and Plasencia, 2011]) in its body, which it gets from eating leaves of the trees of the Salicaceae family (willows, poplars, cottonwoods, and aspens; Lehnert and Scriber, 2011). This makes the caterpillar, and thus the adult viceroy butterfly, bitter and unpalatable to avian (bird) predators. So, both the monarch and the viceroy mimic each other (thus are comimics) and represent an example of Müllerian mimicry (Ritland and Brower, 1991).

However, the classic example of Müllerian mimicry is *Heliconius* butterflies (Figure 10-20; also see Figure 10-16) (family *Nymphalidae*, subfamily Heliconiinae) from the tropics of the Western Hemisphere (Meyer, 2006; Merrill et al., 2015). Figure 10-21, from Joron et al. (2006, Figure 1, p. 1383) more explicitly illustrates both the geographic divergence of some *Heliconius* (and related) species and their dramatic examples of Müllerian mimicry (also see Benson, 1972; Brown, 1981; Flanagan et al., 2004; Meyer, 2006; Counterman et al., 2010; Mallet, 2010; Merrill et al., 2015). In these butterflies, as Meyer (2006, p. 1676) states "...several equally unpleasantly tasting species share a color pattern, and all species benefit mutually, not only the mimic." In other words, these butterflies of the genus *Heliconius* are comimics that converge on similar wing color patterns. The diversity of mimicry color patterns and geographic mimicry associations has been experimentally determined to be primarily under the control of a shared genetic locus (Joron et al., 2006; Meyer, 2006).

Heliconius butterflies are parasitic on, and lay their eggs only on, genera of *Passiflora* (passion-flower) vines. The eggs hatch into larvae that feed on the leaves of the *Passiflora* (Gilbert, 1982). Although species of *Passiflora* have chemical defenses, in the form of cyanogenic glycosides, that make them poisonous to most insects, *Heliconius* butterfly larvae have evolved the ability to circumvent these defenses by sequestering and storing the chemicals in their bodies (Gilbert, 1982; Nishida, 2002; Cardoso and Gilbert, 2013). The adult *Heliconius* butterflies are thus unpalatable to their predators (mostly birds) because they retain the toxic chemicals sequestered and stored by their larval stages (Gilbert, 1982; Nishida, 2002; Cardoso and Gilbert, 2013).

Figure 10-20. A copy of photographs of *Heliconius* butterflies that illustrate Müllerian mimicry. The plates were taken from an article by Meyer (2006). All the butterflies above display aposematic coloration patterns of black/brown, yellow/brown, and orange/red. The original caption in the publication by Meyer (2006) is the following: "Mimicry in Butterflies Is Seen here on These Classic 'Plates' Showing Four Forms of *H. numata*, Two Forms of *H. melpomene*, and theTwo Corresponding Mimicking Forms of *H. erato*. This highlights the diversity of patterns as well as the mimicry associations, which are found to be largely controlled by a shared genetic locus [15]." The citation [15] at the end of the quote is the following: Merrill, R. M.; Dasmahapatra, K. K.; Davey, J. W.; Dell'Aglio, D. D.; Hanly, J. J.; Huber, B.; Jiggins, C. D.; Joron, M.; Kozak, K. M.; Llaurens, V.; Martin; S. H.; Montgomery; S. H.; Morris; J.; Nadeau; N. J; Pinharanda, A. L.; Rosser, N.; Thompson, M. J.; Vanjari, S.;. Wallbank, R. W. R; and Yu, Q.; 2015; The Diversification of *Heliconius* Butterflies: What Have We Learned in 150 Years?: *Journal of Evolutionary Biology*; Vol. 28, Issue 8, p. 417–1438. (Source: Repeating Patterns of Mimicry. Meyer A, PLoS Biology, Vol. 4/10/2006, e341 doi:10.1371/journal.pbio.0040341. The above figure appears in an article that has the following Copyright: © 2006 Axel Meyer. This is an open-access article (including the above associated figure) distributed under the terms of the Creative Commons Attribution License CC BY 2.5, which permits unrestricted use, distribution, and reproduction in any medium, provided the original author and source are credited.) (This image has been modified from color to shades of gray.)

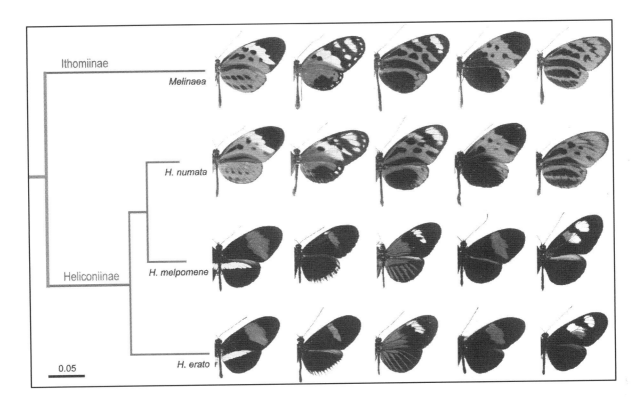

Figure 10-21.   An evolutionary tree (phylogram) of some Heliconiinae and Ithomiinae butterflies (of the genera *Heliconius* and *Melinaea*) from South America that illustrate Müllerian mimicry. The diagram was taken from an article by Joron et al., 2006. All the butterflies above display aposematic coloration patterns of black/brown, yellow/brown, yellow/white and orange/red. The original caption in the publication by Joron et al. (2006) is the following: "Colour Pattern Diversity of *H. numata*, *H. melpomene*, and their Respective Co-Mimics. The upper half of the figure shows five sympatric forms of *H. numata* from northern Peru (second row, left to right: *H. n. f. tarapotensis*, *H. n. f. silvana*, *H. n. f. aurora*, *H. n. f. bicoloratus*, and *H. n. f. arcuella)* with their distantly related comimetic *Melinaea* species (Nymphalidae: Ithomiinae) from the same area (first row: *M. menophilus* ssp. nov., *M. ludovica ludovica*, *M. marsaeus rileyi*, *M. marsaeus mothone*, and *M. marsaeus phasiana*) [20]. The lower half of the figure shows five colour pattern races of *H. melpomene*, each from a different area of South America (third row: *H. m. rosina*, *H. m. cythera*, *H. m. aglaope*, *H. m. melpomene*, and *H. m. plesseni*) with their distantly related comimetic *H. erato* races from the same areas (fourth row: *H. e. cf. petiveranus*, *H. e. cyrbia*, *H. e. emma*, *H. e. hydara*, and *H. e. notabilis*). *H. m. aglaope* and *H. e. emma* are known as rayed forms, whereas *H. m. rosina*, *H. m. melpomene*, and co-mimics are known as postman forms. *H. melpomene* and *H. erato* are from divergent clades of *Heliconius* and are identified in the field using minor morphological characters, such as the different form of the red rays on the hindwing between *H. m. aglaope* and *H. e. emma* (third from left) or the arrangement of red versus white patches in *H. m. plesseni* and *H. e. notabilis* (first from right). Co-mimics *H. numata* and *Melinaea* spp. belong to different subfamilies of the Nymphalidae and have very different body morphology and wing venation. The phylogram on the left is a maximum likelihood tree based on 1,541 bases of mitochondrial DNA (scale bar in substitutions per site, all bootstrap values over 99)." The citation [20] in the quote is the following: Joron, M; Wynne, I. R.; Lamas, G; Mallet, J; 1999; Variable selection and the coexistence of multiple mimetic forms of the butterfly *Heliconius numata. Evolutionary Ecology*; Vol. 13; Issue 7; p. 721–754. (Source: Joron M, Papa R, Beltran M, Chamberlain N, Mavarez J, et al. (2006) A conserved supergene locus controls colour pattern diversity in Heliconius butterflies. PLoS Biol 4(10): e303. DOI: 10.1371/journal.pbio.0040303. The above figure appears in an article that has the following Copyright: © 2006 Joron et al. This is an open-access article (including the associated figure above) distributed under the terms of the Creative Commons Attribution License, which permits unrestricted use, distribution, and reproduction in any medium, provided the original author and source are credited.) (This image has been modified from color to shades of gray.)

An interesting coevolutionary relationship, parasite/host, has also evolved by natural selection between *Heliconius* butterflies and many passion-flower vines (genus *Passiflora*). A study by Gilbert (1982) has identified several factors that may be under strong selection pressure in different species of passion-flower vines (about 500 species of *Passiflora* may be found in the New World Tropics). One of those factors, and the most important in his study, is that certain *Passiflora* species have evolved fake *Heliconius* eggs (i.e., egg mimicry). *Heliconius* larvae (caterpillars) are cannibalistic, thus adult *Heliconius* butterflies have evolved the tendency to not deposit eggs on passion-flower vines that already have the bright yellow eggs deposited. Thus, the fake eggs (as nectar glands, also bright yellow) of certain *Passiflora* species fool the adult *Heliconius* butterflies into avoiding vines that appear to already have eggs deposited on them. Gilbert (1982), and his colleagues, have demonstrated this experimentally.

HERBIVORE/PLANT COEVOLUTION

Plants often develop spines (cactus) or thorns (rose or acacia trees) to discourage herbivores from eating them. However, some herbivores evolve tough mouths and tongues to be able to eat the plants. Also, phytophagous (plant eating) insects can move around the thorns to get to the fleshy plant tissue that they eat (Kardong, 2008). Indeed, although mammals and other vertebrates are certainly important herbivores, the most abundant herbivores are invertebrates, particularly insects, such as grasshoppers, caterpillars, beetles, etc. (Urry et al., 2014).

Most plants evolve chemical warfare to combat animals eating them. This is especially the case against phytophagous insects. Plants synthesize secondary chemical compounds other than chemical compounds that are needed for metabolism (Kardong, 2008; Futuyma, 2009). We have talked about many of these chemical compounds that plants synthesize in the previous section. We discussed toxic chemicals that end up sequestered in prey organisms, such as cardiac glycosides synthesized in milkweeds that end up being sequestered into the body of a monarch butterfly. Some other examples of secondary chemical compounds synthesized by plants to protect themselves from herbivores are nicotine in tobacco plants, d-lysergic acid (chemical cousin of LSD; with hallucinogenic properties) in morning glories, glucosinolates (as mustard oil) in mustards (the family Brassicaceae) that are toxic to many insects, and the poison strychnine produced by the tropical vine *Strychnos toxifera* (Kardong, 2008; Futuyma, 2009; Urry et al., 2014).

However, as expected, many herbivores evolve resistance to these secondary chemical compounds produced by plants. In fact, this was also discussed in the last section. For example, the red milkweed beetle (*Tetraopes tetraopthalmus*) and the monarch butterfly caterpillar have evolved the ability to sequester the toxins from milkweed plants into parts of the body where they can do little harm to the organisms.

PARASITE-HOST RELATIONSHIPS

Parasitism occurs when one organism, the parasite, benefits from the harming of another organism, the host. The parasite typically derives its nourishment by feeding on the tissue or fluids of the host (of course, any activity by one organism that benefits itself at the expense of another organism is parasitism). Parasites may live and feed on the external surfaces of an organism as ectoparasites, such as ticks, fleas, and lice on mammals; or they may live within the body as endoparasites, such as tapeworms or harmful bacteria within the guts of vertebrates (of course, there are also good bacteria that live in the guts of vertebrates in mutualistic relationships). Some parasites are big enough to be seen with the naked eye, these are referred to as macroparasites (for example, many intestinal worms, some leeches, and some caterpillars), whereas protozoa, bacteria, and viruses are examples of microparasites. Usually the parasite does not kill its host, but the parasite may indirectly result in the death of its host due to weakening of its defenses. The weakened state of the host may make it more vulnerable to illness, metabolic malfunctions, lessened competitive ability, or increased potential for predation (Lang and Benbow, 2013). For example, ticks living on moose weaken them by sucking blood and causing hair loss and breakage, such that they may die from cold stress or predation by wolves (Urry et al., 2014).

As stated above, parasites generally do not kill their host. The typical parasite will infect its host, acquire resources from the host to replicate, and then undergo transmission to a new host (Zimmer and Emlen, 2013). In fact, there is a tendency for parasites to become less virulent or even avirulent (virulence is the death rate due to infection by a parasite), so as to not cause death of their host before they can infect new hosts (Ridley, 1996; Futuyma, 2009; Zimmer and Emlen, 2013). However, virulence varies drastically in different parasites (even for the same host). For example, the Ebola virus results in a rapid, terrible death in its human victims, but the common cold virus in humans can replicate rapidly and produce billions of copies of itself in a short time to make us feel bad for only a few days (Zimmer and Emlen, 2013). Also, virulence may vary widely in the same parasite/host system and because virulence is controlled by genetically encoded variation it can evolve in response to natural selection (Kerr et al., 2012; Zimmer and Emlen, 2013). However, often within a particular infected host there may be several different strains of a specific pathogenic parasite (such as different strains of a bacterium or virus), due to mutations within a single ancestor or resulting from independent infection of the host. The more virulent strains reproduce faster and tend to dominate the host, thus competition between strains results in natural selection for increased virulence. But, if the host dies before transmission of the parasite to a new host, the pathogen will also die. So, natural selection for transmission of the pathogen to a new host favors decreased virulence (Zimmer and Emlen, 2013). But remember, because different strains of the pathogen can be present within a host, competition may again lead to increased virulence of the pathogen; these two opposing agents of natural selection, competition and transmissibility, affect the optimal level of virulence of a pathogenic parasite (Zimmer and Emlen, 2013).

The classic example of the coevolution of a parasitic pathogen and host unfolded when the myxoma virus (family Poxviridae, genus *Leporipoxvirus*), which causes myxomatosis, was introduced in 1950 into European rabbit (*Oryctolagus cuniculus*) populations of Australia (Ridley, 1996; Futuyma, 2009; Kerr et al., 2012; Zimmer and Emlen, 2013). Over the past 65 years, or so, the evolutionary attenuation of the initially almost 100% lethal myxoma virus has played out in the rabbit populations of

Australia. The European rabbit was introduced into Australia in 1859, with explosive expansion of the rabbits across the continent due to the lack of predators to control them (Zimmer and Emlen, 2013). To biologically control the pests, agricultural scientists introduced the highly virulent Standard Laboratory Strain (SLS) of myxoma virus, initially isolated in Brazil in 1911 from the jungle rabbits (*Sylvilagus brasiliensis*) of South America, into the European rabbits of Australia (Best and Kerr, 2000). According to Fenner and Marshall (1957, p. 150) this virus strain, although originally from strains taken from *Sylvilagus* in South America, "Almost all laboratory work has been carried out with a South American strain recovered from a naturally infected *Oryctolagus* rabbit by Dr A. Moses of the Oswaldo Cruz Institute in Brazil ..." In the jungle rabbits of South America (*Sylvilagus brasiliensis*), the myxoma virus is passively transmitted by mosquitoes or biting arthropods (such as fleas) and causes a relatively mild disease (a cutaneous fibroma present only at the site of introduction of the virus) (Best and Kerr, 2000). But, when initially introduced into the rabbit populations of Australia, greater than 99% of the rabbits infected developed lethal cases of myxomatosis and, according to Best and Kerr (2000, p. 36), "the virus spread rapidly across the continent dramatically reducing rabbit numbers and becoming endemic in the rabbit population."

Thus, as we have seen above, the myxoma virus was extremely virulent when it first hit the rabbit populations of Australia, killing practically 100% of the rabbits infected. However, the virulence of the myxoma virus in the wild soon decreased (i.e., rapid selection for decreased virulence in the virus) and the mortality rate of rabbits declined (Best and Kerr, 2000). This decline in virulence was demonstrated by inoculating standard rabbit strains in the laboratory with myxoma viruses from wild rabbits in successive years (Ridley, 1996). Also, over time the rabbits were evolving resistance to the myxoma virus. This could be shown by developing other standard laboratory strains of the virus that were used on wild rabbits from a series of times. In this situation, the virus is constant and any decline in mortality (of the rabbits) is due to increased viral resistance in the rabbits (Ridley, 1996). So, as we can now see from this well documented case, both parasitic virulence and host resistance can coevolve via natural selection.

CHAPTER 11

HUMAN EVOLUTION

INTRODUCTION

Humans belong to the order Primates. Primates appear on the scene, at least in terms of anything near their present form, during the Eocene Epoch (from about 55 to 35 million years ago) of the Cenozoic Era (Stringer, 2011). Primates lack the strong specializations found in most other mammalian orders, thus they are difficult to characterize as a group. Although the primates have often been considered to lack clear defining characters (Stringer, 2011), there are several trends in the evolution of primates that help define them (Wicander and Monroe, 2013). These trends are related to their initial arboreal (tree-dwelling) ancestry. These evolving primate characters include changes in the skeleton and mode of locomotion; an increase in brain size; a shift toward smaller, fewer, and less specialized teeth (allowing dietary plasticity); the evolution of stereoscopic vision; the evolution of a grasping hand with opposable thumb; expression of parental investment (a lot of time investment on a few offspring); and others (see Larsen, 2011, p. 158; Stringer, 2011, p. 82). Not all these trends took place in every primate group, nor did they evolve at the same rate in each group (Wicander and Monroe, 2013).

Basal primates, prosimians (the suborder Prosimii), include the lemurs, tarsiers, lorises, and tree shrews; whereas, the suborder Anthropoidea includes the more derived primates. The Anthropoidea includes two infraorders, the Platyrrhini and the Catarrhini. The Catarrhini include the Old World monkeys of Africa and southeast Asia, such as the Macaque of Japan and the baboons of Africa; the apes of Africa, southeast Asia, and Indonesia; and humans. The Platyrrhini include the New World monkeys, such as the white-faced (or white-throated) capuchin monkey (*Cebus capucinus*) (Figure 11-1) and the howler monkey (Figure 11-2) of Central and South American tropical forests (see Larsen, 2011, p. 174-175, for a classification of the order Primates).

Apes and humans differ from all of the other primates in that they lack external tails. They also are more intelligent and more dependent for survival on learned behavior patterns. This is especially true of the great apes (common chimpanzees, bonobos, gorillas, and orangutans) and humans. In the fairly recent past, most anthropologists were satisfied with placing the human clade (present-day humans and extinct fossil taxa that are thought to be more closely related to modern humans than to any other living taxon) in the family Hominidae and the great apes in the family Pongidae (Wood, 2005; Wood, 2010). However, with relatively recent research (primarily genetic and molecular) revealing abundant evidence of a closer relationship between the genera *Homo* and *Pan* (common chimpanzees and bonobos) than between *Pan* and *Gorilla*, many researchers now place *Homo* and *Pan* and their related ancestral genera in the subfamily Homininae (in the new classification the family Hominidae [i.e. hominids] includes the great apes and the human clade). In addition, it appears that most anthropologists are now placing the genus *Homo* (both present-day humans and fossils of extinct *Homo* genera) and all fossil extinct genera that are thought to be more closely related to present-day humans than to present-day *Pan* in the tribe Hominini (i.e., hominins; and they place the genus *Pan* and all those extinct fossil genera thought to be more closely related to present-day *Pan* than to present-day *Homo* in the tribe Panini (i.e., panins) (see Wood, 2005; Wood, 2010).

So, to summarize the above, the lineage (clade) leading to modern humans is placed in the tribe Hominini (hominin in the vernacular), which includes all fossil species leading to modern humans after the split with the ancestors of the modern genus *Pan* (chimpanzees, both the common chimpanzees and bonobos), or all fossil species more closely related to modern humans than to any other primate (Wood and Richmond, 2000). As pointed out by Wood and Lonergan (2008), "It is these extinct taxa plus modern humans that make up the hominin clade; ..." According to the National Academy of Sciences (2010, p. 1, footnote), "The term hominin is used for any member of the evolutionary group of bipedal species most closely related to *Homo sapiens* that evolved following the split between humans and chimpanzees. The term hominid includes all great apes, encompassing chimpanzees, gorillas, orangutans, and humans." This terminology is of relatively recent use, and more correctly reflects the phylogeny of our ancestors and relatives (Kimbel, 2005; Wood, 2005; Wood, 2010). (Note: However, some anthropologists still use the term hominid [family Hominidae, which has now been modified to include the great apes and the hominin clade] to refer to our bipedal ancestors and related kin after the split with the lineage leading to modern chimpanzees; but, this older usage of hominid is confusing and should not be continued.)

Figure 11-1. White-faced (or white-throated) capuchin monkey (*Cebus capucinus*) in the Rincon de la Vieja area of northwestern Costa Rica. (Photo by E. L. Crisp, August 2008.)

Figure 11-2. Mantled howler monkey in a mangrove swamp biome along the Pacific Coast of northwestern Costa Rica. (Photo by E. L. Crisp, August 2008.)

EVOLUTION OF HUMANS AND THEIR BIPEDAL ANCESTORS

Hominins separated from our common ancestor with modern chimpanzees (including common chimpanzees, *Pan troglodytes*, and bonobos, *Pan paniscus*) somewhere between 5 to 8 million years ago (Wood, 2002; Kimbel, 2005; Wood, 2010; Wood, 2014). There are currently about 23 species (there is not universal agreement on this) within the tribe Hominini belonging to 7 genera, with about 9 species within the genus *Homo* (Wood, 2014). According to Tattersall (2012), most anthropologists would use the term human for all members of the genus *Homo* (but this is not true for some, who restrict the use of human to only *Homo sapiens*). The primary trait that separates the chimp lineage (tribe Panini) from the hominin lineage is an upright (bipedal) posture in the Hominini, as opposed to the knuckle walking, quadrapedal posture of chimps (Gorillas and other great apes also have a quadrapedal posture) (Wicander and Monroe, 2010).

The oldest hominin fossil (based on present knowledge) appears to be the late Miocene circa 7 million-year-old (7 Ma) *Sahelanthropus tchadensis* found in the desert savannah of northern Chad in 2001 (Brunet, et al, 2002; Wood, 2002;

Freeman and Heron, 2004; Kimbel, 2005; Gibbons, 2006; Wicander and Monroe, 2010; Wood, 2010; Tattersall, 2012; Papagianni and Morse, 2013). This recent find by a team led by Michael Brunet in the Djurab Desert in Chad "could be a close relative of the last common ancestor (with chimps) – or even, in principle, the last common ancestor itself" (Freeman and Herron, 2004). In fact, some, including Michel Brunet, believe that *S. tchadensis* is an early hominin and descended from the last common ancestor with chimpanzees (Brunet, et al, 2002; Wood, 2002; Freeman and Huron, 2004; Gibbons, 2006). This lone fossil skull (after reconstruction) indicates a foramen magnum that is placed more forward (closer to the underneath central portion of the skull) than in great apes, thus suggesting a bipedal posture and placement within the lineage leading towards modern humans (Kimbel, 2005; Tattersall, 2012).

There are several other early members of the hominin lineage in the 4 to 6 million year time period that have some evidence of possessing an upright, bipedal posture (Kimbel, 2005; Wood, 2010; Tattersall, 2012; Wood, 2014). In 2011, molecular clock dating and fossil primate dating (using DNA sequences in modern monkeys, chimps, and humans, combined with the fossil record itself) by Richard Wilkinson and colleagues of the University of Nottingham indicate that there should be an apelike common ancestor of the hominin and chimp lineages in the time interval from about 5 to 9 Ma that ultimately gave rise to modern chimps and modern humans (but, according to them, the common ancestor of chimps and humans probably lived about 7 Ma) (Zimmer and Emlen, 2013). The geologic dating of *Sahelanthropus tchadensis* places its age at about 7 Ma and many paleoanthropologists currently think that it was on the hominin branch (Gibbons, 2006), as were three other species that have been found in southern and eastern Africa, *Orriorin tugenensis* (at about 6 Ma), *Ardipithecus kadabba* (between 5.8 and 5.2 Ma), and *Ardipithecus ramidus* at circa 4.2 to 4.4 Ma (Kimbel, 2005; Wood, 2010).

From about 4.2 Ma to about 1.4 Ma, a group of hominins consisting of ten or eleven species belonging to one genus, *Australopithecus*, (or some say three genera, *Australopithecus*, *Paranthropus* for some of the more robust members, and *Kenyanthropus*), evolved on the scene (see Wood, 2010, Fig. 1, p. 8903; Wood,

2014, p. 40-41). This group is collectively referred to as australopithecines (southern apes) (Kimbel, 2005). The australopithecines were mostly chimp-sized, with relatively small brains (400 to 500 cubic centimeters). The australopithecines were a diverse group, particularly in terms of dental morphologies and body size. Also, the larger brain development of later australopithecines may have allowed them to produce stone tools. One lineage of robust australopithecines (often referred to the genus *Paranthropus*, although some paleoanthropologists use *Australopithecus* as the genus name for this lineage) split from the others about 3 Ma and developed a masticatory system (large molar teeth and strong jaw musculature) particular adapted to chewing hard brittle fruits, nuts, and underground roots and tubers (Kimbel, 2005). These megadont australopithecines are now thought to be a side branch that became extinct before 1 Ma, but may be the sister group to the first *Homo* species (i.e. they shared a common ancestor with the first *Homo species*) (Kimbel, 2005).

By about 2.5 to 2.0 Ma, two hominins evolve from a common ancestor with the australopithecine sister group that we consider to be what we would call human (i.e. members of the genus *Homo*) (Kimbel, 2005; Wood, 2010). *Homo habilis* and *Homo rudolfensis* showed a significant increase in brains size over the australopithecines, ranging from 610 cubic centimeters in *H. habilis* to 788 cubic centimeters in *H. rudolfensis* (Kimbel, 2005). They made more elaborate stone tools and there is some evidence that these species may have consumed meat protein (Kimbel, 2005).

By 1.9 to about 1.66 Ma another big leap in brain power occurs in the *Homo* lineage (Kimbel, 2005), with the appearance of *Homo ergaster* in Africa (maybe an African form of *Homo erectus,* see discussion in Kimbel, 2005, p. 62-63) and *Homo erectus* in Asia (Wood, 2010; Tattersall, 2012; Papagianni and Morse, 2013; Wood, 2014). Also, the 1.8 million-year-old hominin fossils from Dmanisi in the Republic of Georgia have been assigned to *H. erectus* (Kimbel, 2005; Tattersall, 2012). Most of the fossils of *Homo erectus* had a brain size of about 750 cubic centimeters to 1000 cubic centimeters, with an average of about 850 cubic centimeters (except the Dmanisi skulls, which have cranial volumes of only 600 to 775 cubic centimeters, with an average towards the lower end of the range [see Tattersall, 2012, p. 121]).

Thus with *H. erectus,* by about 1.7 Ma, we are seeing a movement out of Africa, the first migration out of Africa referred to as the Out of Africa 1 model, to colonize other parts of the world, primarily tropical and subtropical eastern and southeastern Asia (Stringer, 2012).

It appears from the current fossil record, that after 1 Ma there was another period of adaptive radiation (reminiscent of the earlier australopithecine adaptive radiation) of the *Homo* lineage, resulting in forms such as *H. antecessor, H. heidelbergensis, H. neanderthalensis,* and early *H. sapiens* with brain capacities more similar to present-day humans. According to Papagianni and Morse (2013), many anthropologists think that our species and *Homo neanderthalensis* were derived from *Homo heidelbergensis* (Figure 11-3) or a close relative, this most likely occurring in Africa for our species (Kimbel, 2005; Papagianni and Morse, 2013) and Europe or the Middle East for *Homo neanderthalensis* (Papagianni and Morse, 2013). Neanderthals appeared between 300,000 and 400,000 years ago (Pääbo, 2014) and early *Homo sapiens* about 200,000 years ago (Tattersall, 2012; Papagianni and Morse, 2013; Seddon, 2014). Although Neanderthals and modern humans share 99.84% identical DNA gene sequences (Gibbons, 2010), the genetic evidence indicates that they diverged from a common ancestor (probably *H. heidelbergensis* or a close relative) about 400,000 to 500,000 years ago. The fossil record of the appearance of *H. heidelbergensis* (with a range of 600,000 years ago to about 250,000 years ago), Neanderthals, and *H. sapiens* is approximately consistent with this interpretation (Papagianni and Morse, 2013). The evidence strongly supports that *Homo sapiens* first evolved on the African continent, prior to spreading out to the rest of the world. And in fact, this happened fairly recently with *Homo sapiens* reaching Eurasia by 90,000 to 100,000 years ago, but did not penetrate into Europe until about 40 thousand years ago (Kimbel, 2005). Thus, *Homo sapiens* eventually arose in Africa about 200,000 years ago (Wood, 2010; Papagianni and Morse, 2013) and ultimately spread to the rest of the world in an Out of Africa 2 event (or maybe 3 or more?) (or, as perhaps more properly referred to by Stringer [2012a], as the Recent African Origin [RAO] model), out competing and perhaps confrontationally combating other hominin species, and perhaps even hybridizing with them to a limited extent (see below for a discussion of this between Neanderthals and *Homo sapiens*) (Gibbons, 2010; Green et al., 2010; Pääbo, 2014). Until today, the approximately 7.5 billion individuals of *Homo sapiens* on Earth remain as the only *Homo* (human) species left on the planet (Stringer, 2012a; Tattersall, 2012).

According to Dr. Ian Tattersall, Curator Emeritus in the Division of Anthropology of the American Museum of Natural History, it is very unusual for *Homo sapiens* to be the only hominin species remaining (Tattersall, 2012). There were many times in the past one million to four million years when more than one to several hominins lived at the same time in eastern and southern Africa (Wood, 2014). Even within the last million years, and until very recently (perhaps until the last 18,000 years [see Gibbons, 2015, p. 365, relative to *Homo floresiensis*]), there have been two, three, or four hominins living contemporaneously on Earth (see Wood, 2010). Tattersall (2012) suggests that for *Homo sapiens* to be the only hominin species left is most likely due to a combination of violent confrontation and indirect economic competition between Neanderthals and *Homo sapiens*. One thing that anthropologists did accept until recently was that Neanderthals (*Homo neanderthalensis*) and other *Homo* species were not species that hybridized with early *Homo sapiens*. However, the recent work of Dr. Svante Pääbo, and his colleagues, at the Max Planck Institute for Evolutionary Anthropology in Leipzig, Germany has challenged that view with a 2010 draft sequence of the Neanderthal nuclear genome (Green et al., 2010; Pääbo, 2014). Based on their study of comparisons of the sequenced nuclear genomes of Neanderthals, present-day Africans (sub-Saharan), and present-day non-African Eurasians, the data indicate that between 1 and 4% of the genomes of people in Eurasia (and other non-African populations) are derived from Neanderthals (Green et al., 2010; Pääbo, 2014). According to the researchers, the limited interbreeding between *Homo neanderthalensis* and *Homo sapiens* most likely occurred about 80,000 years ago in caves within the Middle East (Israel) prior to the spread of *Homo sapiens* into Europe and across Asia (Gibbons, 2010). Thus, while this presents somewhat of a challenge to the simple Recent African Origin model (with no interbreeding) for the origin of modern *Homo sapiens*, "it continues to support the view that the vast majority of genetic variants that exist at appreciable frequencies outside Africa came from Africa with the spread of anatomically modern humans." (Green et al., 2010, p. 721)

Figure 11-3. A sketch of the Broken Hill skull (also called the Kabwe skull) found in Broken Hill, Northern Rhodesia (now Kabwe, Zambia) and first described and named by the British palaeontologist, Arthur Smith Woodward (1864-1944) as *Homo rhodesiensis*, but now assigned to the species *Homo heidelbergensis* (see Stringer, 2012b for a discussion of the status of *Homo heidelbergensis*). The skull has been dated to between 300,000 and 125,000 years old. (Sketch by the author, made from a cast of the original skull housed at the Natural History Museum, London. The author purchased the cast.)

THE DIVERGENCE OF HOMININS FROM CHIMPANZEES

What was the selection pressure for the divergence of the Hominini and chimp (tribe Panini) lineages? What caused the two major adaptive radiations in the hominin lineage? How are we different from our closest relatives, the chimps? There were a series of morphologic changes in the hominin line. What environmental factors may have influenced this? What makes us so much different than our closest relatives, the chimps, even though we share 98.8% of our DNA base pairs in common with them?

What caused the divergence of populations of chimp-like ancestors with a knuckle walking, quadrapedal gait from the first members of the lineage leading to humans? Members of the hominin branch walked (and still walk) with an upright, bipedal stance. In fact, bipedality was the first major step in becoming human. To answer this question, we have to look at how the climate was changing in Africa during the latter part of the Miocene Epoch (23 Ma to 5.3 Ma) and into the Pliocene Epoch (5.3 Ma to 2.6 Ma). In fact, temperatures have been cooling all over the world since the Eocene Epoch thermal maximum (Hanson, 2009), with glaciations initiated near the end of the Pliocene and into the Pleistocene Epoch (2.6 Ma to about 11,700 years ago) (Wicander and Monroe, 2013). During the Pleistocene, there were several major glaciations, with interglacial episodes in between that were warmer. The environment in Africa by late Miocene, 8 to 5 million years ago, and by the beginning of the Pliocene was becoming significantly cooler and drier, this continued into the Pliocene and Pleistocene (Wicander and Monroe, 2010; Skybreak, 2006). Thus, with declining rainfall and a higher degree of seasonality came more open forests and eventually isolated forest patches with savannahs in between, rather than dense forests (particularly in areas outside of the Congo region) (Friedman, 2006). This would result in jungles giving way to drier, less forested regions (open forests), and savannah type habitats with isolated forested areas. It would have been a definite advantage to lift the head up (with its sensory receptors for smell, hearing, and seeing) to survey the terrain for food or predators that may be hidden by brush and grasses (Kardong, 2008). Also, freeing the arms would allow for carrying food and infants and other things (Kardong, 2008). In fact, in this changing environment, natural selection would certainly favor the chimp-hominin ancestor that possessed better adaptations for bipedality (i.e. there would be strong selection pressure for bipedality in this changing environment).

During this time, late Miocene to early Pliocene (five to eight million years ago), the ancestors of chimps and humans were tree dwellers that may have been forced to survive in more open forests and savannahs with isolated forested areas. The break-up of their habitat may have caused isolation of groups of these tree dwellers, and because of the change in climate and food supply many of these groups may have gone extinct. However, some anthropologists think that the ones that could adapt by bipedality in this changing environment, to more efficiently and safely travel from one forested region to another, survived. Others think that this changing climate may have forced these tree dwellers to change their food source to a diet based on insects, nuts, eggs, small reptiles, and fish found on the forest floor and at lake and stream edges, rather than foliage and hanging fruit (Mayell, 2004). Bipedal locomotion during daytime forages for food would free up their hands to carry the food back to the trees at night, where they were safer from predators. According to Mayell (2004; also see discussion in Tattersall, 2015, p. 131-132), Owen Lovejoy, an anthropologist at Kent State University in Ohio, thinks that this may have given rise to division of labor where one parent could search for food and return with a meal for the whole family. The other parent could stay in the forested area ("home") and take care of the offspring. Lovejoy thinks that this would increase survival rates for offspring, since the quadrapedal alternative is for the whole family to forage together, eating as they go. This could have further developed into more cooperation between males and females. Lovejoy states this as follows: "Both males and females would benefit from intensified cooperative care of offspring under such conditions, as this would reduce the burden on females and enhance infant survival rates," Lovejoy said. "Over time, females began to choose males who regularly offered food" (Mayell, 2004).

The above article by Hillary Mayell in National Geographic News online discusses the fossil hominin *Orrorin tugenensis* that was found in Kenya in 2000 and dated at about 6 million years old (Mayell, 2004). CAT scans of the head of the femur showed an internal bone structure that suggests an upright, bipedal posture. As discussed previously, *Sahelanthropus tchadensis*, found in Chad and dated at circa 7 Ma, may also have evolved bipedality. Thus, by about 7 Ma, perhaps bipedal hominins had split from the chimp lineage and had made the first step in becoming human.

However, continued improvements in the efficiency of bipedalism occurred slowly. Fully efficient bipedality may not have evolved until the change in diet of hominins to red meat and to hunting, with the appearance of *Homo erectus* by about

1.9 million years ago (Friedman, 2006). But this may have occurred earlier in later australopithecines or with *Homo habilis* and set the stage for further refinements in bipedality and enlargement of the brain (the second big step that makes us human). The larger brained (than australopithecines) *Homo habilis* is associated with stone tools of Oldowan industry, which may have been used for cutting meat from bones of carnivore prey (Kimbel, 2005). But scavenging of dead carcasses for fat and protein may have also been common in these early members of *Homo* (Kimbel, 2005).

The next big step in the evolution of humans was the enlargement of the brain. This resulted in another wave of adaptive radiation (the first being the various australopithecine species that evolved from a bipedal ancestor). There are many hypotheses for the evolution of a larger brain in hominins, but it appears that one criteria that supplied selection pressure for this was again climate change. The increase in brain size of hominins started with *Homo habilis* and *Homo rudolfensis* between about 2.5 and 2.0 Ma, but increased dramatically in the *Homo* genus over the past 2 million years (Kimbel, 2005). However, this was not a simple, steady increase (Carroll, 2005). The bursts in brain size appear to have taken place in two spurts, one in early Pleistocene (about 1.8 million years ago) and another during the middle Pleistocene (about 600,000 to 150,000 years ago), separated by about a million years of stasis (Carroll, 2005). By early Pleistocene, most of the forests in Africa had been replaced with savannahs due to the much cooler and drier climate. The apes stayed in the more stable equatorial rainforest habitats, but the evolving hominins of the cooler and drier portions of Africa were evolving to changing and highly variable habitats. The changing climate affected food supply, water, hunting, and migration. Thus, there was great selection pressure on these hominins to survive and be reproductively fit. The ones that were best adapted to such constantly changing conditions were the ones that survived and reproduced more effectively. These may have been the ones with slightly larger brains. Under these changing climatic conditions, brain size roughly doubled in a million years, about 50,000 generations (rapid, but not instantaneous) (Carroll, 2005), from *Homo habilis* at an average of around 600 cubic centimeters in early Pleistocene to *Homo erectus* and *Homo heidelbergensis* between about 1000 to 1200 cubic centimeters by middle

Pleistocene, and finally by later Pleistocene at about 250,000 to 30,000 years ago, *Homo neanderthalensis* with an average brain size of 1512 cubic centimeters and *Homo sapiens* at about 200,000 years ago with an average brain size of about 1355 cubic centimeters (Carroll, 2005; Clinton, 2015; Kimbel, 2005; also see Tattersall, 2015).

In conclusion, it appears that climate change has been the main environmental force supplying the selection pressure for the evolution of both bipedality and large brain size in the hominin lineage. Hominins with larger brains could better adapt to the changing climatic conditions (but see Finlayson, 2009 in his book *Humans Who Went Extinct – Why Neanderthals Died Out and We Survived*, and deMenocal, 2014, Climate Shocks in *Scientific American*, for a more detailed discussion of climate change and the evolution of hominins). Of course, further refinement of humans occurred with the evolution of language and culture, which required the larger brain.

CHAPTER 12

WHY EVOLUTION MATTERS AND THE DANGERS OF CREATIONISTIC THINKING

INTRODUCTION

One of the primary reasons for teaching evolutionary theory in our public schools is to educate a new generation in the practical value of the knowledge of evolutionary science in solving problems in biology, environmental science, medicine, agriculture, forensics, conservation, and several other areas. Evolutionary theory is also a shining example of the nature of scientific inquiry. Evolution is an example of a theory that illustrates the empirical methodology of science in its fullest: collecting and observation of initial data, asking questions, formulation of hypotheses, making predictions, testing (by experimentation or observation of nature), collecting more data relative to predictions, then accepting, modifying, or rejecting hypotheses – back to asking more questions, making more predictions, formulating new hypotheses, and the whole process repeated over and over (Berra, 1990). Sound scientific theories assimilate factual evidence (empirical data) that may appear to be unrelated, into a coherent explanation of natural processes; evolutionary theory does this. Evolutionary theory, from Darwin to the present, has withstood the test of time, of course with modifications based on new discoveries and more advanced technologies. But the essence of the theory, as Darwin presented, descent with modification from a common ancestor and natural selection as the defining mechanism, has withstood the many challenges to falsify the theory (Eldredge, 2005). Evolutionary theory stands today as one of the strongest and most robust theories in science. Certainly, there are still arenas of investigation relative to evolutionary theory and debates continue about the patterns, rates, and modes of evolution, but the basic concepts of evolutionary theory are sound and supported by a tremendous amount of factual data. As Eldredge (2005) points out, in addition to giving us the basics of this revolutionary theory, Darwin (1859) exemplified in his formulation of this theory "the fundamentals of the very nature of science."

So, when antievolutionists (primarily creationists, so called creation "scientists", and advocates of Intelligent Design "Theory") attack evolutionary theory, they are really attacking the very nature of science. Teaching evolution in our public school science classrooms has been under a renewed aggressive attack across the nation for the last several decades, since the 1987 Supreme Court decision, *Edwards v. Aguillard*, struck down a Louisiana law requiring that creation science be taught in public school science classrooms along with evolution. The Intelligent Design (ID) Creationism movement has led this renewed attack (of course, the proponents of ID would say "Intelligent Design Theory"). The Discovery Institute of Seattle, Washington, which has served as the primary source of the ID Creationism movement, is a religiously and culturally (and sometimes politically) motivated conservative think tank that has as its primary goal the overthrow of scientific materialism (or scientific naturalism) (Forrest, 2001; Forrest and Gross, 2004; Johnson, 2007). Evolutionary theory is one of the main targets of ID advocates. Their primary strategy is to confuse the public into thinking that there is a crisis in evolutionary theory and much controversy amongst evolutionary biologists about the facts and mechanisms of evolution (which, of course, is not the case). In fact, one of their documents, called The Wedge, leaked out onto the Internet in early 1999 (written in 1998), but it was not until 2002 that the Discovery Institute finally admitted ownership (Forrest and Gross, 2004; Johnson, 2007). The Wedge (see Discovery Institute, 1998) document had statements like the one that follows:

> ...Debunking the traditional conceptions of both God and man, thinkers such as Charles Darwin, Karl Marx, and Sigmund Freud portrayed humans not as moral and spiritual beings, but as animals or machines that inhabited a universe ruled by purely impersonal forces and whose behavior and very thoughts were dictated by the unbending forces of biology, chemistry, and

environment. This materialistic conception of reality eventually infected virtually every area of our culture; from politics and economics to literature and art...Discovery Institute's Center for the Renewal of Science and Culture seeks nothing less than the overthrow of materialism and its cultural legacies.

The Wedge document further stated the goals of the Intelligent Design movement as follows:

"Governing Goals

- To defeat scientific materialism and its destructive moral, cultural and political legacies.
- To replace materialistic explanations with the theistic understanding that nature and human beings are created by God."

As you can see, this is not just an attack on evolutionary theory, but also an attack on all of science. Even after the 2005 federal court decision in Pennsylvania, *Kitzmiller et al. v. Dover Area School District et al.*, relative to the teaching of ID in the Dover Area School District and the ruling by the judge that ID is a form of religion and violates the Establishment Clause of the First Amendment of the U.S. Constitution (and thus cannot be taught in the science classroom), ID creationists are still active and seeking stealth methods to emplace their views into science standards in school districts across the nation (see the excellent book, *Creationism's Trojan Horse – The Wedge of Intelligent Design* by Barbara Forest and Paul R. Gross, 2004; and several articles in Scott and Branch [eds.], 2006).

Evolutionary theory is representative of real and sound science without biased interpretations of data (or presentation of no data), as opposed to all forms of creationism (including ID creationism) that are culturally and religiously (and sometimes politically) motivated. The importance of teaching evolutionary theory in our public school system, including public colleges and universities that train future public school teachers, is stressed by the following statement from Crisp (2008): "With the dangers to science education inherent in the intelligent design creationism movement, explaining evolutionary theory and its strong basis in sound scientific inquiry in introductory college science classes is extremely important if we are going to prevent the teaching of pseudoscience in our public school science classrooms."

There have been numerous legal battles between proponents of evolutionary theory and creationists of all stripes. The following statement from Pennock (2001, p. 758) addresses some of the ways that creationists attempt to get around the legal issues involved in antievolutionary activity:

> The 1925, antievolutionary Butler Act in Tennessee led to the first legal battle over creation and evolution in the schools--the famous *Scopes* trial. Such antievolutionary laws remained in effect until the U.S. Supreme Court finally overturned them in 1968. Creationists countered at first by passing laws in the early 1970s to give "equal emphasis" to the Biblical account. Since these and similar state laws were struck down in the 1980s, creationist activists have turned to other tactics and other venues, getting laws passed that require, for example, "disclaimers" to be read before biology classes in which evolution would be covered. Alabama public school students found a disclaimer pasted in their biology textbooks that began: "This textbook discusses evolution, a controversial theory some scientists present as a scientific explanation for the origin of living things, such as plants, animals and humans. No human was present when life first appeared on earth. Therefore, any statement about life's origins should be considered as theory, not fact." In other states, creationist "stealth candidates" got themselves elected to local School Boards and to State Boards of Education and then worked to change science curriculum standards to include creationism or to gut any evolution component. A few go further and require that "evidence against evolution" be presented. Some creationist teachers sometimes simply ignore the law and go ahead and teach their views in their individual classrooms. These and other examples of creationist activism in the public schools keep this venue at the center of the controversy.

More recently, the attempted introduction of Intelligent Design "Theory" (in reality, Intelligent Design Creationism) into the classroom in Pennsylvania failed (at least in terms of constitutionality, for the federal Middle District of Pennsylvania) as a result of the 2005 court battle *Kitzmiller et al. v. Dover Area School District et al.* Except for the *Scopes* trial, modern evolutionary theory (the modern version of Darwin's theory of evolution) has been the big winner in the major court decisions. The next section will discuss the major courtroom cases of evolution versus creationism (in all its forms); these cases include the following:

State of Tennessee v. Scopes, 1925.

Epperson v. Arkansas, 1968, 393 U.S. 97.

McLean v. Arkansas Board of Education, 1982, 529 F. Supp. 1255.

Edwards v. Aguillard, 1987, 482 U.S. 578.

Kitzmiller et al. v. Dover Area School District et al., 2005, 400 F. Supp. 2d 707 (M.D. Pa 2005).

THE HISTORY OF MAJOR LEGAL BATTLES BETWEEN EVOLUTIONARY SCIENCE AND CREATIONISM

State of Tennessee v. John Thomas Scopes - The 1925 Scopes Trial

The theory that extant (living) species evolved from ancestral species that lived in the past had been around for about a century by the beginning of the 1900s (Larson, 1997). During the late 1700s and early 1800s, hypotheses by Enlightenment scholars of Europe, such as the Frenchmen Cuvier, Buffon, and Lamarck (and others), were proposing that Earth was much older than previously envisioned and that older rocks contained fossils of extinct organisms. Lamarck, as discussed in Chapter 1, published the first comprehensive theory of "transmutation of species" (biologic evolution) in 1802 (Larson, 2004). By a couple of decades after Darwin's publication, *On the Origin of Species* in 1859, the concept of biologic evolution was fairly well established into the accepted scientific knowledge of Europe and America for scientists and naturalists, and also for many of the educated populace of Europe and America (Larson, 1997) (although the mechanism, natural selection, was still hotly debated by some until the early years of the twentieth century). But during the 1800s and into the early 1900s (and even today) many fundamentalist Christians in Europe and the United States still did not (and do not) accept evolutionary theory, in particular that modern humans evolved from past species. According to Edward J. Larson in his Pulitzer Prize winning book, *Summer of the Gods - The Scopes Trial and America's Continuing Debate Over Science and Religion* (Larson, 1997, p. 14): "During the first quarter of the twentieth century, scientists in western Europe and the United States accumulated an increasingly persuasive body of evidence supporting a Darwinian view of human origins, and the American public began to take notice." This statement, in particular, applies to fundamentalist Christians in the

United States. In fact, as the following quote from Larson (1997, p.14) illustrates: "The scientific developments helped set the stage in the early 1920s for a massive crusade by fundamentalists against teaching evolution in public schools, which culminated in the 1925 trial of John Scopes."

The Scopes Trial (sometimes referred to as the "Scopes Monkey Trial") was the first major court battle between proponents of evolutionary theory and biblical fundamentalists, in this case, primarily young-Earth creationists (YECs). According to Scott (2016, p. 289), "YECs are currently the most numerous creationists in the United States." YECs believe that Earth is only 6,000 to 10,000 years old and reject the concepts of modern physics, chemistry, astronomy, and geology, especially with respect to the dating of Earth (at 4.6 billion years old) (Scott, 2004). They further reject Darwinian evolutionary biology, believing that earlier forms of life are not ancestral to later forms, and they do not accept that descent with modification takes place as explained by Darwin (and also modern evolutionary theory) (Scott, 2004; also see Scott, 2004 and Scott, 2016 for a further discussion of the beliefs of YEC and other types of creationism). YECs believe in special (divine) creation of all organisms on Earth, with each "kind" being created suddenly, all at once, and that Earth and life on Earth were created in six 24-hour days (Scott, 2004; Scott, 2016). Thus, they believe in a literal interpretation of the Genesis account as stated in the Bible and in the inerrancy of the Biblical account.

During the early years of the twentieth century, students attending high schools in the U.S. jumped dramatically, from 200,000 in 1890 to more than two million in 1920 (Ruse, 2005; Scott, 2006). In 1910 the state of Tennessee enrolled less than 10,000 students in high schools, but by 1925 (the year of the Scopes trial) the state enrolled more than 50,000 students in secondary education (primarily in public schools) (Larson, 1997). Because of the advancement of biological concepts, including Darwinian evolution (descent with modification and natural selection) and genetics by the 1920s, these topics were included in biology classes in the rapidly growing public high schools (Larson, 1997). Further, finds of human fossils in the early part of the twentieth century provided more evidence of human evolution. Thus, the scene was set for the clash of fundamentalist Christian concepts of YECs and other antievolution Christian creationists [such as old Earth creationists, including gap creationists, day-age creationists, and progressive creationists [see Scott, 2004; Scott, 2016]) with the teaching of biological evolutionary theory in public high schools.

The lead-up to the clash of creationists with the teaching of biological evolution in the high schools was hastened by the crusade against evolution (and its teaching to high school youths) by William Jennings Bryan (1860-1925). Bryan was a lawyer and three-time democratic candidate for President of the United States. He served as a representative in the U. S. House of Representatives from Nebraska, and served as Secretary of State from 1913 to 1915 under President Woodrow Wilson. Bryan was also an ardent defender of the Christian faith and a staunch opponent of Darwinian evolution, which he thought was undermining the faith of students in public high schools (and colleges). He encouraged states to legislate laws prohibiting the teaching of biologic evolution in public schools (Larson, 1997). In March of 1925 the Tennessee legislature passed the Butler Act, prohibiting public school teachers from teaching human evolution (the first law in the U. S. to ban the teaching of evolution in public schools) (Larson, 1997), in particular (according to the State of Tennessee, 1925) the law stated the following relative to the teaching of evolution:

> *Be it enacted by the General Assembly of the State of Tennessee,* That it shall be unlawful for any teacher in any of the Universities, Normals and all other public schools of the State which are supported in whole or in part by the public school funds of the State, to teach any theory that denies the story of Divine Creation of man as taught in the Bible, and to teach instead that man had descended from a lower order of animal.

After the bill had been enacted (March 21, 1925), the American Civil Liberties Union (ACLU) was anxious to challenge the constitutionality of the bill and was looking for a test case. By May of 1925, the ACLU "issued a press release in New York City offering to defend any Tennessee schoolteacher willing to challenge the validity of the new statute in state court" (Larson, 1997; Larson, 2003; Larson, 2004). The offer was promptly accepted by a 24 year-old general science teacher at Rhea County Central High School in the east Tennessee town of Dayton, John Thomas

Scopes, at the instigation of prominent townspeople (Hannon, 2010). Scopes said that he had probably already broken the law by teaching human evolution in the biology class (he had been substituting in biology class for the regular teacher, who was ill) from the textbook Hunter's *A Civic Biology* (*A Civic Biology: Presented in Problems* by George W. Hunter), the state-approved biology textbook for Tennessee high schools (Hannon, 2010). Leading townspeople in Dayton thought that this challenge to the state antievolution law would put the town on the map (Larson, 1997; Larson, 2003; Larson, 2004). Hannon (2010, p. 25) states the following about what happened next:

> On May 9, Scopes attended his preliminary hearing, where he was represented by John Neal and John Godsey. He was bound over for a grand jury set for August. Scopes soon left Tennessee and went back to Kentucky to visit his family. On May 25, a special grand jury was called by local judge John T. Raulston to speed up the legal process - so that no other Tennessee city could create a test case and steal Dayton's chance at fame. During the grand jury hearing, three of Scopes' students testified that he had taught evolution (Scopes recalled that seven of his students testified). The grand jury indicted Scopes for violating the Butler Act. He was arrested as a formality but was not detained. Scopes' bail was set at $500 and paid by the owner of the *Baltimore Sun* newspaper.

So, John Scopes went to trial, in *State of Tennessee v. John Thomas Scopes*, on July 10, 1925 for violating the Butler Act. William Jennings Brian volunteered his services as one of the prosecutors for the case and religious agnostic Clarence Darrow (1857-1938), the famous trial lawyer (and "America's leading defense Lawyer" [Larson, 1997]) from Chicago (who had just the previous year defended two confessed murderers in the Leopold-Loeb trial and kept them from receiving the death penalty), volunteered to join the ACLU's defense team (Larson, 1997; Ruse, 2005). Of course, the ACLU did not expect to win the initial constitutional test case against this antievolution law; they expected that the constitutionality of the antievolution law would be upheld and that Scopes would be found guilty in the Dayton trial of violating the Butler Act, then they would appeal the decision to a higher court, preferably all the way to the United States Supreme Court (Linder, 2008; Hannon, 2010). This case was important for the ACLU to set a precedent against the establishment of religion in public schools and to protect the rights and civil liberties of individuals, in this case, the right of an individual to have free speech and academic freedom rights to teach the accepted science of the time.

On Monday July 13 (day 2 of the trial; day 1 on Friday July 10 was taken up in selecting a jury) the lawyers for the defense challenged the constitutionality of the Butler Act with a motion to "quash the indictment" (Larson, 1997). The jury was asked to leave the courtroom because only the Judge could rule on the constitutionality of the law (the jury did not return to the courtroom until Wednesday afternoon, July 15). Larson (1997, p. 158-159) states the following relative to the primary objections to the law (as delivered to the court by one of the defense attorneys, John R. Neal [former professor of law and dean of the law school at the University of Tennessee]):

> Most of the serious objections invoked provisions of the Tennessee state constitution because the federal Bill of Rights then had limited application against states. The key constitutional provisions included express guarantees of individual freedom of speech and religion, requirements of clearly understandable indictments and titles for legislation, and a clause directing the legislature to cherish science and education. Furthermore, both the Tennessee and United States constitutions barred the state from depriving any person of liberty without due process of law, which courts then interpreted as precluding patently unreasonable state laws and actions.

Larson (1997, p. 159) further describes the arguments by Neal as follows: "During his presentation, Neal returned most often to the constitutional bar against the establishment of religion in public schools, asserting that 'the legislature spoke for the majority of the people of Tennessee [in passing the antievolution law], but we represent the minority, the minority that is protected by this great provision in our constitution.'" Larson (1997, p. 163) also highlights this argument for the defense with the following statement by Clarence Darrow as he addressed the court: "Darrow's opening introduced his main point. The antievolution statute was illegal because it established a particular religious viewpoint in the public schools. Darrow presented this defense in state constitutional terms because the U. S. Supreme Court had not yet interpreted the Constitution's establishment clause to limit state laws – but otherwise both state and federal constitutions offered similar protections."

Despite the arguments presented by the defense for the unconstitutionality of the Butler Act, Judge Raulston denied the motion to quash the indictment and rejected each of the defense objections to the statute, one by one (Larson, 1997). With the issue of the constitutionality of the Butler Act settled, the prosecution presented witness testimony by students of Scopes and school officials that verified that Scopes had indeed taught evolution in biology class (Larson, 2004).

Afterwards, the defense called its first witness, Maynard Metcalf, zoology professor from John Hopkins University, an expert witness on evolutionary theory. After Darrow establishes Metcalf's qualifications as an expert on evolution and that Metcalf has been brought up as a protestant Christian that also accepts and teaches evolutionary theory, Darrow asked him to tell the court the meaning of evolution in the origin of man. At this point the prosecution objected to the line of questioning that Darrow had used with Metcalf. According to Larson (1997, p. 174), "The prosecution maintained that the statute outlawed any teaching about human evolution regardless of what evolution meant or whether it conflicted with the Bible." Judge Raulston excused the jury from the court and allowed Darrow to question Metcalf the remainder of the afternoon, which was concentrated on the meaning of evolution and the strong evidence supporting evolutionary theory. Dr. Metcalf did a superb job of defining evolution and giving support to evolutionary theory. However, the next day, after much argument by the defense and prosecution, on the issue of allowing expert witnesses to testify before the jury, Judge Raulston rules that the testimony of Dr. Metcalf will not be allowed, nor will the remaining experts be allowed to give testimony, and thus the defense will not be allowed to submit expert testimony relative to evolutionary theory or its relationship to the Biblical account of creation to the jury (Larson, 1997; Linder, 2008). Judge Raulston did not view the testimony of the expert witnesses by the defense as pertinent to whether John Scopes did or did not violate the Butler Act. The judge did allow the defense to "read into the record," for the purpose of review by an appellate court, summaries of prepared statements by eight expert scientists on evolution and four religious experts who had attended the trial to testify (Linder, 2008). Linder (2008) states the following: "The statements of the experts were widely reported by the press, helping Darrow succeed in his efforts to turn the trial into a national biology lesson."

On the next to last day of the trial, July 20, Darrow called Bryan as a witness for the defense to show that the Bible cannot be taken literally and that Earth is much older than 6,000 to 10,000 years. Bryan agreed to take the stand, against concerns by his fellow prosecutors. It was a very hot day and the judge had moved the court to the courthouse lawn for fear that the court flooring would collapse under the weight of the thousands (perhaps as many as 3000 people) attending the court session (Larson, 1997). Darrow mercilessly poured questions about the Bible upon Bryan. The following statement by Linder (2008) illustrates the type of questions Darrow was using to attack the literal interpretation of the Bible:

> Darrow began his interrogation of Bryan with a quiet question: "You have given considerable study to the Bible, haven't you, Mr. Bryan?" Bryan replied, "Yes, I have. I have studied the Bible for about fifty years." Thus, began a series of questions designed to undermine a literalist interpretation of the Bible. Bryan was asked about a whale swallowing Jonah, Joshua making the sun stand still, Noah and the great flood, the temptation of Adam in the garden of Eden, and the creation according to Genesis. After initially contending that "everything in the Bible should be accepted as it is given there," Bryan finally conceded that the words of the Bible should not always be taken literally. In response to Darrow's relentless questions as to whether the six days of creation, as described in Genesis, were twenty-four hour days, Bryan said "My impression is that they were periods."

The press reported that Bryan had been defeated and humiliated in the Biblical confrontation with Darrow. The next day (July 21, the last day of the trial), Judge Raulston ruled that Bryan's testimony from the previous day would be stricken from the record and he would not be allowed to return to the stand (Larson, 1997; Linder, 2008). The jury returned to the courtroom a little before noon (after having heard only two hours of testimony in the entire trial) and was charged by the judge to render a verdict; they left the room and returned with the verdict of guilty nine minutes later. Judge Raulston, rather than the jury, fined Scopes $100.00 and declared the trial adjourned about lunchtime (Larson, 1997).

Even though Scopes lost the trial, the defense appealed the decision to the highest court in the state. On January 17, 1927, the Tennessee Supreme Court overturned the conviction of Scopes by the lower court based on a technicality; it was the duty of the jury in the lower court, rather than the judge, to levee the minimum fine ($100.00) that was imposed upon Scopes (Tennessee law required that any fine above $50.00 had to be levied by the jury, rather than the judge) (Larson, 2003; Linder, 2008). Thus, even though the penalty to the defendant was overturned, the Butler Act still remained as law (the Tennessee Supreme Court ruled that the law was constitutional). Because no punishment was imposed upon Scopes for violating the Butler Act, there was no recourse to appeal the decision to a higher court. Therefore, the ACLU failed to get a real test case of the Butler Act as they had planned, with an appeal all the way to the United States Supreme Court (Larson, 1997). However, the Butler Act was repealed by the Tennessee Legislature in 1967, and the ACLU did get their test case of an antievolution law 43 years later in Arkansas, in the case *Epperson v. Arkansas*, with the United States Supreme Court ruling that the state law prohibiting the teaching of evolution violates the Establishment Clause of the First Amendment of the U. S. Constitution (Larson, 2003; Linder, 2008). William Jennings Bryan died in his sleep in Dayton, Tennessee five days after the *Scopes* trial ended (July 26, 1925). Many considered Bryan a martyr who died defending the Bible against the agnostic and atheistic evolutionists (exemplified by Clarence Darrow and H. L. Mencken, reporter for the *Baltimore Sun* during the trial) (Larson, 1997; Larson, 2004).

The *Scopes* trial did result in growing resistance to antievolution forces, but certainly did not stop the controversy between fundamentalist Christians and evolutionists. Of the fifteen states that were attempting to pass antievolution legislation in 1925, only Mississippi (in 1926) and Arkansas (in 1928) passed antievolution laws similar to the Butler Act (Larson, 2003; Larson, 2004; Linder, 2008; Scott, 2006). However, there were local objections to the teaching of evolution by school boards, parents, and teachers across the nation (particularly in the deep south), so that textbook publishers tended to slight the inclusion of evolution (or drop it completely) in high school biology textbooks. Because of this, evolution disappeared or was severely slighted in biology classrooms nationally. According to Scott (2006, p. 4): "By 1930, only five years after the Scopes trial an estimated 70 percent of American classrooms omitted evolution,[6] and the amount diminished even further thereafter. From the 1930s until the late 1950s, there was no need for a creationist movement because evolution was not being taught." (Note: The [6] refers to Larson, 2003, as used in this book.)

Everything changed in the late 1950s. With the cold war between the U.S. and the Soviet Union escalating and the successful launching of sputnik in 1957 by the Soviet Union, the U.S. surveyed its scientific establishment, including public school science education, and found it to be wanting and in serious need of an overhaul (Scott, 2004). The following statement from Scott (2006, p. 4-5) makes this clear:

> The late 1950s saw a boom in federal expenditure for science education. Model textbooks in physics, chemistry, geology, and biology were commissioned. The scientists and master teachers at the National Science Foundation-funded Biological Science Curriculum Study (BSCS) ignored tradition and composed textbooks that reflected science as it was taught at the university level, including evolution, ecology, and human reproduction. In 1963 the first three textbooks were released, and all of them included evolution as a prominent theme.

Since the BSCS textbooks carried the approval of the National Science Foundation (NSF), school boards and textbook selection committees were keen to adopt them. Commercial textbook publishers quickly adopted similar content as the BSCS textbooks (Larson, 1997; Larson, 2003; Scott, 2004; Scott, 2006). So now, the teaching of evolution in high school biology classes was escalating at a rapid pace and alarmed creationists were gearing-up to challenge this new wave of evolution education that would challenge their Biblical account.

Epperson v. Arkansas – U.S. Supreme Court, 1968

In the spring of 1965, Susan Epperson was in her second semester as a first year tenth grade biology teacher at Little Rock's Central High School in Arkansas (Berkman and Plutzer, 2010). She was born in Arkansas and had graduated from a local private religious college, but had received a master's degree in zoology from the University of Illinois (Larson, 2003). She had recently married a young Air Force officer and would most likely leave Arkansas when he was transferred to another base (Larson, 2003). That spring Susan Epperson served on a committee of local biology teachers that would choose the biology textbook for the following school year (1965-1966). The committee adopted the biology textbook *Modern Biology* by James H. Otto and Albert Towle, a biology text in the mold of a BSCS textbook (Larson, 2003; Berkman and Plutzer, 2010). The textbook chosen by this committee would be put into use the following fall (of 1965) and "in accordance with school district procedures, the Little Rock Board of Education supplied the text to all its high schools" (Larson, 2003).

Since the biology textbook chosen for the Little Rock, Arkansas high school was in the mold of BSCS texts, it contained material on evolution, including human evolution. However, since Arkansas was one of the states that passed an antievolution law after the Scopes trial, these books were in conflict with Arkansas law. The 1928 Arkansas antievolution law was not passed by the Arkansas legislature (in fact, the state legislature narrowly defeated an antievolution law), but the editors of two state Baptist denominational journals initiated a campaign to place an initiative act ("a law proposed by citizens and subject to approval by a majority of voters in a general election" [Berkman and Plutzer, 2010, p. 18]) on the state ballot in the fall of 1928 (Larson, 2003). The initiative act (known as Initiative Act No. 1) was almost identical to the 1926 Mississippi antievolution law passed by the Mississippi legislature and also similar to Tennessee's Butler Act (Larson, 2003). The initiative passed by 63% of Arkansas voters and became state law (Berkman and Plutzer, 2010). Thus, it was plain that a majority of the citizens of Arkansas in 1928 did not want evolution taught in the public schools.

According to Randy Moore in his book *Evolution in the Courtroom* (2002, p. 45) relative to the Arkansas antievolution law:

> Initiated Act No. 1 – The only antievolution law ever approved by a popular vote – read as follows:
>
> 1. It shall be unlawful for any teacher or other instructor in any University, College, Normal, Public School, or other institution of the State, which is supported in whole or in part from public funds derived by State and local taxation to teach the theory or doctrine that mankind ascended or descended from a lower order of animals and also it shall be unlawful for any teacher, textbook commission, or other authority exercising the power to select textbooks for above mentioned educational institutions to adopt

or use in any such institution a textbook that teaches the doctrine or theory that mankind descended or ascended from a lower order of animals.

2. Any teacher or other instructor or textbook or commissioner who is found guilty of violation of this act by teaching the theory or doctrine mentioned in Section 1 hereof, or by using or adopting any such textbooks in any such educational institution shall be guilty of a misdemeanor and upon conviction shall vacate the position thus held in any educational institutions of the character above mentioned or any commission of which he may be a member.

The textbook that had been adopted by the Little Rock Board of Education, *Modern Biology* by James H. Otto and Albert Towle (1965), for the 1965-1966 academic year at Little Rock's Central High School contained material that conflicted with the Arkansas antievolution law (as stated above) (Berkman and Plutzer, 2010; Greenburg, 1983; Larson, 2003; Moore, 2002). In particular, as stated by Larson (2003, p. 98), the *Modern Biology* textbook had statements such as the following: "In 1871 the English biologist Charles Darwin published his famous book entitled *The Descent of Man*...Darwin proposed that the same forces operating to bring about changes in plants and animals, could also effect man and his development." Larson (2003, p. 98) continues to quote from the text the conclusion of the section on human evolution: "It is believed by many anthropologists that, although man evolved along separate lines from the primates, the two forms may have had a common, generalized ancestor in the remote past."

Obviously, based on the Arkansas antievolution law as previously stated, any teacher instructing students about evolution using this textbook would be in violation of the law. The Arkansas Education Association (AEA – a professional teachers' association affiliated with the National Education Association [NEA]) wanted to challenge the Arkansas antievolution statute in court to determine its constitutionality. In early December of 1965, prior to biology teachers in Little Rock having taught the material on evolution in the illegal textbook, the AEA asked Susan Epperson if she would serve as plaintiff in a suit to test the constitutionality of the statute (Larson, 2003). As stated by Larson (2003, p. 99): "...seeing the action as 'my responsibility both as a teacher of biology and as an American citizen,' Epperson agreed to bring the test case..."

Epperson filed her complaint on December 6, 1965 and asserted that her First Amendment right of freedom of speech (and other constitutional rights) were violated by the antievolution law (Larson, 2003). The trial began and ended on April 1, 1966, lasting only two hours. The judge presiding over the case (without a jury) was Chancellor Murray O. Reed. Chancellor Reed issued his opinion two months later and found for the plaintiff, Susan Epperson. As stated by Berkman and Plutzer (2010, p. 18) in their book *Evolution, Creationism, and the Battle to Control America's Classrooms*: "Epperson was victorious in Arkansas's Chancery Court, which held that the law violated the First Amendment because it 'tends to hinder the quest for knowledge, restrict the freedom to learn, and restrain the freedom to teach' (quoted in *Epperson v. Arkansas*, 100)." According to Moore (2002, p 54), Chancellor Reed also ruled that the Arkansas antievolution law "violated the Fourteenth Amendment because it was arbitrary and vague, thereby making interpretation difficult."

In less than one month after the ruling by Reed, Arkansas' Attorney General's Office appealed the decision to the Arkansas Supreme Court (Moore, 2002). On June 5, 1967, the Arkansas Supreme Court basically reversed the decision of the lower court and gave a ruling similar to the *Scopes v. Tennessee* decision by the Tennessee Supreme Court. The following statement in Moore (2002, p. 55) summarizes the result of the decision:

> ...the Arkansas Supreme Court issued a bizarre, unsigned, two-sentence per curiam ruling. Unlike most decisions by the court, this decision was not written by any one justice and did not include the usual published opinion... The first sentence of the decision ruled that the state law "is a valid exercise of the state's power to specify the curriculum in its public schools." The second sentence puzzled everyone: "The court expresses no opinion on the question whether the Act prohibits any explanation of the theory of evolution or merely prohibits teaching that the theory is correct..."

So, the Arkansas Supreme Court ruled that the antievolution statute was constitutional because the majority rule in Arkansas allowed the voters to decide what could be taught in the public schools (and since this law was passed by the initiative, the majority rule meant the actual approval of the law by the people, rather than the legislature) (Moore, 2002; Larson, 2003; Berkman and Plutzer, 2010).

Epperson appealed the decision of the Arkansas Supreme Court to the U.S. Supreme Court and that court agreed to hear the case on March 4, 1968 (Moore, 2002). In a decision issued November 12, 1968, the justices voted 9-0 to overturn the Arkansas Supreme Court ruling and declared the Arkansas antievolution law unconstitutional based on religious establishment (Larson, 2003). The liberal Justice Abe Fortas wrote the majority opinion for the court. The following statement from Larson (2003, p. 114-115) summarizes the position taken by the court as written by Justice Fortas:

> ...his majority opinion overturned the law solely for violating the Establishment Clause. In reaching this result, Fortas's decision expressly passed over the primary objections raised by other justices in conference, undue vagueness and interference with free speech. "Today's problem is capable of resolution in the narrower terms of the First Amendment's prohibition of laws respecting establishment of religion," Fortas wrote. The 1963 *Schempp* decision against school Bible reading interpreted the Establishment Clause as barring any statute having solely a religious purpose or primarily a religious effect. Invoking the religious-purpose part of this test for the first time to void a statute, Fortas concluded, "The overriding fact is that Arkansas' law selects from the body of knowledge a particular segment which it proscribes for the sole reason that it is seemed to conflict with a particular religious doctrine..."

With this decision by the U. S. Supreme Court, the Arkansas antievolution law was dead. Now only one antievolution law remained in the U. S., the Mississippi antievolution law. This antievolution law had almost identical wording as the Arkansas antievolution law, thus would not survive if seriously challenged. However, the Mississippi law did not go away easily; but as stated by Larson (2003, p. 120) "...even though the Epperson decision almost certainly doomed its statute as well." The Mississippi law was challenged in courts in the state beginning in 1969, but the charges were dismissed for several reasons. The Mississippi legislature also had members that tried to kill the law, but failed to do so. Federal district court would not take the case until the state courts made a decision on the law, so eventually the case went to the Mississippi Supreme Court on appeal (Larson, 2003). The state's highest court finally bit the bullet and made the unavoidable decision. The following statement according to Larson (2003, p. 122), announces how the court ended the reign of the Mississippi antievolution law on December 21, 1970: "'...we are constrained to follow the decisions of the Supreme Court of the United States.' After quoting at length from *Epperson*, the Mississippi justices unanimously concluded, 'It is clear to us from what was said in *Epperson* that the Supreme Court of the United States has for all practical purposes already held that our anti-evolution statutes are unconstitutional.'" The Mississippi Supreme Court finally ruled the Mississippi antievolution law "void and of no effect," thus after 45 years since the *Scopes* trial, the last antievolution law in the U. S. was ruled invalid (Larson, 2003).

However, creationists were already working on ways to get creationism into the public schools. Rather than try to keep evolution out of the classroom, creationists throughout the 1970s and into the early 1980s were trying to establish empirical scientific evidence for creationism to avoid the violation of the First Amendment's Establishment Clause. Several laws and court battles ensued during the 1970s and early 1980s to test the validity of scientific creationism with balanced (or equal time) teaching of both evolution and creation science. This ultimately led to equal time, or balanced treatment, laws in several states, including Tennessee with a 1973 creationism law, which was declared unconstitutional by a federal appellate court in 1975 due to violation of the Establishment Clause (Larson, 2003). These laws mandated that if evolution is taught in the public schools then creation science must also be taught. Both Arkansas and Louisiana passed balanced treatment creationism laws in 1981, responding to the strong popular biblical creationistic sentiment in both of these states. However, the ACLU moved quickly to challenge these laws (Moore, 2002; Larson, 2003; Moore, Decker, and Cotner, 2010).

McLean v. Arkansas Board of Education – Federal District Court, Little Rock, Arkansas, 1982

In March of 1981, the Arkansas legislature passed, and Arkansas Governor Frank White enacted by signing, a law known as Act 590, which required balanced treatment of teaching evolution and creation science (Moore, 2002; Berkman and Plutzer, 2010). The law included in the first two sections the following statements (see Berkman and Plutzer, 2010, p. 19-20):

BE IT ENACTED BY THE GENERAL ASSEMBLY OF THE STATE OF ARKANSAS:

SECTION 1. Requirement for Balanced Treatment. Public schools within this state shall give balanced treatment to creation-science and to evolution-science. Balanced treatment to these two models shall be given in classroom lectures taken as a whole for each course, in textbook materials taken as a whole for each course, in library materials taken as a whole for the sciences and taken as a whole for the humanities, and in other educational programs in public schools, to the extent that such lectures, textbooks, library materials, or educational programs deal in any way with the subject of the origin of man, life, the earth, or the universe.

Section 2. Prohibition against Religious Instruction. Treatment of either evolution-science or creation-science shall be limited to scientific evidences for each model and inferences from those scientific evidences, and must not include any religious instruction or references to religious writings.

In addition, Act 590 also defined both evolution-science and creation-science in Section 4 as follows (see Moore, 2002, p. 311):

(a) "Creation-science" means the scientific evidences for creation and inferences from those scientific evidences. Creation-science includes the scientific evidence and related inferences that indicate: (1) Sudden creation of the universe, energy, and life from nothing; (2) The insufficiency of mutation and natural selection in bringing about development of all living kinds from a single organism; (3) Changes only within fixed limits of originally created kinds of plants and animals; (4) Separate ancestry for man and apes; (5) Explanation of the earth's geology by catastrophism, including the occurrence of a worldwide flood; and (6) A relatively recent inception of the earth and living kinds.

(b) "Evolution-science" means the scientific evidences for evolution and inferences from those scientific evidences. Evolution-science includes the scientific evidence and related inferences that indicate: (1) Emergence by naturalistic processes of the universe from disordered matter and emergence of life from nonlife; (2) The sufficiency of mutation and natural selection in bringing about development of present living kinds from simple earlier kinds; (3) Emergence by mutation and natural selection of present living kinds from simple earlier kinds; (4) Emergence of man from a common ancestor with apes; (5) Explanation of the earth's geology and the evolutionary sequence by uniformitarianism; and (6) An inception several billion years ago of the earth and somewhat later of life.

The Little Rock, Arkansas chapter of the American Civil Liberties Union (ACLU) filed a lawsuit on May 27, 1981 on behalf of the plaintiffs, who were opposed to the teaching of creation science in the public schools (Moore, 2002; Numbers, 2006). The plaintiffs charged, "that Act 590 was an attempt to establish religion in the public schools and therefore violated the first amendment..." (Moore, 2002, p. 83). The lead plaintiff was Reverend William S. McLean, a United Methodist minister. According to the Memorandum Opinion of Judge William R. Overton (1939-1987) (*McLean v. Arkansas Board of Education*, 529 F. Supp. 1255, 1258-1264 [ED Ark. 1982]), the Federal trial judge for the United States District Court for the Eastern District of Arkansas, Western Division, the other plaintiffs included:

...the resident Arkansas Bishops of the United Methodist, Episcopal, Roman Catholic and African Methodist Episcopal Churches, the principal official of the Presbyterian Churches in Arkansas, other United Methodist, Southern Baptist and Presbyterian clergy, as well as several persons who sue[d] as parents and next friends of minor children attending Arkansas public schools. One plaintiff is [was] a high school biology teacher. All are [were] also Arkansas taxpayers. Among the organizational plaintiffs are [were] the American Jewish Congress, the Union of American Hebrew Congregations, the American Jewish Committee, the Arkansas Education Association, the National Association of Biology Teachers and the national Coalition for Public Education and Religious Liberty, all of which sue[d] on behalf of members [then] living in Arkansas... (Words within brackets added by the present author.)

So, as the above statement illustrates, many of the plaintiffs in this trial were associated with religious denominations. According to Numbers (2006, p. 278), "The Arkansas trial also shattered myths about the so-called warfare between science and religion." Numbers (2006, p. 279) further states, "There may have been conflict in Little Rock, but it scarcely conformed to any simplistic science-versus-religion formula." In fact, the religions involved on the plaintiff side of this trial may be described as Modernist Christian religions (more liberal and scientific leaning forms of Christianity) as compared to the more Fundamentalist Christian religions (and primarily young Earth creationist with catastrophic flood geology beliefs), associated with the defense.

The antievolution defendants in this trial "were the Arkansas Board of Education and its members, the director of the Arkansas Department of Education, and the State Textbooks and Instructional Materials Selecting Committee" (Moore, 2002, p.

83). However, there were "...No religious groups that appeared on the list of defendants" (Numbers, 2006, p. 278). But, of course, Act 590: Balanced Treatment for Creation Science and Evolution Science, was legislated by the state of Arkansas and became law when signed by the governor of Arkansas in 1981, based on claims that creation science represented real science; even though it was supported and pushed through the Arkansas legislative body by Fundamentalist Christians intent upon balancing the supposed evils of teaching evolution in the public schools. Thus Act 590 was weighted-down with Fundamentalist religion from its beginnings. Certainly, this was in part, thanks to the creationistic scientific claims of Henry Morris and the legal theories of Wendell Bird, both (at the time) associated with the Institute for Creation Research (ICR), then located near San Diego, California (Moore, 2002; Larson, 2003; Numbers, 2006). Henry M. Morris (1918-2006), a Virginia Tech (then Virginia Polytechnic Institute) engineering professor (until the early 1970s), was a Fundamentalist young Earth creationist who believed in the literal truth of the Genesis account for the origin of Earth and life in six 24-hour days and the inerrancy of the Bible (Moore, 2002; Larson, 2003; Numbers, 2006). In 1961 he coauthored, with John C. Whitcomb, Jr. (another creationist), *The Genesis Flood: The Biblical Record and Its Scientific Implications,* "a biblically orthodox book that revived George McCready Price's flood geology, ignited the creationist movement in the late twentieth century, and became a foundation for creationists' demands for 'equal time' in classrooms" (Moore, 2002, p. 135). As stated by Moore (2002, p. 115), Wendel Bird (1954-) was the "Leading architect of 'creationists' demands for 'equal time' and 'balanced treatment' in public schools. Bird devised his strategy in 1978 and subsequently worked at the Institute for Creation Research as a legal adviser." Bird earned a J. D. from Yale Law School in 1978. While at the Institute for Creation Research (ICR) serving as a legal advisor and staff attorney, Bird updated the ICR equal time resolution that was originally written by Morris in 1975 (Moore, 2002; Larson, 2003;). Bird's revised resolution "was designed for adoption by school boards wishing to include scientific creationism in their curriculum" (Larson, 2003, p. 149) and was the model for the creation science/evolution science resolution adopted (with minor modifications) by both Arkansas and Louisiana for their 1981 creationism laws (Moore, 2002; Larson, 2003). Later, as the trial was about to get underway, Wendell Bird requested that he be allowed to assist the Arkansas Attorney General, Steve Clark, in defending Act 590. But as pointed out by Larson (2003, 159-160), "...the court refused to let Wendell Bird intervene in the lawsuit on behalf of four creationist organizations and fifteen like-minded individuals. The state 'Attorney General will defend the suit with adequate vigor and diligence' without outside help, the court ruled in rejecting Bird's petition."

The Arkansas trial lasted about a week, from December 7, 1981 to December 17, 1981. Judge Overton issued his thirty-eight-page decision on January 5, 1982 (Moore, 2002). His decision primarily focused on whether Act 590 violated the First Amendment Establishment Clause of the U. S. Constitution. During the trial, lawyers from the American Civil Liberties Union (ACLU) had concentrated their assault for the plaintiffs on the Establishment Clause of the First Amendment (Larson, 2003). According to Larson (2003, p. 160-161):

> ..., the ACLU gathered an impressive array of leading experts to testify that the Arkansas statute had a purely religious purpose and effect. "Consistent with this approach," one of the plaintiff's lawyers explained, "the expert testimony at the trial was intended to offer the trial judge an understanding of the history and social context of the 'creation science' movement, of the consideration and conclusions of the scientific and philosophical communities regarding the status of 'creation science' as science, of the relationship of the 'two model approach' enshrined in the Act and the history and theology of Christian Fundamentalism, and of the impact of Act 590 on the educational system within Arkansas."

With the following quote, Larson (2003, p. 161) continues to stress the importance of the expert witnesses in this trial and how their testimony showed that Act 590 violated the Establishment Clause of the First Amendment:

> The plaintiff's ten expert witnesses were divided into two teams. "The ACLU first presented expert witnesses from their 'religious team,'" reported one of those witnesses, Cornell University sociologist Dorothy Nelkin. "These witnesses argued that, historically, philosophically, and sociologically, creationism is a religious movement of fundamentalists who base their beliefs on the inerrancy of the Bible and that creation science is no more than religious apologetics." Nelkin went on, "The ACLU then presented its 'scientific team': a geneticist, a paleontologist, a geologist, and a biophysicist. They documented the absence of scientific evidence for the creationist beliefs" and the affirmative case for evolution. "Here then is the significance

of science in the case," ACLU attorney Jack Novik explained. "For if, as we contended, creationism was not science at all, then whatever else the Arkansas legislature thought it was doing, meaningful science education could not provide a legitimate secular purpose for enacting the creationism statute." ...At last, expert testimony on the creation-evolution controversy had made it into court. (Note [added by current author]: The paleontologist mentioned above was the popular science writer, Harvard paleontologist and evolutionary biologist, Stephen Jay Gould [1941-2002]. He is well known for the significant contribution to evolutionary biology of the theory of punctuated equilibrium [with Niles Eldridge of the American Museum of Natural History].)

Judge Overton, based on the testimony by the plaintiff's expert witnesses and his own research, that creation science was not science but instead represented the establishment of religion in the classroom, relied on a court decision from the 1971 U. S. Supreme Court, *Lemon v. Kurtzman* to make his ruling (*Lemon v. Kurtzman*, 403 U.S. 602 (1971); Moore, 2002; Scott, 2004). *Lemon v. Kurtzman* (1971) established a procedure to determine whether a legislated statute violated the Establishment Clause of the First Amendment of the U. S. Constitution. This procedure has become known as the Lemon Test and consists of three prongs, (1) the "purpose" prong requires that the statute must have a secular legislative purpose, (2) the "effect" prong means that the primary effect of the statute must be one that neither advances nor inhibits religion, and (3) the "entanglement" prong states that the statute must not foster excessive entanglement of the government with religion (*Lemon v. Kurtzman*, 403 U.S. 602 [1971]; Moore, 2002; Scott, 2004). If a statute violates any of these prongs of the Lemon Test, it is unconstitutional (Moore, 2002). Judge Overton agreed with the plaintiffs that Act 590 violated all three of the prongs of the Lemon Test, and thus was unconstitutional (*McLean v. Arkansas Board of Education*, 1982).

Relative to prong 1 of the Lemon Test, Judge Overton determined that there was no secular purpose of Act 590, and in fact there was an obvious sectarian purpose because "...the legislative history of the law clearly demonstrated that the legislators intended to promote a religious view" (Scott, 2004). Overton wrote the following in his Memorandum Opinion (see Larson, 2003, p. 163): "The State failed to produce any evidence which would warrant an inference or conclusion that at any point in the process anyone considered the legitimate educational value of the Act" (*McLean v. Arkansas Board of Education*, 1982).

With respect to prong 2 (the effect prong), Larson (2003, p. 164) states the following: "...the concept and tenets of creation science, as defined in the Arkansas law, were inescapably religious and identical to the Genesis account." Further, Judge Overton, in his Memorandum Opinion (*McLean v. Arkansas Board of Education*, 1982), made the following statement:

> The conclusion that creation science has no scientific merit or educational value as science has legal significance in light of the Court's previous conclusion that creation science has, as one major effect, the advancement of religion. The second part of the three-pronged test for establishment reaches only those statutes as having their *primary* effect the advancement of religion. Secondary effects which advance religion are not constitutional fatal. Since creation science is not science, the conclusion is inescapable that the *only* real effect of Act 590 is the advancement of religion. The Act therefore fails both the first and second portions of the test in *Lemon v. Kurtzman*, 403 U. S. 602 (1971).

With respect to prong 3 of the Lemon Test (the law must not foster an excessive entanglement of government with religion), Judge Overton noted in his Memorandum Opinion (*McLean v. Arkansas Board of Education*, 1982) the following relative to Act 590:

> Act 590 mandates "balanced treatment" for creation science and evolution science. The Act prohibits instruction in any religious doctrine or references to religious writings. The Act is self-contradictory and compliance is impossible unless the public schools elect to forego significant portions of subjects such as biology, world history, geology, zoology, botany, psychology, anthropology, sociology, philosophy, physics and chemistry. Presently, the concepts of evolutionary theory as described in 4(b) permeate the public textbooks. There is no way teachers can teach the Genesis account of creation in a secular manner.
>
> The State Department of Education, through its textbook selection committee, school boards and school administrators will be required to constantly monitor materials to avoid using religious references. The school boards, administrators and teachers face an impossible task. How is the teacher to respond to questions about a creation suddenly and out of nothing? How will a teacher explain the occurrence of a worldwide flood? How will a teacher explain the concept of a relatively recent age of the earth? The answer is obvious because the only source of this information is ultimately contained in the Book of Genesis.

References to the pervasive nature of religious concepts in creation science texts amply demonstrate why State entanglement with religion is inevitable under Act 590. Involvement of the State in screening texts for impermissible religious references will require State officials to make delicate religious judgments. The need to monitor classroom discussion in order to uphold the Act's prohibition against religious instruction will necessarily involve administrators in questions concerning religion. These continuing involvements of State officials in questions and issues of religion create an excessive and prohibited entanglement with religion. *Brandon v. Board of Education*, 487 F.Supp 1219, 1230 (N.D.N.Y.), *aff'd.*, 635 F.2d 971 (2nd Cir. 1980).

Judge Overton's decision in *McLean v. Arkansas Board of Education* was so soundly reasoned and strongly worded that Act 590 was a violation of the Establishment Clause of the First Amendment of the U. S. Constitution, and thus unconstitutional, that the state, under the leadership of Arkansas Attorney General J. Steven Clark, decided not to appeal the decision to the Court of Appeals (Moore, 2002; Larson, 2003; Scott, 2004). Judge Overton's decision in this trial (even though it technically only applied to the Eastern District of Arkansas), including an analysis and statements of "what science is and is not", strongly questioned the legality of creation science in the public schools, and established that evolution is not religion (as claimed by creationists), but is science (Moore, 2002). This was the case despite the fact "...that most Americans favored teaching creationism" (Larson, 2003, p. 164).

According to Moore (2002, p. 84), "...a 1981 poll by *NBC news* found that 76 percent of Americans favored equal presentations of evolution and creationism in public schools." However, Judge Overton had made it clear before the trial "that his decision would not be affected by public opinion" (Moore, 2002). As quoted by Moore (2002, p. 86), he later wrote in his Memorandum Opinion the following statement relative to this:

> The application and content of First Amendment principles are not determined by public opinion polls or by a majority vote. Whether the proponents of Act 590 constitute the majority or minority is quite irrelevant under a constitutional form of government. No group, no matter how large or small, may use the organs of government, of which the public schools are the most conspicuous and influential, to foist its religious beliefs on others.

As Larson (2003, p. 165) points out relative to Judge Overton's statement above:

> Public opinion, acting through the legislature, still had its place in fixing the school curriculum, but its free rein to prescribe creationism or to proscribe evolution was checked by the Establishment Clause. Yet public opinion exerted some influence even within these limits. Overton appealed to it in refuting the defense contention that the religious nature of evolution justified restricting evolution teaching. Citing *Epperson*, *Willoughby*, and *Wright*, Overton observed, "it is clearly established in the case law, and perhaps also in common sense, that evolution is not a religion and that teaching evolution does not violate the Establishment Clause." ...Thanks to the Constitution and common sense, in Arkansas, evolutionary teaching was legal and creationist instruction was illegal.

So, with the failure of equal time (or balanced treatment) of creation science and evolution in Arkansas, a strong precedent had been established against the teaching of creation science or creationism in the public schools. However, a very similar law had been enacted in Louisiana just a few months before the *McLean v. Arkansas Board of Education* decision (Scott, 2004). But the Louisiana Creationism Act, "Balanced Treatment for Creation Science and Evolution Science in Public School Instruction," had purposely been framed by the founders of the law to avoid defining creation science in explicitly religious language and to attempt to cloak creation science in a mask of scientific integrity, as a ploy to avoid the fate of Act 590 in Arkansas (Moore, 2002; Larson, 2003; Scott, 2004). However, as we will see in the next section, this thin disguise did not fool the Federal District Court, Court of Appeals, and U.S. Supreme Court judges that would evaluate the constitutionality of Louisiana's Creationism Law. And, as stated by Scott (2004, p. 196), relative to the importance of the *McLean v. Arkansas Board of Education* decision in evaluating Louisiana's Creationism Law, "...the McLean decision was not appealed, so its conclusions were not extended beyond its district. However, its reasoning was highly influential in the subsequent Supreme Court case that struck down 'equal time' laws nationwide."

Edwards, Governor of Louisiana, et al. v. Aguillard et al. – U. S. Supreme Court, 1987

Louisiana's Creationism Act (Balanced Treatment for Creation Science and Evolution Science in Public School Instruction Act), signed into law by Louisiana Governor David Treen on July 19, 1981, prohibited the teaching of evolutionary theory in the public elementary and secondary schools, unless "creation science" was also taught (Moore, 2002). Of course, the Act did not require that either had to be taught, only that if evolution is taught, so must it be balanced (or given equal-time) by the teaching of creation science. Although the Creationism Act was in place as law by mid-summer 1981, it was "not scheduled to take effect until the fall of 1983" (Larson, 2004). However, Wendell Bird, who had recently been excluded from participating in the *McLean v. Arkansas Board of Education* trial, wanted the Louisiana balanced-treatment law to take effect immediately and before the ACLU could file suit to challenge it (Moore, 2002). So, on December 2, 1981, Bird filed suit (*Keith v. Louisiana*) on behalf of Louisiana Senator Keith and fifty-four other plaintiffs in federal court (Federal District Court) in Baton Rouge to force educational agencies in Louisiana to comply with the new act (Moore, 2002; Larson, 2003). The very next day, December 3, 1981, the ACLU filed its own lawsuit in the New Orleans federal court (Federal District Court) that challenged the constitutionality of the Louisiana Creationism Act (Larson, 2003; Moore, 2002). According to Larson (2003, p. 166-167):

> The second action, *Aguillard v. Treen*, named twenty-six organizational and individual plaintiffs, beginning with Louisiana educator [biology teacher at Acadiana High School in Lafayette, Louisiana] Donald Aguillard and including such familiar participants as the National Association of Biology Teachers [NABT], the National Science Teachers Association [NSTA], and the American Association for the Advancement of Science [AAAS].

The Louisiana state attorney general deputized Wendell Bird of Atlanta, Georgia as a special assistant attorney general for both trials *(Keith v. Louisiana* and *Aguillard v. Treen)* (Moore, 2002; Larson, 2003). The creationists were placing all their hope in Bird to "…give creationists their day in court to defend the last surviving equal-time statute" (Larson, 2003).

The *Keith v. Louisiana* lawsuit was dismissed in June 1982 because it did not raise a federal question, even though it had been filed in U.S. District Court in Baton Rouge (Moore, 2002). This left *Aguillard v. Treen* as the lawsuit remaining in U.S. District Court in New Orleans for a trial to challenge Louisiana's Creationism Act. Because of jurisdictional issues, the case was batted around between state court and federal district court for several years (Scott, 2004). The following statement from Scott (2004, p. 109) summarizes the path to the ultimate decision by the U.S. Supreme Court in 1987:

> ...Finally the case was heard by the Federal District Court. Rather than holding a full trial, as in Arkansas, the District Court tried the case by "summary judgment": the judge accepted written statements from both sides and decided the outcome of the case based on these documents.
> In 1985, the Federal District Court decided that the law was unconstitutional because it advanced a religious view by prohibiting the teaching of evolution unless creationism - a religious view - was also taught. The Court of Appeals agreed, and finally the case made its way to the Supreme Court in 1987.

By the time the case appeared before the U.S. Supreme Court, it was known as *Edwards, Governor of Louisiana, et al. v. Aguillard et al.* The Supreme Court's 7-2 decision for the appellees (Aguillard et al.) affirmed the decisions of the lower courts, as the following statements from the Opinion by Justice Brennan demonstrate:

> The preeminent purpose of the Louisiana Legislature was clearly to advance the religious viewpoint that a supernatural being created humankind... The term "creation science" was defined as embracing this particular religious doctrine by those responsible for the passage of the Creationism Act. Senator Keith's leading expert on creation science, Edward Boudreaux, testified at the legislative hearings that the theory of creation science included belief in the existence of a supernatural creator... The Louisiana Creationism Act advances a religious doctrine by requiring either the banishment of the theory of evolution from public school classrooms or the presentation of a religious viewpoint that rejects evolution in its entirety. The Act violates the Establishment Clause of the First Amendment because it seeks to employ the symbolic and financial support of government to achieve a religious purpose. The judgment of the Court of Appeals therefore is *Affirmed* (*Edwards, Governor of Louisiana, et al. v. Aguillard et al.,* No. 85-1513 Supreme Court of the United States; 482 U.S. 578; 107 S. Ct. 2573, 1987 U.S. Lexis 2729; 96 L. Ed. 2d 510; 55 U.S.L.W. 4860).

Thus, equal time or balanced treatment for teaching "Creation Science" (i.e.,

Creationism) in the science classes of the public schools in the U.S. was no longer a legal choice (Scott, 2004). But, as always, creationists were already thinking of new methods for getting creationism into the public school system.

The Rise of Neocreationism and Intelligent Design

In *Edwards v. Aguillard*, Judge Brennan, in his majority opinion, left some room open for the possibility of teaching various scientific theories about origins in public schools, with the following statement: "...teaching a variety of scientific theories about the origins of humankind to schoolchildren might be validly done with the clear secular interest of enhancing the effectiveness of science instruction" *(Edwards v. Aguillard*; 1987; No. 85-1513; 482 U.S. 578). Also, relative to the *Edwards v. Aguillard* decision, Moore (2002, p. 103) states that creationists "were also encouraged by Justice Scalia's dissenting opinion (which Justice William H. Rehnquist joined) ... that accepted creationists' claims when concluding that 'the people of Louisiana, including those who are Christian fundamentalists, are quite entitled, as a secular matter, to have whatever scientific evidence there may be against evolution presented in their schools.'"

After the filing of the decision in *Edwards v. Aguillard*, creationists immediately went to work to find new schemes for getting antievolution propaganda into public school classrooms. As Scott (2004, p. 113) contends: "Antievolution strategies subsequently were developed that avoided the use of any form of the words 'creation,' 'creator,' or 'creationism.' In effect, proponents shifted their strategy from proposing to balance evolution with Creation Science to proposing to balance evolution with the teaching of Creation Science-like 'scientific alternatives to evolution' or 'evidence against evolution' - avoiding referring to such purported disciplines as 'creationism.' Scott (2004, p. 113) further states that "The avoidance of Creation Science terminology and the development of Creation Science-like 'alternatives to evolution,' plus the repackaging of Creation Science content into 'evidence against evolution,' constitute what I have called neocreationism, which continues into the twenty-first century."

Wendell Bird started work immediately after the *Edwards v. Aguillard* decision to develop a scientific alternative to evolution that would not be construed by the courts as having a religious purpose. He proposed a concept that he called "Abrupt Appearance Theory" and which he claimed was quite distinct from Creation Science (Scott, 2004). However, as Scott (2004, p. 115) points out in the following quote, this tactic was not going to fool anyone because this was nothing more than Creation Science with a different name:

> The phrase "abrupt appearance" was part of the definition of Creation Science in literature presented by the creationist side in the *Edwards v. Aguillard* case. Creation Science in fact was defined in *Edwards* as including "the scientific evidences for creation and inferences from those scientific evidences," but also including "origin through abrupt appearance in complex form." Bird reworked his brief for the *Edwards* case into *The Origin of Species Revisited*, published in 1987. Abrupt Appearance Theory was held to be the scientific evidence for the sudden appearance of all living things--in fact, the entire universe--in essentially its present form. No material or transcendent agent was identified as causing this event; Bird was meticulous in avoiding any references that could be interpreted as religious and would therefore expose Abrupt Appearance Theory to the same First Amendment challenges as Creation Science.

Bird's Abrupt Appearance Theory and his 1987 book, *The Origin of Species Revisited* (two volumes), did not make a lasting impression on creationists, and certainly did not on evolutionary scientists (Scott, 2004). In her own words, Scott (2004, p. 115) states the following: "..., *The Origin of Species Revisited* is rarely cited today in creationist literature. It was, and remains, ignored in the scientific literature, and after the mid-1990s virtually disappeared from the political realm as well. It has been supplanted by another 'alternative to evolution' that was evolving in parallel to it."

The other "alternative to evolution" referred to in the above paragraph is "Intelligent Design Theory" (ID). ID, as stated in the introduction to this chapter, is really a form of creationism. However, it is a more sophisticated type of special creationism, whose proponents claim that they can empirically determine by studying nature that the universe, Earth, and life (including humans) was designed by an intelligent designer. The proponents of ID are of the opinion that they can distinguish ID from design that is produced through purely natural processes (for example, natural selection) (Scott, 2004). The Wedge Strategy as outlined in the

Wedge Document and other statements initiated from the Discovery Institute's Center for Renewal of Science and Culture (CRSC; later changed to Center for Science and Culture [CSC]) in the1990s (and afterward), made it clear that ID was interested in abolishing methodological materialism (or methodological naturalism; the way that science is practiced by most working scientists [regardless of their religious beliefs], without an appeal to supernatural causes as an explanation for natural phenomena) (Forrest and Gross, 2004; Scott, 2004; Numbers, 2006; Ruse, 2009). Although ID advocates avoid the use of the term God in their statements and writings, most of the founders of the ID movement would personally admit that the Intelligent Designer is God (and most of the founders and fellows of the Center for Science and Culture [of the Discovery Institute] lean towards evangelical Christianity in the form of old Earth creationism or progressive creationism [see Scott, 2004 for a discussion of these forms of creationism]). The following quote from Scott (2004, p. 124) elaborates on this theme:

> Intelligent Design supporters are hostile to methodological materialism and propose a new kind of science, "Theistic Science." This is an alleged subclass of science that deals with the class of scientific problems dealing with "origins" ("origins science"), which are unrepeatable. Such phenomena as the origin of life and the evolution of living things (unspecified) constitute origins science. Although the majority of science may be performed in a methodologically materialistic fashion, explaining only with reference to natural causes, origins science allows (indeed, requires) the occasional invocation of "intelligence" - by which is meant the direct hand of God. Theistic science, then, is a proposal to radically change how we do science by abandoning methodological materialism in favor of allowing a supernatural cause - and still calling the process science.

(Note: Metaphysical [or philosophical] materialism [or metaphysical {or philosophical} naturalism], unlike methodological materialism [or methodological naturalism], denies the existence of a supernatural God; of course, ID is hostile to both types, but most Christian scientists practice methodological materialism.)

Contemporary ID had its beginning in 1984 with the publication of *The Mystery of Life's Origin: Reassessing Current Theories* (copyright by the Dallas, Texas Christian based conservative Foundation for Thought and Ethics [FTE]) by the chemist Charles B. Thaxton (1930-), the mechanical engineer Walter L. Bradley

(1943-), and the geochemist Roger L. Olsen (1950-) (Scott, 2004; Numbers, 2006). These authors stated that natural laws could not explain the origin of life and "they attributed the complex process of originating life to a divine creator" (Numbers, 2006; also see Scott, 2004). Thaxton and Bradley are currently fellows of the Discovery Institute's Center for Science and Culture. In 1985, Michael Denton (1943-), a British/Australian medical doctor, biochemist, and geneticist published the book *Evolution: Theory in Crisis*, strongly criticizing the neo-Darwinian view of evolution and arguing that there was evidence of divine design in nature (Numbers, 2006). Denton is currently a senior fellow with the Discovery Institute's Center for Science and Culture. Both of these books, *The Mystery of Life's Origin* and *Evolution: Theory in Crisis*, "helped to lay the intellectual foundation for the ID movement of the 1990s" (Numbers, 2006).

Then, in 1989, the book *Of Pandas and People*: *The Central Question of Biological Origins*, the first high school level supplemental biology textbook dealing with ID, was published by the Foundation for Thought and Ethics (FTE) with authors Percival Davis and Dean H. Kenyon (currently a CSC Discovery Institute fellow) (Charles B. Thaxton, Academic Editor) (see Scott, 2004, p. 116; see Numbers, 2006, p. 375-376;). FTE now has a close relationship with the Discovery Institute relative to the publication of books associated with ID. The authors, Davis and Kenyon (biologists, and creationists), with their editor, Jon A. Buell (1939-) (Founder and President of FTE, and director of FTE Books, an imprint of Discovery Institute Press [see http://www.discoveryinstitutepress.com/fte/]), prior to publication, thought that their book, *Of Pandas and People*, would both serve God and generate a considerable monetary reward for them (Numbers, 2006). According to Numbers (2006, p. 375-376):

> Kenyon and Davis had originally conceived their book as a scientific brief for creationism. In the wake of the Supreme Court's negative decision on creation science, however, they quickly revised their manuscript, substituting the phrases "intelligent design" and "design proponents" for the legally suspect terms "creation" and "creationists." A work they had initially called *Biology and Creation* now became *Of Pandas and People*.

Even with the above publications, the ID movement was relatively unnoticed, until it received a big boost with the 1991 publication of *Darwin on Trial* by University of California - Berkeley law professor Phillip E Johnson (1940-) (Larson, 2003; Forrest and Gross, 2004; Scott, 2004; Numbers, 2006). From reading Richard Dawkin's (1986) *The Blind Watchmaker: Why the Evidence of Evolution Reveals a Universe without Design*, Johnson came to the erroneous conclusion that the argument for evolutionary theory and natural selection was not based on factual information, but was more rhetorical in nature (Numbers, 2006). If he had taken the time to read some of the popular textbooks on evolution and/or some of the more technical peer-reviewed journal articles of the time, he would have seen that a wealth of empirical factual information supports evolutionary theory. However, one of Johnson's major criticisms of evolutionary theory was the premise that science must be practiced according to naturalistic methods (methodological naturalism), without resorting to supernatural explanations. As stated by Numbers (2006, p. 376), Johnson was of the opinion that "This bias, he argued, unfairly limited the range of possible explanations and ruled out, a priori, any consideration of theistic factors" (internal references deleted). Of course, as soon as supernatural explanations (miracles) enter into the methods of science, there is no longer an objective and rational approach to the study of nature. If the methods of science allowed supernatural explanations of natural phenomena, it would result in the opening of a "Pandora's box." The use of supernatural explanations to explain nature results in something that is no longer science, thus anyone can propose explanations based on their faith in a certain religion, their ideology, or their philosophy.

So, is the belief in ID (intelligent design) a form of creationism? According to ID proponents, they are not creationists. The following quote from Scott (2004, p. 128) discusses Phillip Johnson's position relative to ID and evolution:

> Phillip Johnson contends that the scientific data and theory supporting evolution are weak, and that evolution persists as a scientific idea only because it reinforces philosophical materialism. To him and most other ID proponents, the most important issue in the creation/evolution controversy is whether the universe came to its present state "through purposeless, natural processes known to science" (Johnson 1990: 30) or whether God had meaningful involvement with the process. Intelligent Design proponents

clearly believe that God is an active participant in creation, though they are divided as to whether this activity takes the form of front-loading all outcomes at the big bang, episodic intervention of the progressive creationism form, or other, less well-articulated possibilities. *Theistic evolution, however, is ruled out.* (Emphasis on last sentence by the present author.) (The reference cited above is the following: Johnson, Phillip E.; 1990; *Evolution as Dogma: The Establishment of Naturalism*: Haughton Publishing Company; Dallas, Texas; 37 p.)

(Note: The majority of working Christian scientists practice methodological naturalism in their work, but many accept Theistic evolution as their belief system. Theistic evolution basically is the belief that God established the laws of nature and allows creation and evolution of organisms to proceed according to these laws. There is a range of differences among Theistic evolutionists in accepting the amount of "tinkering" that God has done (or does) in the process of biologic evolution (Scott, 2004). According to Scott (2004, p. 64): "...TE [Theistic Evolution] is the view of creation taught at the majority of mainline Protestant seminaries, and it is the position of the Catholic Church. In 1996, Pope John Paul II reiterated the Catholic version of the TE position, in which God created, evolution happened, humans may indeed be descended from more primitive forms, but the Hand of God was required for the production of the human soul (John Paul II, 1996).")

To fight the evil of evolution and methodological naturalism (as well as metaphysical naturalism), Phillip Johnson, during the 1990s devised a movement and a strategy referred to as the Wedge (the Wedge strategy was briefly discussed in the introduction to this chapter). The following statement in Johnson's *Darwin on Trial*, quoted from Numbers (2006, p. 377), explains Johnson's Wedge metaphor:

> A log is a seeming solid object, but a wedge can eventually split it by penetrating a crack and gradually widening the split. In this case the ideology of scientific materialism is the apparently solid log. The widening crack is the important but seldom-recognized difference between the facts revealed by scientific investigation and the materialist philosophy that dominates scientific culture.

Johnson's Wedge strategy consisted of a series of stages or phases, within a five year plan and a twenty year plan, that involved academic scientific research and publication; and political and social maneuvering involving federal, state, and local legislative bodies, boards of education, parents of school children, and religious groups and individuals (especially those that were large monetary sources for the ID Movement via the Discovery Institute's CRSC or CSC), that would establish (he thought) ID as a likely alternative to teaching "materialistic evolutionism" in the science classrooms of the public schools. To date, no real scientific publications or scientific research programs have been established by the academic phase of the Wedge strategy. Although the *Kitzmiller et al. v. Dover Area School District et al.* legal decision in 2005 against ID has somewhat tarnished the use of "Intelligent Design Theory" as a viable alternative to evolution in public schools, the political and social aspects of the Wedge strategy has been more successful in getting "teaching the controversy" and "teaching critical analysis" of evolution into the public school science classrooms as alternatives that allow teaching ID without using the name Intelligent Design. But as stated by Scott (2004, p. 117):

> ...educated conservative Christians, for whom Creation Science was unacceptable because of its often outlandish scientific claims, found Johnson's message very attractive indeed. Largely because of its more respectable pedigree, Intelligent Design obtained considerably more coverage in the popular media than Creation Science – though the latter boasts many more organizations and activists than ID.
>
> ...As is the case with Creation Science, ID combines a scholarly focus with an effort to promote a sectarian religious view.

Kitzmiller et al. v. Dover Area School District et al. – Intelligent Design Trial, U.S District Court, Harrisburg, Pennsylvania, 2005

As indicated above, 2005 was the first test of ID in a court of law. Several popular books have been written that describe this challenge to the teaching of Intelligent Design in the rural public school district of Dover, York County, Pennsylvania. These books include the following: *Monkey Girl: Evolution, Education, Religion, and the Battle for America's Soul* by Edward Humes (2007); *The Devil in Dover: An Insider's Story of Dogma V. Darwin in Small-Town America* by Lauri Lebo (2008); *The Battle Over the Meaning of Everything: Evolution, Intelligent Design, and a School Board in Dover, PA* by Gordy Slack (2007); and *40 Days and 40 Nights: Darwin, Intelligent Design, God, Oxycontin®, and Other Oddities on Trial in*

Pennsylvania by Matthew Chapman (2007). In addition to these books, and other popular accounts, Judge John E. Jones III, federal district judge for the U.S. District Court for the Middle District of Pennsylvania, Harrisburg, PA, and the trial judge for this case, wrote a 139-page Memorandum Opinion that was made public on December 20, 2005 for the decision in this case.

This test case involved eleven distressed and irate parents of school children within the Dover Area School District as plaintiffs, including Tammy Kitzmiller as the lead plaintiff and mother of a ninth-grade daughter and an eleventh grade daughter at Dover Area High School in Dover, Pennsylvania (*Kitzmiller v. Dover*, 2005; Numbers, 2006). The plaintiffs were distressed and irate because the defendants in this case, the Dover Area School District and Dover Area School District Board of Directors had implemented in October and November 2004 an ID Policy that would require the reading of a statement relative to the teaching of ninth-grade biology (*Kitzmiller v. Dover*, 2005; Numbers, 2006). The plaintiffs asked the American Civil Liberties Union (ACLU) and Americans United for Separation of Church and State (AU) to intervene on their behalf and the ACLU and AU did on December 14, 2004 by filing suit challenging the constitutionality of the implemented ID Policy (*Kitzmiller v. Dover*, 2005; Numbers, 2006). The large Philadelphia law firm Pepper Hamilton also worked with the ACLU and AU to assist the plaintiffs in this case (Slack, 2007).

The ID Policy as referred to here is in two parts; the first part was a resolution adopted by a vote of 6-3 by the Dover Area School Board of Directors on October 18, 2004, stating the following (*Kitzmiller v. Dover*, 2005):

> Students will be made aware of gaps/problems in Darwin's theory and of other theories of evolution including, but not limited to, intelligent design. Note: Origins of Life is not taught.

The second part of the ID policy was announced by the Dover Area School District by press release on November 19, 2004 and required that, commencing on January 2005, teachers would be required to read the following statement to ninth grade biology students (*Kitzmiller v. Dover*, 2005):

> The Pennsylvania Academic Standards require students to learn about Darwin's Theory of Evolution and eventually to take a standardized test of which evolution is a part. Because Darwin's Theory is a theory, it continues to

be tested as new evidence is discovered. The Theory is not a fact. Gaps in the Theory exist for which there is no evidence. A theory is defined as a well-tested explanation that unifies a broad range of observations.

Intelligent Design is an explanation of the origin of life that differs from Darwin's view. The reference book, *Of Pandas and People*, is available for students who might be interested in gaining an understanding of what Intelligent Design actually involves.

With respect to any theory, students are encouraged to keep an open mind. The school leaves the discussion of the Origins of Life, to individual students and their families. As a Standards-driven district, class instruction focuses upon preparing students to achieve proficiency on Standards-based assessments.

The trial lasted about six weeks, beginning on September 26, 2005 and ending on November 4, 2005. The federal judge in the Harrisburg, Pennsylvania U.S. District Court for the Middle District of Pennsylvania was (as mentioned above), John E. Jones III, a conservative Republican appointed by President George W. Bush "and a practicing Lutheran" (Numbers, 2006, p. 394). The following statement by Singham (2009, p. 145-146) stresses the conservative nature of Judge Jones:

> ...a Republican and long-time member of the Lutheran church, the judge was sponsored for this post by then U.S. Senator Rick Santorum (himself a strong supporter of intelligent design) and was nominated to the bench by then-President George W. Bush (who has argued that "both sides" of the evolution issue, whatever that means, should be taught). His assignment to the Dover case was praised by Tom Ridge (former Republican governor of Pennsylvania and then head of the Department of Homeland security), who said, "I can't imagine a better judge presiding over such an emotionally charged issue."

The plaintiffs in the suit claimed that the ID Policy violated the First Amendment to the U.S. Constitution by an establishment of religion (and thus defied the Establishment Clause of the First Amendment). The Fourteenth Amendment makes the First Amendment relevant and applicable to the states. The plaintiffs further claimed that the ID Policy also violated, by an establishment of religion, the Constitution of the Commonwealth of Pennsylvania (*Kitzmiller v. Dover*, 2005).

The defendants (the Dover Area School District and the Dover Area School Board of Directors) were represented by the pro bono services of the Thomas More

Law Center of Ann Arbor, Michigan. The Thomas More Law Center is a conservative, Christian-based law firm that was "eager to promote intelligent design and to challenge the American Civil Liberties Union in court" (Foster, Clark, and York, 2008, p. 14). The Thomas More Law Center was established in 1999 by Thomas Monaghan (1937-), "conservative Catholic pizza baron ... of Domino's fame" (Numbers, 2006, p. 392) and states on its website the mission and purpose of the law firm: "The Law Center's purpose is to be the sword and shield for people of faith, providing legal representation without charge. We achieve this goal principally through litigation, seeking out significant cases consistent with our mission" (Thomas More Law Center, accessed 04/03/2017; also see Singham, 2009, p. 128).

The Dover Area School District Board wanted to present an alternative theory of biologic origins to Dover Area High School students that would not be construed as an establishment of religion. The defendants (the Dover Area School District and the Dover Area School District Board) asserted and argued that the Santorum explanatory note (non-binding) that had been attached to George W. Bush's No Child Left Behind Act of 2001 justified their actions (Numbers, 2006). The Santorum note was initially proposed by U.S. Senator Rick Santorum (1958-) (a conservative Catholic Republican from Pennsylvania, and "who also sat on the board of directors of the Thomas More Law Center" [Humes, 2007. p. 251]) as an amendment to the No Child Left Behind Act and was authored by none other than Phillip Johnson himself (Numbers, 2006). According to Numbers (2006, p. 389-390) the proposal stated the following:

> It is the sense of the Senate that –
> (1) good science education should prepare students to distinguish the data or testable theories of science from philosophical or religious claims that are made in the name of science; and
> (2) where biological evolution is taught, the curriculum should help students to understand why this subject generates so much continuing controversy and should prepare the students to be informed participants in public discussions regarding the subject.

The amendment, which Johnson hoped, would "make it very difficult for the public school authorities to justify firing or disciplining a teacher who informs students of the weaknesses of the Darwinian theory, rather than teaching it in the authoritarian and dogmatic manner that Darwinians have been able to enforce up until now" (as spoken by Johnson in 2001 and quoted by Numbers, 2006, p. 390). The amendment was approved by a 91-8 vote in the Senate, but met resistance in committee negotiations between the House and Senate ("the presidents of eighty science societies sent a letter asking Congress to delete Santorum's resolution from the bill's final text" [Larson, 2003, p. 208]), thus it was downgraded to just a non-binding explanatory note (report) attached to the bill (Numbers, 2006). Larson (2003, p. 208-209) further explains relative to the No Child Left Behind Act with amendments and explanatory notes: "Santorum and Johnson hoped that the non-binding report language would embolden school boards to add dissenting views to the biology curriculum, but in its overall effect, the new law [No Child Left Behind Act] strengthened the force of existing (mostly pro-evolution) state science standards by requiring that, as a condition of federal education funding, schools test student achievement under their state's standards" (internal citations deleted).

The ninth-grade biology teachers of Dover High School refused to read the ID Policy statement that the board of directors had approved because they felt that doing so violated their ethical duty to their students to teach sound and accepted science, so the school administrators agreed to read the ID Policy statement. Students that did not want to have the ID Policy statement read to them were allowed to wait in the hall outside the classroom while the statement was being read. Sixty copies of the ID text *Of Pandas and People* had been placed in the library for interested students to read. The copies of the ID textbook *Of Pandas and People* was donated "anonymously" (supposedly) to the Dover Area High School library. It was later determined that the money to purchase the ID reference texts was obtained by Bill Buckingham, chair of the school board curriculum committee, by donations from area churches. The following statement from (Singham, 2009, p. 127) is revealing:

> As another example of the religious motivation behind the school board's actions that would cause problems during the trial, Buckingham raised money in churches to buy sixty copies of the creationist textbook *Of Pandas and People*, then gave the money to fellow board member Alan Bonsell's father, who then donated the books "anonymously" to the school's library to be made available as "reference" books for biology students. Later, Buckingham and

Bonsell both denied, under oath in their depositions, any knowledge of where the books had come from.

During the trial, this and other falsehoods were revealed in open court and clearly angered the judge, who said in his ruling, "It is ironic that several of these individuals, who so staunchly and proudly touted their religious convictions in public, would time and again lie to cover their tracks and disguise the real purpose behind the ID Policy."

Both sides, plaintiffs and defendants, were allowed to have expert witnesses for the trial. There were six expert witnesses for the plaintiffs. The lead expert witness for the plaintiffs was Kenneth R. Miller, Ph.D. (1948-). Dr. Miller is an endowed Professor of Biology (Royce Family Professor for Teaching Excellence) at Brown University where he teaches both introductory biology classes and upper level courses in cell biology and molecular biology; he is also well known as an author of high school biology textbooks. Prior to this trial he had often debated on the side of evolution against proponents of creationism (including intelligent design creationism), even though he is a devout Roman Catholic (Humes, 2007). The other five expert witnesses for the plaintiffs were the following: Brian J. Alters, Ph.D., Associate Professor of Education, McGill University, an expert on teaching and education; Barbara C. Forrest (1952-), Ph.D., Professor of Philosophy, Southeastern Louisiana University, an expert on the history and philosophy of the ID creationist movement; John F. Haught (1942-), Ph.D., Theology Professor, Georgetown University, an expert on religion; Kevin Padian (1951-), Ph.D., Professor of Integrative Biology, University of California, Berkeley and Curator of the Museum of Paleontology, University of California, Berkeley, an expert on evolution and paleontology; and Robert T. Pennock , Ph.D., Associate Professor of Science and Technology Studies and Associate Professor of Philosophy, Michigan State University, an expert on the philosophy of science. (Note: The above affiliations for the expert witnesses were their affiliations at the time of the trial.)

The expert witnesses above, during much of the first nine days of the trial, put on a spectacular presentation of "cutting-edge science" that show-cased evolutionary biology and revealed what it says about the natural world of living organisms (both past and present) and their environments (see Humes, 2007, p. 254-278). The testimony of the expert witnesses for the plaintiffs would also clearly illustrate that ID is a form of creationism that is deceitfully "dressed-up" to appear to be scientific by officially not mentioning God or the Bible and insisting that there is evidence in nature for design of living organisms, but proponents of ID require that a supernatural intelligence is essential for the origin and development of life on Earth (*Kitzmiller v. Dover*, 2005; Humes, 2007;). The testimony of these witnesses would show that ID is not science at all and has no real peer-reviewed scientific journal articles to justify its status as a science. These witnesses would show during the trial, in particular the testimony of Dr. Barbara Forrest, that the book *Of Pandas and People* was intended (based on old manuscripts subpoenaed from the publisher) to be a textbook of creation science until the *Edwards v. Aguillard* decision was handed-down in 1987, then became an ID textbook when it was first published in 1989; Forrest showed that "...every reference to 'creation' and 'creationism' in the book was simply changed to 'intelligent design,' and 'creator' was changed to 'designer' or 'intelligent agency' " (Humes, 2007, p. 284-285).

On October 17, 2005, Day 10 of the trial, the defense began their presentation of witnesses. The attorneys for the defendants originally had six expert witnesses lined-up to testify, with two more rebuttal experts available if needed to offset plaintiff expert witnesses late in the trial. However, with the leaders of the Center for Science and Culture of the Discovery Institute becoming nervous about the antics of the School Board during 2004 and 2005 (leading up to the trial), to push an apparent religious agenda to get God back into the classroom, particularly by William Buckingham (chair of the curriculum committee) and Alan Bonsell (president of the School Board), expert witnesses for the defense began to drop out of the trial. The following statement from Humes (2007, p. 270) expounds upon this:

> The departure of the defense experts in the middle of the trial was an embarrassment to the defense, and it could not be chalked up to a perception that the plaintiffs' case was weak, requiring little response. On the contrary, the presentations that followed Ken Miller's began to seem like a scientific juggernaut to most observers of the trial. Nick Matzke [then with the National Center for Science Education; he assisted the plaintiffs' attorneys and expert witnesses with the *Dover* trial] began referring to the departed ID experts as jumping off a sinking ship. Even Richard Thompson [President and Chief Counsel of the Thomas More Law Center], who appeared on C-Span while the case was still being tried, complained bitterly about the Discovery

Institute's witnesses fleeing the case, "victimizing" the Dover school board, and ruining his own ability to counter the plaintiffs' experts. (Note: Text in brackets added by the current author.)

There were originally six primary expert witnesses that agreed to serve as witnesses for the defense, however, four of those dropped out before the trial started; two of those four were fellows of the Discovery Institute. In addition, there were two rebuttal witnesses that had been lined up for the trial to rebut the testimony of certain plaintiff expert witnesses, but one of those also dropped out before the trial began. The rebuttal witness that dropped out prior to the trial beginning was Stephen C. Meyer (1958-), Ph.D., Senior Fellow and Director of the Center for Science and Culture of the Discovery Institute. Thus, only three expert witnesses, including one rebuttal expert witness, participated in the trial, those being the following: Michael Behe (1952-). Ph.D., Professor of Biochemistry, Lehigh University, Senior Fellow of the Discovery Institute (DI), an expert on intelligent design (ID) and biochemistry; Scott A. Minnich, Ph.D., Associate Professor of Microbiology, University of Idaho, Fellow of DI, expert on ID and microbiology; and the rebuttal witness, Steven W. Fuller (1959-), Ph.D., Professor of Sociology, University of Warwick (England), an expert on ID and sociology. (Note: The above affiliations for the expert witnesses were their affiliations at the time of the trial.)

So, why the mass exodus of Discovery Institute fellows as potential expert witnesses for the *Kitzmiller v. Dover* trial? Is it that the Discovery Institute did not want this trial to be a test of their ID program? Yes, it does appear that is the case. The following two paragraphs from Singham (2009, p. 123) makes this conclusion clear:

> The Dover trial was a bad situation from the beginning for the ID people, especially the strategists at the Discovery Institute, because it took events out of their control and put them in the hands of people [such as Buckingham and Bonsell] who did not really understand what ID was all about [and had pushed an agenda for the school board that reeked of creationism]. The ID theorists were trying to implement a carefully crafted stealth strategy, avoiding any taint of religion. The Discovery Institute's Wedge strategy required everyone to be very discreet, carefully avoiding any mention of God or religion or anything remotely connected to them.
> The Dover school board was much too clumsy in its attempts to introduce ID ideas into its curriculum. They had little patience for the subtlety of the

slow, long-range plan envisaged by the Discovery Institute. They wanted God, the Bible, and prayer back in their schools, and they wanted it *now*. As a result, they left their religious fingerprints all over the policy in a way that the sophisticated strategists suspected would be fatal to their case. While ID strategists were walking on eggshells, the Dover school board members were clumping around in thick boots. (Note: Text in italics by Singham, but the text in brackets added by the current author.)

The first and star expert witness for the defense was Michael Behe. As stated previously, Dr. Behe is a professor of biochemistry at Lehigh University, Bethlehem, Pennsylvania. Dr. Behe testified for the first three days of the testimony by the defense, followed later in the trial by the testimony of the two other ID expert witnesses, Dr. Fuller and Dr. Minnich. In 1996, Dr. Behe published a book, entitled *Darwin's Black Box: The Biochemical Challenge to Evolution*, in which he claimed that many molecular structures and processes in living organisms (such as the bacterial flagellum, the blood clotting (coagulation) cascade mechanism, and "the intracellular transport system" within living cells) are "irreducibly complex" and cannot evolve (and could not have evolved in the past) via Darwinian natural selection. Behe contended in his book, and also in his testimony at the *Kitzmiller v. Dover* trial, that living systems consist of irreducibly complex structures intelligently designed and that they could not form by the "incremental additions of natural selection (Behe 1996)" (Scott, 2004, p. 117; also see Behe, 1996). So, according to Dr. Behe these structures could not form one step at a time by natural selection, with each step giving a selective advantage. Thus, irreducibly complex structures would need to be formed in an abrupt act of special creation, with the entire structure consisting of all the required parts forming in one fell swoop by a supernatural intelligence. Here again, the idea of abrupt appearance that is so characteristic of creationism is evident in the creation of irreducibly complex structures and biological processes. This would appear to mean that the organism itself would also have to form abruptly and all at once, presumably created by an intelligent designer. The following statement from the ID textbook, *Of Pandas and People* (Davis and Kenyon, 1993, p. 99-100), that was made available to ninth grade biology students attending the Dover Area High School, makes the abrupt appearance concept apparent and also supports "the conclusion that ID is predicated on supernatural causation"

(*Kitzmiller v. Dover*, p. 67):

> Darwinists object to the view of intelligent design because it does not give a natural cause explanation of how the various forms of life started in the first place. Intelligent design means that various forms of life began abruptly through an intelligent agency, with their distinctive features already intact–fish with fins and scales, birds with feathers, beaks and wings, etc.

Davis and Kenyon (1993, p. 100) continue with their line of reasoning supporting the concept of abrupt appearance as evidence for an intelligent designer:

> Some scientists have arrived at this view since fossil forms first appear in the rock record with their distinctive features intact, and apparently fully functional, rather than gradually developing. No creatures with a partial wing or partial eye are known. Should we close our minds to the possibility that the various types of plants and animals were intelligently designed? This alternative suggests that a reasonable natural cause explanation for origins may never be found, and that intelligent design best fits the data.

(Note: The great majority of evolutionary biologists and paleontologists would not arrive at the views as presented in the above statements. It is also questionable who "some scientists" are in the above statement and whether they are even paleontologists [they are certainly not evolutionary biologists]).

Judge Jones, in his Memorandum Opinion of the *Kitzmiller v. Dover* trial (p. 76-79), bashed the reliability of Dr. Behe's irreducible complexity concept with the following discussion:

> As irreducible complexity is only a negative argument against evolution, it is testable, unlike ID, by showing that here are intermediate structures with selectable functions that could have evolved into the allegedly irreducibly complex systems. ...Professor Behe has applied the concept of irreducible complexity to only a few select systems: (1) the bacterial flagellum; (2) the blood-clotting cascade; and (3) the immune system. Contrary to Professor Behe's assertions with respect to these few biochemical systems among the myriad existing in nature, however, Dr. Miller [Kenneth Miller, Brown University, plaintiff expert witness on evolutionary biology] presented evidence, based upon peer-reviewed studies, that they are not in fact irreducibly complex.
> First, with regard to the bacterial flagellum, Dr. Miller pointed to peer-reviewed studies that identified a possible precursor to the bacterial flagellum, a subsystem that was fully functional, namely the Type-III Secretory System. Moreover, defense expert Professor Minnich [Scott Minnich, University of Idaho, defense expert witness on microbiology] admitted that there is serious scientific research on the question of whether the bacterial flagellum evolved

into the Type-III Secretory System, the Type-III Secretory System into the bacterial flagellum, or whether they both evolved from a common ancestor. None of the research or thinking involves ID. In fact, Professor Minnich testified about his research as follows: "we're looking at the function of these systems and how they could have been derived one from the other. And it is legitimate scientific inquiry.

Second, with regard to the blood-clotting cascade, Dr. Miller demonstrated that the alleged irreducible complexity of the blood-clotting cascade has been disproven by peer-reviewed studies dating back to 1969, which show that dolphins' and whales' blood clots despite missing a part of the cascade, a study that was confirmed by molecular testing in 1998. Additionally, and more recently, scientists published studies showing that in puffer fish, blood clots despite the cascade missing not only one, but three parts. Accordingly, scientists in peer-reviewed publications have refuted Professor Behe's predication about the alleged irreducible complexity of the blood-clotting cascade. Moreover, cross-examination revealed that Professor Behe's definition of the blood-clotting system was likely designed to avoid peer-reviewed scientific evidence that falsifies his argument, as it was not a scientifically warranted redefinition.

The immune system is the third system to which Professor Behe has applied the definition of irreducible complexity. Although in Darwin's Black Box, Professor Behe wrote that not only were there no natural explanations for the immune system at the time, but natural explanations were impossible regarding its origin. However, Dr. Miller presented peer-reviewed studies refuting Professor Behe's claim that the immune system was irreducibly complex. Between 1996 and 2002, various studies confirmed each element of the evolutionary hypothesis explaining the origin of the immune system. In fact, on cross-examination, Professor Behe was questioned concerning his 1996 claim that science would never find an evolutionary explanation for the immune system. He was presented with fifty-eight peer-reviewed publications, nine books, and several immunology textbook chapters about the evolution of the immune system; however, he simply insisted that this was still not sufficient evidence of evolution, and that it was not "good enough."

We find that such evidence demonstrates that the ID argument is dependent upon setting a scientifically unreasonable burden of proof for the theory of evolution…

We therefore find that Professor Behe's claim for irreducible complexity has been refuted in peer-reviewed research papers and has been rejected by the scientific community at large. Additionally, even if irreducible complexity had not been rejected, it still does not support ID as it is merely a test for evolution, not design. (Note: Internal citations have been deleted for clarity. The text in brackets is by the current author.)

The trial was completed on November 04, 2005 and U.S. District Court Judge John E. Jones III announced his ruling on the *Kitzmiller v. Dover* case on December

20, 2005. Judge Jones, in his 139-page decision, ruled overwhelmingly in favor of the plaintiffs. He ruled that the school board had violated the Establishment Clause of the First Amendment to the U.S. Constitution by violating the purpose prong of the Lemon Test (it was obvious from the history of the school board members that the purpose of the Dover ID Policy was not secular, but religious in nature), as well as the effect prong (the effect of the Dover ID Policy was to establish a particular type of religion in the Dover Area High School biology classroom), either of which would render the Dover ID Policy unconstitutional (Singham, 2009). In addition, the ID Policy failed the endorsement test of the Establishment Clause. The endorsement test is in reality a reconceptualization of the effect prong of the Lemon test (Singham, 2009). In this case, a hypothetical, well informed, reasonable observer that knew the history of creationism, the history of the ID movement, the actions of the school board prior to the trial during 2004 and 2005, and the testimony given during the trial, would conclude that the Defendants' conduct and the challenged ID Policy represented an endorsement of religion, and thus violated the Establishment Clause and therefore is unconstitutional (*Kitzmiller v. Dover;* Singham, 2009).

To be complete, the judge also ruled on whether ID was science. Although both the plaintiffs and defendants (in particular the Thomas More Law Center attorneys) wanted the judge to adjudicate a decision on whether ID was science or not, the ID proponents at the Discovery Institute hoped that this would not take place. According to Singham (2009, p. 132): "...The ID strategists at the Discovery Institute had desperately wanted to avoid having a judicial determination on this question, and the Discovery Institute and a group of people sympathetic to their views filed two amicus curiae ("friend of the court') briefs asking him *not* to rule on the question of the scientific status of ID."

Of course, Judge Jones did make a very studied and extensive determination of whether ID is science, and determined that, indeed, ID is not science but represents a form of creationism. Judge Jones (*Kitzmiller v. Dover*, 2005, p. 64) made the following statement relative to his conclusion on whether ID is science:

> After a searching review of the record and applicable caselaw, we find that while ID arguments may be true, a proposition on which the Court takes no position, ID is not science. We find that ID fails on three different levels, any

one of which is sufficient to preclude a determination that ID is science. They are: (1) ID violates the centuries-old ground rules of science by invoking and permitting supernatural causation; (2) the argument of irreducible complexity, central to ID, employs the same flawed and illogical contrived dualism that doomed creation science in the 1980's; and (3) ID's negative attacks on evolution have been refuted by the scientific community. As we will discuss in more detail below, it is additionally important to note that ID has failed to gain acceptance in the scientific community, it has not generated peer-reviewed publications, nor has it been the subject of testing and research.

Expert testimony reveals that since the scientific revolution of the 16th and 17th centuries, science has been limited to the search for natural causes to explain natural phenomena. This entailed the rejection of appeal to authority, and by extension, revelation, in favor of empirical evidence. Since that time period, science has been a discipline in which testability, rather than any ecclesiastical authority or philosophical coherence, has been the measure of a scientific idea's worth. In deliberately omitting theological or "ultimate" explanations for the existence or characteristics of the natural world, science does not consider issues of "meaning" and "purpose" in the world. While supernatural explanations may be important and have merit, they are not part of science. This self-imposed convention of science, which limits inquiry to testable, natural explanations about the natural world, is referred to by philosophers as "methodological naturalism" and is sometimes known as the scientific method. Methodological naturalism is a "ground rule" of science today which requires scientists to seek explanations in the world around us based upon what we can observe, test, replicate, and verify.

...ID is predicated on supernatural causation, as we previously explained and as various expert testimony revealed. ID takes a natural phenomenon and, instead of accepting or seeking a natural explanation, argues that the explanation is supernatural. ...

What better way to summarize the *Kitzmiller v. Dover* trial and this section of Chapter 12 than to quote parts of Judge Jones' conclusion from his Memorandum Opinion of the trial, as follows:

The proper application of both the endorsement and Lemon tests to the facts of this case makes it abundantly clear that the Board's ID Policy violates the Establishment Clause. In making this determination, we have addressed the seminal question of whether ID is science. We have concluded that it is not, and moreover that ID cannot uncouple itself from its creationist, and thus religious, antecedents.

Both Defendants and many of the leading proponents of ID make a bedrock assumption which is utterly false. Their presupposition is that evolutionary theory is antithetical to a belief in the existence of a supreme being and to religion in general. Repeatedly in this trial, Plaintiff's scientific experts testified that the theory of evolution represents good science, is overwhelmingly accepted by the scientific community, and that it in no way conflicts with, nor does it deny, the existence of a divine creator. ...

The citizens of the Dover area were poorly served by the members of the Board who voted for the ID Policy. It is ironic that several of these individuals, who so staunchly and proudly touted their religious convictions in public, would time and again lie to cover their backs and disguise the real purpose behind the ID Policy. ...

... The breathtaking inanity of the Board's decision [relative to the ID Policy] is evident when considered against the backdrop which has now been fully revealed through this trial. The students, parents, and teachers of the Dover Area School District deserved better than to be dragged into this legal maelstrom, with its resulting utter waste of monetary and personal resources.

To preserve the separation of church and state mandated by the Establishment Clause of the First Amendment to the United States Constitution, and Art. I, § 3 of the Pennsylvania Constitution, we will enter an order permanently enjoining Defendants from maintaining the ID Policy in any school within the Dover Area School District, from requiring teachers to denigrate or disparage the scientific theory of evolution, and from requiring teachers to refer to a religious, alternative theory known as ID. We will also issue a declaratory judgment that Plaintiffs' rights under the Constitutions of the United States and the Commonwealth of Pennsylvania have been violated by Defendants' actions. Defendants' actions in violation of Plaintiffs' civil rights as guaranteed to them by the Constitution of the United States and 42 U.S.C. § 1983 subject Defendants to liability with respect to injunctive and declaratory relief, but also for nominal damages and the reasonable value of Plaintiffs' attorneys' services and costs incurred in vindicating Plaintiffs' constitutional rights.

MEDICENE AND MODERN EVOLUTIONARY THINKING

This section will discuss some effects of an understanding of evolutionary theory as related to modern medical practices that affect the quality of human life. This really gets to the heart of why evolution matters. Norman Johnson (2007) in his recent book entitled: *Darwinian Detectives – Revealing the Natural History of Genes and Genomes*, introduces a relevant quote at the beginning of Chapter 1 by Olivia Judson as follows: "Whether it's preventing a flu pandemic or tackling malaria, we can use our knowledge of evolutionary processes in powerful and practical ways, potentially saving the lives of tens of millions of people. So, let's not strip evolution from the textbooks, or banish from the class, or replace it with ideologies born of wishful thinking. If we do, we might find ourselves facing the consequences of natural selection. – Olivia Judson (Judson, 2005)." This quote was taken from an Op-ed in the New York Times, November 6, 2005.

Johnson (2007), in the opening paragraph of the above book, recounts the October 1984 operation on a newborn baby girl, known as Baby Fae, with a defective heart. Dr. Leonard Bailey and his team at Loma Linda University Hospital Medical Center (A Seventh Day Adventist institution in California) replaced Baby Fae's defective heart with the heart of baboon. Baby Fae died three weeks later. As we know, Baby Fae's immune system attacked the baboon heart. When asked why he had used a baboon heart, rather than the heart of a more closely related chimpanzee or gorilla, he replied that he did not believe in evolution and went on to say: "The scientists that are keen on the evolutionary concept that we actually developed serially from subhuman primates to human, with mitochondrial DNA dating and that sort of thing, the differences have to do with millions of years. That boggles my mind somehow. I don't understand it well, and I'm not sure that it means a great deal in terms of tissue homology."

Johnson (2007) goes on to state that Baby Fae's operation perhaps would not have been successful if an animal more closely related to humans had been used for the transplant, but the casual dismissal of evolutionary theory (the central organizing theme of the biological sciences) by a medical doctor are very disturbing when human lives are at stake. As stated by Antolin et al. (2012, p. 1992), "Application of evolutionary principles to medical research, public health, and clinical practice can improve health care, reduce suffering and save lives."

Evolution of Antibiotic Resistance in Bacteria

One example of evolution in action is the evolution of antibiotic resistance in bacteria (see Genereux and Bergstrom, 2005; Perlman, 2013; Anderson, 2016). Because bacteria reproduce rapidly, they can evolve resistant strains to antibiotics in a relatively short period of time. When bacteria are exposed to antibiotics, some individuals are more resistant to the antibiotics than others in the population. Also, as the bacteria reproduce, chance mutations in their genome form resistant individuals within a population, natural selection works to choose those antibiotic resistant individuals for survival and reproduction. Some bacteria may also become resistant by acquiring resistance from other bacteria during horizontal transfer by bacterial conjugation (even different species and strains of bacteria may laterally transfer resistant genes to each other via conjugation). After many generations (which may occur in a relatively short time) the bacteria may evolve into a resistant strain to a specific antibiotic. This is becoming an increasingly serious problem in hospital intensive care units (Genereux and Bergstrom, 2005; Perlman, 2013; Anderson, 2016). Some bacteria are becoming resistant to most of the antibiotics currently available (Perlman, 2013; Anderson, 2016). The seriousness of the situation is emphasized in the following quote from Anderson (2016, p. 147):

> Antiobiotics have revolutionized human and veterinary medicine, and over the last 70 years they have made it possible to efficiently treat most types of bacterial infections. Unfortunately, the extensive use – and frequent misuse – of antibiotics has resulted in the rapid evolution and spread of bacteria that are resistant to antibiotics. Arguably, the global use of antibiotics is one of the largest evolution experiments performed by humans, and the frightening consequence is that we are now at the brink of a postantibiotic era in which antibiotics have lost their miraculous power. This problem originates from the strong selection imposed by the extensive use of antibiotics and the resulting enrichment of resistance mutations and horizontally acquired resistance genes [in bacteria]. Together these factors have generated high-level antibiotic resistance in the majority of significant human and veterinary pathogens. ...The development of new classes of antibiotics, coupled with more prudent use of antibiotics, will be required to maintain antibiotics as efficient agents for treating bacterial infections.

According to Anderson (2016) (and others) the worldwide restraint in using antibiotics will take place only if there is a concerted effort by all involved (from pharmaceutical companies to doctors to patients, etc.) to decrease the use of antibiotics. Anderson (2016) further explains that this will involve the following strategies: 1) avoidance of the use of antibiotics, except for bacterial infection (e.g., rather than infections or diseases caused by viruses); 2) to no longer use large doses of antibiotics to stimulate growth in the production of animals for food; 3) to discontinue the use of antibiotics in aquaculture and plant crop production; 4) to avoid economic situations where there is a significant profit motive for antibiotic prescriptions by doctors (or other prescribers); 5) to control regulation of marketing of antibiotics by the pharmaceutical industry to doctors (and other prescribers), pharmacists, and patients (or other consumers); and 6) to prohibit the sale of antibiotics to the public (via the internet, from pharmacies, or other outlets) without the need for a prescription.

Also, the continued and increased use of hygienic and infection control methods to reduce the transmission of disease from pathogenic bacteria will result in the reduction of antibiotic use (Anderson, 2016). Anderson (2016, p. 159) states the following relative to hygienic methods: "Pathogens for which hygienic measures have been shown to be particularly successful include various food-borne pathogens (e.g., *Salmonella*) [certain strains of *Escherica coli* are other common examples] and nosocomial [hospital acquired] infections such as methicillin-resistant *Staphylococcus aureus* (MRSA). For MRSA infections, screening strategies to track and isolate affected patients, coupled with improved hospital hygiene, have been successful in reducing the transmission of these dangerous bacteria." (Text inside of brackets is by the current author.)

The relatively recent emergence of methicillin-resistant *Staphylococcus aureus* (MRSA) is one of the more worrisome steps involving the evolution of resistance to antibiotics (Perlman, 2013). *S. aureus* commonly inhabits mucus membranes (including the nose, the most common site) and the skin. The bacterium is typically part of the harmless microbiome of humans, usually not causing serious problems other than skin infections and inflammatory and immunological responses (Perlman, 2013). However, *S. aureus* is an opportunistic bacterial pathogen and as stated in the following by Perlman (2013, p. 88) may sometimes cause very serious, even fatal, infections: "The bacterium can infect a large number of sites and so can cause a wide variety of clinical infections, including skin and soft tissue infections, pneumonia, systemic sepsis, septic arthritis, osteomyelitis, and toxic shock syndrome." *S. aureus* can be transmitted directly between humans by contact with skin infections, but can survive outside the body of its host for relatively long periods of time resulting in fomites (objects that carry infection, such as hair, skin cells, hospital clothing, bedding, utensils, etc.) that serve a major role in transmission (Perlman, 2013).

Penicillin, discovered in 1928 by Alexander Fleming (1881-1955) at St. Mary's Hospital in London, was introduced as an antibiotic against staphylococcal and streptococcal bacterial infections in 1942 and soon was produced in commercial quantities to fight these bacterial infections, and other serious bacterial infections (American Chemical Society, accessed 06/26/2017). In fact, the first bacterium studied by Dr. Fleming for its susceptibility to a substance generated by the mold *Penicillium notatum* was *Staphylococcus aureus*. The following statement from Markel, 2013 (accessed online from PBS Newshour 07/14/2017) makes this clear: "Upon examining some colonies of Staphylococcus aureus, Dr. Fleming noted that a mold called Penicillium notatum had contaminated his Petri dishes. After carefully placing the dishes under his microscope, he was amazed to find that the mold prevented the normal growth of the staphylococci." Later, other genera of *Penicillium* were found to generate more volume of penicillin than *P. notatum*. But, eventually, because of overuse and misuse of the antibiotic, many formerly susceptible bacteria, including *S. aureus*, became resistant to penicillin.

By the late 1940s and early 1950s, penicillin-resistant strains of *S. aureus* (both hospital-acquired and community-acquired) were showing up (Chambers, 2001; Perlman, 2013). In order to treat penicillin-resistant *S. aureus*, semisynthetic penicillin derivitives, one of the first being methicillin, were developed that would kill *S. aureus*. Methicillin was introduced for clinical use in 1959, but by 1961 methicillin-resistant *Staphylococcus aureus* (MRSA) had evolved, at first in hospital settings (hospital acquired MRSA) but more recently MRSA strains have been found outside the hospital setting (community acquired MRSA) (Perlman, 2013). Most strains of MRSA have also evolved multidrug resistance to other penicillin-derived antibiotics (such as nafcillin and oxacillin), as well as to macrolides, tetracyclines, and aminoglycosides (see Tenover, Biddle, and Lancaster, 2001). By the 1980s, vancomycin emerged as the preferred treatment for MRSA infections. However, because of the increased use of vancomycin resulting in selection pressure on staphylococcal bacteria, by the middle to late 1990s this "eventually led to the emergence of strains of *S. aureus* and other species of staphylococci with decreased susceptibility to vancomycin and other glycopeptides" (Tenover, Biddle, and Lancaster, 2001). So, as is apparent from this discussion, the spread of staphylococcal bacterial infections (and the spread of other infectious bacteria) is a global health problem that is becoming worse with time. Hopefully, increased health consciousness and hygiene in both healthcare facilities and in communities across the world will slow the spread of MRSA infections, VRSA (vancomycin-resistant *Staphylococcus aureus*) infections, and other bacterial infections.

Very Rapid Evolution of the Human Immunodeficiency Virus (HIV)

There are numerous other examples of using evolutionary principles in medicine, in particular phylogenetic sequence analysis (today, by primarily looking at the DNA nucleotide sequence) to reconstruct the evolutionary history of a particular pathogen (whether they are a virus, a bacterium, or members of another group). Understanding the relationships of different strains of bacteria and viruses (and sometimes just to identify a particular bacterium or virus or determine that it is new to science) allows scientists to develop drug treatments or vaccines that may be effective to combat the diseases caused by these bacteria or viruses (or other pathogens) (see Hillis, 2005 for numerous examples). Also, new breakthroughs in molecular technologies and genetic engineering (including germ cell gene editing) are rapidly affecting evolutionary medical procedures and our ability to actually direct the evolutionary process (see the next section which will discuss some of these methods, including CRISPR [clustered regularly interspaced short palindromic repeats; see Doudna and Sternberg, 2017, p. 40-41; Losos, 2017, p. 309]). However, the remainder of this discussion on evolution and medicine will concentrate on the relatively recent AIDS (acquired immune deficiency syndrome) epidemic caused by the human immunodeficiency virus (HIV), which so far has taken 25-30 million lives (McNeil, 2010; Perlman, 2013; Quammen, 2012). Other statistics relative to HIV/AIDS, as presented by Perlman (2013, p. 98-99), are staggering:

> ...some recent estimates from the U.N. Program on HIV/AIDS (UNAIDS) indicate the magnitude of the burden of human disease caused by HIV. Some 30-35 million people are thought to be infected with HIV and another 2.5-3 million people per year are becoming newly infected. Of the 30-35 million

infected individuals, approximately two-thirds live in sub-Saharan Africa. Up to 30% of adults in some urban areas of sub-Saharan Africa carry the virus. India, with 3-5 million HIV-infected people, is the country with the greatest prevalence, but this is less than 1% of the Indian population. In the 30 years since AIDS was identified [in 1981], it has been responsible for some 25-30 million deaths and, ...it is currently causing close to 3 million deaths per year.

Viruses consist basically of a capsid of protein surrounding a viral genome of DNA or RNA, but some also have a membraneous envelope of phospholipids and membrane proteins derived from the host cell. In addition, these viral envelopes also contain proteins and glycoproteins that are derived from the virus itself (Urry et al., 2014). Viruses are extremely small, from about 15-20 nannometers (nm) up to about 300 nm, but, unless inside a host cell, cannot metabolise or reproduce (Quammen, 2012; Urry et al., 2014). "Most biologists studying viruses today would likely agree that they are not alive but exist in a shady area between life-forms and chemicals" (Urry et al., 2014). HIV is a retrovirus and contains two identical single strands of RNA (Freeman and Huron, 2007; Quammen, 2012; Urry et al., 2014). According to Quammen (2012, p. 391), "Retroviruses are fiendish beasts, even more devious and persistent than the average virus. They take their name from the capacity to move backward (retro) against the usual expectations of how a creature translates its genes into working proteins. Instead of using RNA as a template for translating DNA into proteins, the retrovirus converts its RNA into DNA within a host cell; its viral DNA then penetrates the cell nucleus and gets itself integrated into the genome of the host cell, thereby guaranteeing replication of the virus whenever the host cell reproduces itself."

HIV is transmitted from human to human when a body fluid containing the virus from an infected person, usually blood or semen, carries the virus onto a mucus membrane or into the bloodstream of an uninfected person (Freeman and Herron, 2007). The virus, once in the bloodstream, attacks the human immune system by parasitizing white blood cells, specifically dendritic cells, macrophages, and T cells of the immune system. HIV invades these cells by binding to two proteins on the surface of the cells and then discharging the contents of the virus into the cells. The contents include two identical single strains of viral RNA and three proteins that aid the virus to take over the workings of the cell (using the cell's own enzymatic machinery) and to replicate. HIV is a retrovirus that once in the cell uses an enzyme called reverse transcriptase to synthesize viral DNA from the viral RNA template. Another viral enzyme then splices the new viral DNA into the host's DNA. Then the host cell's RNA polymerase transcribes the viral DNA into mRNA. The viral mRNA then uses the cells own ribosomes to make viral protein and new viral units (virions). Once many copies of the virus are made and assembled in the host cell, they bud from the host cell's membrane to invade another of their host's cells or to invade another host. In this process, the host cell is destroyed or dies. Of course, the host's immune system attacks the viruses in the bloodstream and kills its own cells that are infected. But the macrophages and T cells are depleted so quickly by the rapidly reproducing viruses that the host's immune system collapses, thus giving rise to AIDS. Then of course, the host cannot fight off opportunistic diseases (bacteria and fungi) and the AIDS patient usually dies within 2 or 3 years if not treated (Freeman and Hurron, 2004; Freeman and Huron, 2007).

Several drugs have been developed that attempt to block some phase of the viral reproductive cycle. Some of the most popular antiretroviral drugs developed to fight HIV are reverse transcriptase inhibitors (Freeman and Herron, 2007). These drugs were the ones first used for the treatment of HIV in AIDS patients (Perlman, 2013). These drugs tend to kill retroviruses with limited side effects (Freeman and Huron, 2007). One of the first of these drugs used, primarily during the mid-1980s to early 1990s (but still used today), with AIDS patients was azidothymidine (AZT). AZT has a very similar chemical structure as the nucleoside thymidine that is in DNA. Using AZT is an attempt to fool the viral reverse transcriptase into substituting the AZT into the viral DNA, rather than thymidine. This effectively stops the DNA synthesis and thus stops the production of new virions, thus stopping the infection. This treatment is effective in many patients for a short time and often drastically reduces the number of viruses in the bloodstream, but usually only for about 6 months. The problem is that the virus is reproducing so rapidly and the construction of double stranded viral DNA from viral RNA is so error prone that mutations occur at a very high rate. After a short time, variants of the virus are resistant to AZT (they are not fooled by the AZT and pick normal thymidine to construct the viral DNA). These variant forms of the virus then survive and reproduce (via natural selection) and the virus increases in the bloodstream of the patient again (Freeman and Herron, 2007). Other drugs have been developed to try to shut down the replication process at other points in the viral cycle, but used alone they also result in viral resistance after a relatively short time. Multidrug cocktails are somewhat more effective in treating AIDS, because they can provide treatment for several different variants of the virus. However, even with this, HIV evolves so rapidly that the virus is becoming simultaneously resistant to multiple drugs. The other problem with multidrug cocktails is that they are very expensive and patients often experience serious side effects (Freeman and Herron, 2004; Freeman and Huron, 2007).

So, as is obvious from the above discussion, AIDS is a disease that is very hard to combat. Phylogenetic analysis offers great hope of conquering this disease. Based on phylogenetic analyses that have already been completed, we are learning much about the evolutionary history of HIV and where the virus originated. Based on phylogenetic analysis, we know that HIV arose from simian immunodeficiency viruses (SIV) in Africa. HIV occurs as two major types HIV-1 and HIV-2. HIV-2 is pretty much restricted to western Africa and is less virulent than HIV-1. HIV-2 arose from the sooty mangabey (an old world monkey) due to mutations in a form of SIV that jumped to a human host (infections in humans resulting from pathogens jumping from an animal host to a human host are referred to as zoonoses (pl.; zoonosis, sing.; AIDS is a zoonotic disease) (Freeman and Huron, 2004; Freeman and Huron, 2007; Quammen, 2012). HIV-1 arose from chimpanzees. Both HIV-1 and HIV-2 occur as several strains and probably have made the transitions from their chimp and monkey hosts several times over centuries (Hillis, 2005). The M-subgroup of HIV-1 (the strain that is most prevalent in North America and western Europe) has been dated (via molecular clock dating) to between 1915 and 1941 (Hillis, 2005; Freeman and Huron, 2004; Freeman and Huron, 2007). According to Freeman and Huron (2007, p. 27), "There is considerable uncertainty, but the best estimate is that the last common ancestor of the group M HIV-1 viruses lived in the 1930s." However, the following statement from Quammen (2012, p. 477) suggests a significantly earlier date for the initiation of the AIDS pandemic and the most probably location where HIV-1 group M made its debut:

> Here's what you have come to understand. That the AIDS pandemic is traceable to a single contingent event. That this event involved a bloody interaction between one chimpanzee and one human. That this occurred in southeastern Cameroon, around the year 1908, give or take. That it led to the proliferation of one strain of virus, now known as HIV-1 group M. That this virus was probably lethal in chimpanzees before the spillover occurred, and that it was certainly lethal in humans afterward. That from southeastern Cameroon it must have traveled downriver, along the Sangha and then the Congo, to Brassaville and Léopoldville. That from these entrepôts it spread to the world.

Evidently, even though the transfer of HIV from simians to humans has occurred repeatedly over centuries, AIDS was restricted primarily to Africa. However, according to Hillis (2005), because of the modern major urbanization of Africa starting in the 1950s and 1960s and continuing today (and the associated increase in prostitution, illegal drug use, reuse of hypodermic needles, etc.) and the movement of people within and out of Africa to the rest of the world, AIDS has changed from local epidemics in Africa to a global epidemic today. Phylogenetic analyses are critical to track the spread of HIV around the world and to identify the routes along which the virus has spread and is spreading. These studies are necessary for slowing the transmission of HIV by identifying the risk factors in different geographic regions and understanding the growing divergence of the viruses to help control and treat HIV (Hillis, 2005; Freeman and Herron, 2004; Freeman and Herron, 2007; Quammen, 2012).

There is much more that could be discussed here, such as, the difficulty in developing a vaccine for AIDS because of the rapid mutation rate of HIV, what preventive measures to take to slow the incidence of infection, etc. However, the following statement from Perlman (2013, p. 102) broadly sums up the current status of the treatment for HIV to try to prevent the development of AIDS:

> The development of combination therapies was motivated at least in part by evolutionary considerations. The use of two different types of reverse transcriptase inhibitors, which inhibited the enzyme by different mechanisms, was based on the premise that resistant mutants would require at least two mutations in reverse transcriptase and would likely have decreased fitness.

The development of combination anti-retroviral therapy has had a dramatic impact on the natural history of HIV infections. Drug therapy controls the disease in most patients and has turned HIV from a progressive, lethal disease into a chronic, manageable disease. Drug treatment is not without its problems. It is expensive, it depends on patients actually taking their medication, and it has side effects. But in addition to prolonging the healthy lives of people who are infected with HIV, drug therapy appears dramatically to reduce the risk of transmission of the virus. Combination drug therapy has created new optimism that we may finally be able to curb the spread of HIV and reduce the toll from AIDS. [internal citations deleted]

GENETIC ENGINEERING AND GENE EDITING USING CRISPR

Certainly, there are other areas (as we have discussed throughout this book) than medicine where understanding evolution is of utmost importance to the human race, such as agriculture (and thus, feeding the billions of humans on Earth), conservation (for example, the preservation of species and understanding what constitutes a species), understanding how species adapt to (or do not adapt to) climate change, etc. For sure, one area that is of utmost importance in studying and understanding the process of evolutionary change, as with all the sciences, is basic research. One field of research that has come to the forefront most recently in biomolecular technology (due to basic research in molecular biology, biochemistry, and bacteriology) is genomic editing using CRISPR (clustered regularly interspaced short palindromic repeats), or similar methods (Doudna and Sternberg, 2017; Losos, 2017). CRISPR has ushered in the very real possibility of genetically modifying germ (sex) cells in wild populations of organisms and according to Losos (2017) "...of basically being able to direct genetic evolution in the wild." Of course, CRISPR technology also has the potential to be used to modify both somatic (body) cells and germ (sex) cells (including fertilized eggs and embryos) in humans for the benefit of humans for medical purposes, and also purposes of supposed enhancement of human characteristics. Obviously, germline cell modification of human embryos must be approached very cautiously due to both practical and ethical concerns (Travis, 2015; Specter, 2016; Doudna and Sternberg, 2017; Losos, 2017). I will discuss more on this later, but now an explanation of the nature of CRISPR is in order.

The 2015 Breakthrough of the Year in *Science* magazine (published by the American Association for the Advancement of Science, AAAS) was CRISPR, a new genome editing method (Travis, 2015). According to Travis (2015, p. 1456),

> CRISPR has appeared in Breakthrough sections twice before, in 2012 and 2013, each time as a runner-up in combination with other genome-editing techniques. But this is the year it broke away from the pack, revealing its true power in a series of spectacular achievements. Two striking examples—the creation of a long-sought "gene drive" that could eliminate pests or the diseases they carry, and the first deliberate editing of the DNA of human embryos—debuted to headlines and concern.

The August 2016 *National Geographic* issue states on its cover the following: "The DNA Revolution – With new gene-editing techniques, we can transform life---but should we?" The article inside by Specter (2016) discusses the tremendous potential of CRISPR to "quickly and precisely alter, delete, and rearrange the DNA of nearly any living organism, including us."

CRISPR was first noted during sequencing research on the single DNA molecules of prokaryotic cells in bacteria and archaea. Most of these cells contained the CRISPR array in one portion of the circular DNA molecule. CRISPRs are present in about half of all bacterial genomes and almost all archaeal genomes that have been sequenced (Doudna and Sternberg, 2017). The CRISPR array consists of repeating sequences of DNA (the "clustered regularly interspaced short palindromic repeats" – here palindromic means that the DNA base-pair sequence reads the same [or nearly so] backward as forward) consisting of about 29 base-pairs, with unique sequences of DNA consisting of about 32 base-pairs sandwiched between the repeat sequences; these unique sequences of DNA are referred to as spacers (Jinek et al., 2012; Zimmer, 2015; Doudna and Sternberg, 2017). The earlier research on the unique sequences of DNA that made up the spacers in CRISPR "also showed that many of the interspaced sequences in the CRISPR arrays matched viral DNA sequences in the environment" (Doudna and Sternberg, 2017, p. 43). This provided "evidence suggesting that CRISPR was likely to be part of an archaeal and bacterial immune system, an adaptation that allowed microbes to fight off viruses" (Doudna and Sternberg, 2017, p. 44). Viruses are the primary pathogens that attack organisms with prokaryotic cells (bacteria and archaea). Viral pathogens that infect and destroy bacteria are called bacteriophages (or often referred to as simply phages). Bacteriophages are the most abundant biological entity on Earth (Doudna and Sternberg, 2017). According to Doudna and Sternberg (2017, p. 57), researchers determined that "CRISPR functioned like a molecular vaccination card: by storing memories of past phage infections in the form of spacer DNA sequences buried within the repeat-spacer arrays, bacteria could use this information to recognize and destroy those same invading phages during future infections."

It also became apparent to researchers early on in studying CRISPR that there were other genes that almost always occurred near the CRISPR sequence of repeats and spacers. These became known as CRISPR-associated (Cas) genes. Soon it was determined that these Cas genes coded for enzymes called endonucleases that could be used to slice up or cut DNA molecules (Doudna and Sternberg, 2017). To make a long story short, it was also determined by research teams that RNAs associated with the CRISPR/Cas system and containing short DNA sequences from the CRISPR spacers are used to guide a Cas protein (endonuclease) to a specific site where foreign (viral) DNA is located and then to cut and destroy that DNA. Once this was accomplished, researchers realized that they could cut out sections of DNA in any organism's genome and replace it with DNA of their choice, thus the CRISPR/Cas system was programmable for genome editing (Jinek et al., 2012; Doudna and Sternberg, 2017).

One of the team leaders that was instrumental in developing the programmable CRISPR/Cas 9 genome editing tool, Jennifer A. Doudna, a professor of biochemistry at the University of California, Berkeley, and her former graduate student, Samuel H. Sternberg, also a biochemist, have very recently written an outstanding book, *A Crack in Creation: Gene Editing and the Unthinkable Power to Control Evolution*, on how the system was developed, its various uses, and praise for its value for mankind, but with a gentle warning of the ultimate power of this gene editing tool. Carl Zimmer, in an excellent article in Quanta Magazine (2015), explains, in the following quote, how the CRISPR/Cas 9 system works:

To create a DNA-cutting tool, Doudna and her colleagues picked out the CRISPR-Cas system from *Streptococcus pyogenes*, the bacteria that cause strep throat. It was a system they already understood fairly well, having worked out the function of its main enzyme, called Cas9. Doudna and her colleagues figured out how to supply Cas9 with an RNA molecule that matched a sequence of DNA they wanted to cut. The RNA molecule then guided Cas9 along the DNA to the target site, and then the enzyme made its incision. Using two Cas9 enzymes, the scientists could make a pair of snips, chopping out any segment of DNA they wanted. They could then coax a cell to stitch a new gene into the open space. Doudna and her colleagues thus invented a biological version of find-and-replace — one that could work in virtually any species they chose to work on.

So, the CRISPR-Cas9 system has evolved in the bacterium *Streptococcus pyogenes* as an immune system to fight invading viruses (bacteriophages) (similar CRISPR-Cas immune systems have evolved in other prokaryotic cells). This realization by researchers doing basic research on bacteria and archaea has led to a tremendously powerful gene editing tool; as illuminated in the following statement by Doudna and her colleagues in the last sentence of the abstract of their 2012 ground-breaking paper called "A Programmable Dual-RNA-Guided DNA Endonuclease in Adaptive Bacterial Immunity" in *Science* magazine (see Jinek, 2012, p. 816): "Our study reveals a family of endonucleases that use dual-RNAs for site-specific DNA cleavage and highlights the potential to exploit the system for RNA-programmable genome editing."

The above CRISPR-Cas9 genome editing system is the most powerful, the most accurate and efficient, and the cheapest system that has been developed by scientists to date and offers great promise for editing the genomes of organisms, including humans (Doudna and Sternberg, 2017). Specter (2016, p. 40) describes this gene editing tool in the August 2016 issue of National Geographic as follows:

> CRISPR-CAS9 has two components. The first is an enzyme—Cas9—that functions as a cellular scalpel to cut DNA. (In Nature, bacteria use it to sever and disarm the genetic code of invading viruses.) The other consists of an RNA guide that leads the scalpel to the precise nucleotides—the chemical letters of DNA—it has been sent to cut...
>
> The guide's accuracy is uncanny; scientists can dispatch a synthetic replacement part to any location in a genome made of billions of nucleotides. When it reaches its destination, the Cas9 enzyme snips out the unwanted

DNA sequence. To patch the break, the cell inserts the chain of nucleotides that has been delivered in the CRISPR package.

The CRISPR-Cas9 system is already being used to modify genomes and offers great promise for correcting human genetic disorders, such as muscular dystrophy, cystic fibrosis, even HIV/AIDS and some cancers. In addition to directly correcting human genetic disorders, CRISPR-Cas9 can be used to alter genes in animals for the benefit of human health. Research is underway with CRISPR to grow pig organs that will be more compatible for transplantation into humans, as illustrated in the following quote from Doudna and Sternberg (2017, p. 141): "Gene editing is now being harnessed to shut down pig genes that might provoke the human immune response and to eliminate the risk that porcine viruses embedded in the pig genome could hop over and infect humans during transplantation." Also, research is being conducted on the use of CRISPR-Cas9 to edit the genes of *Anopheles* mosquitoes such that they can no longer transmit the malaria parasite when they bite a person (Specter, 2016). It may also be possible to use the CRISPR-Cas9 system to prevent other types of mosquitoes from transmitting pathogens that cause diseases such as yellow fever and dengue fever, as well as carrying pathogens such as chikungunya virus, West Nile virus, and Zika virus to human hosts (Specter, 2016). CRISPR-Cas9 will, no doubt, have a tremendous number of applications in agriculture (for the benefit of both plant crops and livestock for food), as well as opportunities for bringing back organisms that have relatively recently gone extinct (such as the woolly mammoth – of which we have sequenced the genome) or those that will go extinct in the near future due to climate change (see Specter, 2016; Doudna and Sternberg, 2017).

Of course, the other area of continued research and potential treatment with CRISPR-Cas9 genomic engineering will involve editing the germline of organisms, including humans. According to Doudna and Sternberg (2017), *Science* magazine published an article resulting from a January 2015 meeting in Napa, California--of top scientists and experts on bioethics, relative to genomic engineering, gene editing, and ethics of biological experimentation--on March 19, 2015 entitled "A Prudent Path Forward for Genomic Engineering and Germline Gene Modification."

The article, after explaining how the CRISPR tool works and the applications presently being conducted, discussed the topic of germline editing (also see Baltimore et al., 2015). The following statement by Doudna and Sternberg (2017, p. 211; also see Baltimore et al., 2015) summarizes (from the article) the need for great caution when experimenting with heritable alterations of the genomes of organisms, particularly with human germline cells:

> … We asked experts from the scientific and bioethics communities to create forums that would allow interested members of the public to access reliable information about new gene-editing technologies, their potential risks and rewards, and their associated ethical, social, and legal implications. We called on researchers to continue testing and developing the CRISPR technology in cultured human cells and in nonhuman animal models so that its safety profile could be better understood in advance of any clinical applications. We called for an international meeting to ensure that all the relevant safety and ethical implications could be openly and transparently discussed – not just among scientists and bioethicists, but also among the many diverse stakeholders who would surely want to weigh in: religious leaders, patient- and disability-rights advocates, social scientists, regulatory and governmental agencies, and other groups.
>
> Last, and perhaps most significant, we asked scientists to refrain from attempting to make heritable changes to the human genome. Even in countries with lax regulations, we wanted researchers to hold off until governments and societies around the world had a chance to consider the issue. Although we ultimately avoided using the word *ban* or *moratorium*, the message was clear: for the time being, such clinical applications would be off-limits.

However, it was learned soon after this meeting that some genomic alterations of human embryos had already taken place in experiments by one researcher in China, but these embryos were not viable and did not grow to full term; in fact, these human embryos were triploblastic [the cells contained three sets of chromosomes] and would never have been implanted into a potential mother (see Doudna and Sternberg, 2017, p. 214-218). In addition, the procedures in these gene editing experiments, showed that the CRISPR tool was only about 5% efficient, and in some cases CRISPR even edited DNA sequences other than the one's targeted (Doudna and Sternberg, 2017). According to Doudna and Sternberg (2017), these problems will most likely be fixed in the near future, but this only highlights the need for further research before gene editing of human germline cells.

Nevertheless, there is a high probability that CRISPR-Cas9 (or some other variant of CRISPR) will be used to edit germline cells for genetic diseases and (most likely also in some countries and perhaps under some conditions in all countries) to enhance the genomes of newborn babies. China is continuing to be the most active nation on CRISPR experimentation, on both somatic cells for many genetic diseases (such as HIV/AIDS and several cancers – where they are currently recruiting patients for clinical trials on several types of cancer) and continued research on human embryos (Normile, 2017). It may be unfortunate that germline modification of human embryos has already begun, but not all researchers are discouraged by this. Hopefully (and most likely in all developed countries, at least), great care will be taken to make sure the technique is honed to precision and accuracy before CRISPR babies are born in large numbers.

So, as we have seen throughout this chapter, and throughout this entire book, EVOLUTION IS REAL SCIENCE and EVOLUTION MATTERS! The understanding that we get from a knowledge of how evolution works and how much we depend upon this knowledge to make our modern world a better and more livable place, free of the demons of superstition and ignorance, is inspiring. Darwin's Theory, as modified by Modern Evolutionary Theory, is an example of one of the parts of the candle's flame in Carl Sagan's *The Demon-Haunted World: Science as a Candle in the Dark* (Sagan, 1996).

REFERENCES CITED

Agenbroad, Larry D.; 2009; *Mammuthus exilis* from the California Channel Islands: Height, Mass, and Geologic Age: in Damiani, C.C. and Garcelon, D.K. (eds.); 2009; *Proceedings of the 7th California Islands Symposium*; Institute for Wildlife Studies; Arcata, California; p. 15-19.

Alexandrino, João; Baird, Stuart J. E.; Lawson, Lucinda; Macey, J. Robert; Moritz, Craig; and Wake, David B.; 2005; Strong Selection Against Hybrids at a Hybrid Zone in the Ensatina Ring Species Complex and Its Evolutionary Implications: *Evolution*; Vol. 59; No. 6; p. 1334–1347.

Alvarez, Walter; 1997; *T. rex and the Crater of Doom*: Princeton University Press (Princeton Science Library, Paperback Edition, 09/15/2015); 208 p.

American Chemical Society; International Historic Chemical Landmarks: Discovery and Development of Penicillin; at http://www.acs.org/content/acs/en/education/whatischemistry/landmarks/flemingpenicillin.html; accessed 06/26/2017.

AmphibiaWeb; 2014; Information on amphibian biology and conservation; *Rana clamitans*: Berkeley, California: AmphibiaWeb; Available at: http://amphibiaweb.org/; accessed 08/31/2014.

Anderson, Dan I.; 2016; Chapter 10 – Evolution of Antibiotic Resistance: in Losos, Jonathan B. and Lenski, Richard E. (eds.); 2016; *How Evolution Shapes Our Lives – Essays on Biology and Society*; Princeton University Press; Princeton, New Jersey and Oxford, United Kingdom; p. 147-164.

Antolin, Michael F.; Jenkins, Kristen P.; Bergstrom, Carl T.; Crespi, Bernard J.; De, Subhajyoti; Hancock, Angela; Hanley, Kathryn A., Meagher, Thomas R.; Voreno-Estrada, Andres; Nesse, Randolph M.; Omenn, Gilbert S.; and Stearns. Stephen C.; 2012; Evolution and Medicine in Undergraduate Education: A Prescription for all Biology Students: *Evolution*; Vol. 66; No. 6; p. 1991-2006.

Arnold, J. R. and Libby, W. F.; 1951; Radiocarbon Dates: *Science*; V. 113; No. 2927; p. 111-120.

Arny, Thomas T.; 2006; *Explorations – An Introduction to Astronomy, Fourth Edition*: McGraw-Hill Higher Education; New York, New York; 578 p.

Ausich, William I. and Lane, N. Gary; 1999; *Life of the Past (Fourth Ed.):* Prentice Hall; Upper Saddle River, New Jersey; 321 p.

Bada, Jeffrey L. and Lazcano, Antonio; 2009; The Origin of Life: in Ruse, Michael and Travis, Joseph (eds.); *Evolution – The First Four Billion Years*: The Belknap Press of Harvard University Press; Cambridge, Massachusetts and London, England; p. 49-79.

Baker, Andrew C.; 2003; Flexibility and Specificity in Coral-Algal Symbiosis: Diversity, Ecology, and Biogeography of *Symbiodinium*: *Annual Review of Ecology, Evolution, and Systematics;* Vol. 34; p. 661-689.

Baltimore, David; Berg, Paul; Botchan, Michael; Carroll, Dana; Charo, R. Alta;

Church, George; Corn, Jacob E.; Daley, George Q.; Doudna, Jennifer A.; Fenner, Marsha; Greely, Henry T.; Jinek, Martin; Martin, G. Steven; Penhoet, Edward; Puck, Jennifer; Sternberg, Samuel H.; Weissman, Jonathan S.; Yamamoto, Keith R.; 2015; A Prudent Path Forward for Genomic Engineering and Germline Gene Modification: *Science*; published online at the following website: http://science.sciencemag.org/content/early/2015/03/18/science.aab1028, accessed 10/23/2017.

Bates, Henry W.; 1862; Contributions to an Insect Fauna of the Amazon Valley, Lepidoptera: Heliconidae: *Transactions of the Linnean Society of London*; Vol. 23; p. 495-566.

Baxter, Simon W.; Papa, Riccardo; Chamberlain, Nicola; Humphray, Sean J.; Joron, Mathieu; Morrison, Clay; ffrench-Constant, Richard H.; McMillan, W. Owen; and Jiggins, Chris D.; 2008; Convergent Evolution in the Genetic Basis of Müllerian Mimicry in *Heliconius* Butterflies: *Genetics*; Vol.180; p.1567-1577.

Behe, Michael J.; 1996; *Darwin's Black Box – The Biochemical Challenge to Evolution*: Published by Simon & Schuster (First Touchstone Edition 1998); New York, NY; 307 p.

Bennett, Jeffrey; Donahue, Megan; Schneider, Nicholas; and Voit, Mark; 2010; *The Cosmic Perspective – Fundamentals*: Pearson Education, Inc., Publishing as Pearson Addison-Wesley; San Francisco, California; 263 p.

Benton, Michael; 1993; Chapter Three, Four Feet on the Ground: in Gould, Stephen J. (ed.), 1993, *The Book of Life*: W. W. Nortan & Company, New York and London, 256 p.; Chapter Three; p. 76-125.

Benton, Michael J.; 2003; *When Life Nearly Died – The Greatest Mass Extinction of All Time*: Thames & Hudson Ltd.; London; 336 p.

Benton, Michael J.; 2014; How Birds Became Birds: Sustained Size Reduction was Essential for the Origin of Birds and Avian Flight: *Science*; 1 August 2014; Vol. 345; Issue 6196; p. 508-509.

Benson, Woodruff W.; 1972; Natural Selection for Müllerian Mimicry in *Heliconius erato* in Costa Rica: *Science*; Vol. 176; 26 May 1972; p. 936-939.

Bergstrom, Carl T. and Dugatkin, Lee Alan; 2012; *Evolution*: W. W. Norton & Company; New York and London; 677 p.

Berkman, Michael and Plutzer, Eric; 2010; *Evolution, Creationism, and the Battle to Control America's Classrooms*: Cambridge University Press; Cambridge, England and New York, New York; 287 p.

Berner, Robert A.; 1971; *Principles of Chemical Sedimentology*: McGraw-Hill Book Company; New York; 240 p.

Berra, Tim M.; 2009; *Charles Darwin -The Concise Story of an Extraordinary Man*: The John Hopkins University Press, Baltimore; 114 p.

Best, Sonja M. and Kerr, Peter J.; 2000; Coevolution of Host and Virus: The Pathogenesis of Virulent and Attenuated Strains of Myxoma Virus in Resistant and Susceptible European Rabbits: *Virology*; Vol. 267, p. 36-48.

Bierema, Andrea M.-K. and Rudge, David W.; 2014; Using David Lack's Observations of Finch Beak Size to Teach Natural Selection and the Nature of Science: *The American Biology Teacher*; Vol. 76; No. 5; p. 312-317.

Blumberg, Roger B.; 1997; MendelWeb Notes (Notes to accompany Gregor Mendel's 1865 paper "Experiments in Plant Hybridization"): located at http://www.mendelweb.org/MWNotes.html ; accessed 10/27/2010.

Boag, Peter T.; 1983; The Heritability of External Morphology in Darwin's Finches (*Geospiza*) on Isla Daphne Major, Galapagos: *Evolution*; Vol. 37; No. 5; p. 877-894.

Boag, Peter T. and Grant, Peter R.; 1981; Intense Natural Selection in a Population of Darwin's Finches (Geospizinae) In the Galapagos: *Science*; Vol. 214; No. 4516; p. 82-85.

Boggs, Jr., Sam; 2006; Principles of Sedimentology and Stratigraphy (Fourth Edition): Pearson/Prentice Hall; Upper Saddle River, New Jersey; 662 p.

Bonatti, Enrico; 1968; Fissure Basalts and Ocean-Floor Spreading on the East Pacific Rise: *Science*; Vol. 161; No. 3844; p. 886-888.

Bowler, Peter; 2009; *Evolution - The History of an Idea 25th Anniversary Edition*: University of California Press; Berkeley, Los Angeles, and London; 464 p.

Brodie, Edmund D. Jr; 1968; Investigations on the Skin Toxin of the Adult Rough-Skinned Newt, *Taricha granulosa*: *Copeia*; Vol. 1968; No. 2; p. 307-313.

Brodie, Edmund D. III and Brodie, Edmund D. Jr.; 1999; Predator–Prey Arms Races: Asymmetrical Selection on Predators and Prey May Be Reduced When Prey are Dangerous: BioScience; Vol. 49; No. 7; p. 557-568.

Brodie, Edmund D. III; Feldman, Chris R.; Hanifin, Charles T.; Motychak, Jeffrey E.; Mulcahy, Daniel G.; Williams, Becky L.; and Brodie, Edmund D. Jr.; 2005; Evolutionary Response of Predators to Dangerous Prey: Parallel Arms Races Between Garter Snakes and Newts Involving Tetrodotoxin as the Phenotypic Interface of Coevolution: Journal of Chemical Ecology; Vol. 31; No. 2; p. 343–355.

Brower, Lincoln P. and Glazier, Susan C.; 1975; Localization of Heart Poisons in the Monarch Butterfly: Science; Vol. 188; 4 April 1975; p. 19-25.

Brower, Lincoln P.; Ryerson, William N.; Coppinger, Lorna L.; and Glazier, Susan C.; 1968; Ecological Chemistry and the Palatability Spectrum: Science; Vol.161; 27 September 1968' p.1349-1350.

Brown, Keith S., Jr.; 1981;The Biology of Heliconius and Related Genera: Annual Review of Entomology; Vol. 26; p. 427-456.

Brown, Charles W.; 1974; Hybridization Among the Subspecies of the Plethodontid Salamander *Ensatina eschscholtzi*: University of California Publications in Zoology; Vol. 98; 56 p.

Brown, Jason L.; Maan, Martine E.; Cummings, Molly E.; and Summers, Kyle; 2010; Evidence for Selection on Coloration in a Panamanian Poison Frog: A Coalescent-Based Approach: *Journal of Biogeography*; Vol. 37; p. 891–901.

Brown, Jason L.; Morales, Victor; and Summers, Kyle; 2010; A Key Ecological Trait Drove the Evolution of Biparental Care and Monogamy in an Amphibian: *The American Naturalist*; Vol. 175; No. 4; p. 436-446.

Browne, Janet; 1995; *Charles Darwin – Voyaging: A Biography*: Princeton University Press; Princeton, New Jersey; 605 p.

Browne, Janet; 2002; *Charles Darwin - The Power Of Place: Volume II Of A Biography*: Princeton University Press; Princeton and Oxford; 591 p.

Brunet, Michel; Guy, Franck; Pilbeam, David; Mackaye, Hassane Taisso; Likius, Andossa; Ahounta, Djimdoumalbaye; Beauvilain, Alain; Blondel, Cécile; Bocherens, Hervé; Boisserie, Jean-Renaud; De Bonis, Louis; Coppens, Yves; Dejax, Jean; Denys, Christiane; Duringer, Philippe; Eisenmann, Véra; Fanone, Gongdibé; Fronty, Pierre; Geraads, Denis; Lehmann, Thomas; Lihoreau, Fabrice; Louchart, Antoine; Mahamat, Adoum; Merceron, Gildas; Mouchelin, Guy; Otero, Olga; Campomanes, Pablo Pelaez; Ponce De Leon, Marcia; Rage, Jean-Claude; Sapanet, Michel; Schuster, Mathieu; Sudre, Jean; Tassy, Pascal; Valentin, Xavier; Vignaud, Patrick; Viriot, Laurent; Zazzo, Antoine; and Zollikofer, Christoph; 2002; A New Hominid from the Upper Miocene of Chad, Central Africa: *Nature*; Vol. 418; July 11, 2002; p. 145-151.

Budd, Graham E.; 2003; The Cambrian Fossil Record and the Origin of the Phyla: *Integrative and Comparative Biology*; Vol. 43; p. 157–165.

Caldera, Eric J.; Poulsen, Michael; Suen, Garret; and Currie, Cameron R.; 2009; Insect Symbioses: A Case Study of Past, Present, and Future Fungus-Growing Ant Research: *Environmental Entomology*; V. 38, No. 1; p. 78-92.

Campbell, Neil A.; 1996; *Biology (4ᵗʰ ed.):* The Benjamin/Cummings Publishing Co., Inc.; Menlo Park, California; 1206 p.

Campbell, Neil A.; Reece, Jane B.; and Mitchell, Lawrence; 1999; *Biology (5ᵗʰ ed.)*: Benjamin/Cummings, an imprint of Addison Wesley Longman, Inc.; Menlo Park, California; 1175 p.

Canfield, Donald E.; 2014; *Oxygen – A Four Billion Year History*: Princeton University Press; Princeton, New Jersey and Oxford, England; 196 p.

Cardoso, M. Z. and Gilbert, L. E.; 2013; Pollen Feeding, Resource Allocation and the Evolution of Chemical Defence in Passion Vine Butterflies: *Journal of Evolutionary Biology*; Vol. 26; p. 1254–1260.

Carey, Nessa; 2012; *The Epigenetics Revolution – How Modern Biology Is Rewriting Our Understanding of Genetics, Disease, and Inheritance*: Columbia University Press; New York; 339 p.

Carey, Nessa; 2015a; Junk Matters – Part 1: Although 98 Percent of Human DNA Doesn't Code for Protein, It May Be Vital to Gene Expression: *Natural History*; March 2015; Vol. 123; No. 2; p. 12-17.

Carey, Nessa; 2015b; Junk Matters – Part 2: Telomeres, Built Largely of Junk DNA, Maintain the Integrity of Our Genome: *Natural History*; April 2015; Vol. 123; No. 3; p. 18-23.

Carlson, Sandra J.; 1999; Evolution and Systematics: in Scotchmoor, Judy and Springer, Dale (eds.); *Evolution – Investigating the Evidence*; The Paleontological Society: Vol. 9; 1999; Printed by New Image Press; Pittsburgh, PA; p. 95-117.

Carroll, Sean B.; 2005; *Endless Forms Most Beautiful - The New Science of Evo Devo and the Making of the Animal Kingdom*: W. W. Norton & Company; New York and London; 349 p.

Censky, Ellen J., Hodge, Karim, and Dudley, Judy; 1998; Over-Water Dispersal of Lizards Due to Hurricanes: *Nature*; Vol. 395; No. 6702 (8 October 1998); p. 527-621.

Charles Darwin Research Station Fact Sheet; 2006; Land Iguanas: Charles Darwin Foundation for the Galapagos Islands (AISBL); Galapagos, Ecuador.

Chambers, Henry F.; 2001; The Changing Epidemiology of *Staphylococcus aureus*? Emerging Infectious Diseases; Vol. 7, No. 2, March-April 2001, Special Issue; p. 178-182. (This article may be obtained online at the following web site: https://www.ncbi.nlm.nih.gov/pmc/articles/PMC2631711/pdf/11294701.pdf, last accessed 07/15/2017.)

Chapman, Matthew; 2007; *40 Days and 40 Nights: Darwin, Intelligent Design, God, Oxycontin®, and Other Oddities on Trial in Pennsylvania*: HarperCollins Publishers; New York; 288 p.

Clack, Jennifer A.; 2006; *Acanthostega. Acanthostega gunnari*: Version 13, June 2006: http://tolweb.org/Acanthostega_gunnari/15016/2006.06.13 in The Tree of Life Web Project, http://tolweb.org/; accessed 03/02/2015.

Clack, Jennifer A.; 2012; *Gaining Ground – The Origin and Evolution of Tetrapods (2nd ed.):* Indiana University Press; Bloomington, Indiana; 544 p.

Clary, Renee and Wandersee, James; 2013; Classification - Putting Everything in its Place: *The Science Teacher*; V. 80; No. 9; p. 31-36.

Clarkson, E. N. K.; 1998; *Invertebrate Palaeontology and Evolution (Fourth ed.)*: Blackwell Publishing; Malden, MA, USA; Oxford, UK; Carlton, Victoria, Australia; 452 p.

Clinton, Keely; 2015; Average Cranium/Brain Size of *Homo neanderthalensis* vs. *Homo sapiens*: *The Backbone Journal*; Howard University; December 24, 2015; published online at the following web address: https://www.cobbresearchlab.com/issue-2-1/2015/12/24/average-cranium-brain-size-of-homo-neanderthalensis-vs-homo-sapiens.

Cohen, Tracey; Goodnight, Gary W.; Jensen, Deborah L.; Koballa, Thomas R. Jr.; Rainis, Kenneth; Nassis, George; Talkmitt, Susan Green; and Tocci, Salvatore; 1998; *Biology – Principles & Explorations*: Holt, Rinehart and Winston; Orlando, Florida; 1072 p.

Connell, Joseph H.; 1961; The Influence of Interspecific Competition and Other Factors on the Distribution of the Barnacle *Chthamalus stellatus*: *Ecology*; Vol. 42; No. 4; p. 710-723.

Conway Morris, Simon; 2006; Darwin's Dilemma: The Realities of the Cambrian 'Explosion'; *Philos. Trans. R. Soc. Lond. B Biol. Sci.*; Vol. 361; p. 1069–1083.

Conway Morris, Simon and Caron, Jean-Bernard; 2012; *Pikaia gracilens* Walcott, A Stem-group Chordate from the Middle Cambrian of British Columbia: *Biological Reviews*, Cambridge Philosophical Society; p. 1-33.

Cook, L. M., Grant, B. S., Saccherl, I. J., and Mallet, J.; 2012; Selective Bird Predation on the Peppered Moth: The Last Experiment of Michael Majerus: *Biology Letters*, Evolutionary Biology; Vol. 8; Issue 4; p. 609-612.

Costa, James T.; 2018; *Darwin's Backyard – How Small Experiments Led to a Big Theory*: W. W. Norton & Company; New York and London; 441 p.

Counterman, Brian A.; Araujo-Perez, Felix; Hines; Heather M.; Baxter, Simon W.; Morrison, Clay M.; Lindstrom, Daniel P.; Papa, Riccardo; Ferguson, Laura; Joron, Mathieu; ffrench-Constant, Richard H.; Smith, Christopher P.; Nielsen, Dahlia M.; Chen, Rui; Jiggins, Chris D.; Reed, Robert D.; Halder, Georg;

Mallet, Jim; and Owen, W. McMillan; 2010; Genomic Hotspots for Adaptation: The Population Genetics of Müllerian Mimicry in *Heliconius erato*: *PLoS Genetics*; Vol. 6; Issue 2; p. 1-13.

Coyne, Jerry A.; The Peppered Moth Story is Solid: in *Why Evolution is True* (Blog at WordPress.com); at https://whyevolutionistrue.wordpress.com/2012/02/10/the-peppered-moth-story-is-solid/; accessed 06/24/2015.

Coyne, Jerry A.; 2009; *Why Evolution is True*: Viking, Published by the Penguin Group; New York, New York; 282 p.

Cracraft, Joel; 2005; An Overview of the Tree of Life: in Cracraft, Joel and Bybee, Roger W. (eds.): *Evolution Science and Society: Educating a New Generation*: Revised Proceedings of the BSCS, AIBS Symposium, November 2004, Chicago, Il.; BSCS, Colorado Springs, Colorado; Chapter 5, p. 43-51

Crisp, Edward L.; 2008; Student Attitudes Relative to Scientific Inquiry, Evolution, and Creationism in Introductory Science Courses at a Small College in West Virginia: *Geological Society of America Abstracts with Programs*, V. 40, No. 6, p. 365. (May be accessed online at: https://gsa.confex.com/gsa/2008AM/finalprogram/abstract_147277.htm).

Currie Lab; University of Wisconsin-Madison; at https://currielab.wisc.edu/index.php; accessed 08/27/2015.

Darwin, Charles; 1845; *Voyage of the Beagle Round the World – Journal of Researches into Natural History and Geology (2nd ed.)*: This special republished edition by Tess Press; New York, New York; 490 p.

Darwin, Charles R.; 1859; *On the Origin of Species By Means of Natural Selection Or The Preservation of Favoured Races in the Struggle for Life*: John Murray; London; 502 p.

Darwin, Charles R.; 1861; *On the Origin of Species By Means of Natural Selection Or The Preservation of Favoured Races in the Struggle for Life* (3rd ed.): John Murray; London; 538 p. [Citation: John van Wyhe, editor. 2002-. The Complete Work of Charles Darwin Online. (http://darwin-online.org.uk/): accessed 3/13/2014].

Darwin, Charles R.; 1862; *On the Various Contrivances by Which British and Foreign Orchids are Fertilized by Insects, and on the Good Effects of Intercrossing*: John Murray; London; 417 p. [Citation: John van Wyhe, editor. 2002-. The Complete Work of Charles Darwin Online. (http://darwin-online.org.uk/): accessed 06/10/2014].

Darwin, Charles R.; 1866; *On the Origin of Species By Means of Natural Selection Or The Preservation of Favoured Races in the Struggle for Life* (4th ed.): John Murray; London; 593 p. [Citation: John van Wyhe, editor. 2002-. The Complete Work of Charles Darwin Online. (http://darwin-online.org.uk/): accessed 08/02/2014].

Darwin, Charles; 1871; *The Descent of Man and Selection in Relation to Sex*, Two Volumes: John Murray; London; Vol. 1, 423 p.; Vol. 2, 475 p.

Darwin, Charles; 1872 (this printing, 2004); *The Origin of Species By Means of*

Natural Selection or the Preservation of Favoured Races in the Struggle for Life (6th ed.): Castle Books; Edison, New Jersey; 703 p. (Originally published in1859 (1st ed.) by John Murray; London).

Darwin, Charles R. (Nora Barlow, ed.); 1969 (reissued 2005) (Originally in 1958, Collins: London); *The Autobiography of Charles Darwin 1809-1882:* W.W. Norton & Company; New York, London; 210 p.

Darwin, Charles; 2005 (Originally published 1845, John Murray, London): *Charles Darwin's Voyage of the Beagle Round the World:* Tess Press, an imprint of Black Dog & Leventhal Publishers, Inc.; New York, New York; 490 p.

Davis, Percival and Kenyon, Dean H.; 1989; *Of Pandas and People – The Central Question of Biological Origins:* Haughton Publishing Company (Copyright 1989, 1993 by Foundation for Thought and Ethics; Richardson, Texas); Dallas, Texas; 170 p.

Dawkins, Richard; 1996; *The Blind Watchmaker - Why the Evidence of Evolution Reveals a Universe Without Design:* W. W. Nortan & Company, Inc.; New York, New York; 496 p.

Dawkins, Richard; 2004; *The Ancestor's Tale – A Pilgrimage to the Dawn of Evolution:* Houghton Mifflin Company; Boston and New York; 673 p.

deMenocal, Peter B.; 2014; Climate Shocks: *Scientific American;* Vol. 311; No. 3; p. 48-53.

de Queiroz, Alan; 2014; *The Monkey's Voyage – How Improbable Journeys Shaped the History of Life:* Basic Books (A Member of the Perseus Books Group); New York; 360 p.

DeSalle, Rob and Yudell, Michael; 2005; *Welcome to the Genome - A User's Guide to the Genetic Past, Present, and Future:* John Wiley & Sons, Inc., in association with the American Museum of Natural History; Hobeken, New Jersey; 215 p.

Desmond, Adrian and Moore, James; 1991; *Darwin - The Life of a Tormented Evolutionist:* W. W. Norton & Company; New York and London; 808 p.

Diamond, Jared M.; 1987; Did Komodo Dragons Evolve to Eat Pygmy Elephants?: *Nature,* Vol. 326, April 30 1987; p. 832.

Dingus, Lowell and Rowe, Timothy; 1998; *The Mistaken Extinction – Dinosaur Evolution and the Origin of Birds:* W. H. Freeman and Company; New York; 332 p.

Discovery Institute; 1998; The Wedge: Center for the Renewal of Science and Culture, Discovery Institute: at http://www.antievolution.org/features/wedge.pdf; last accessed 03/02/2017. (A more legible copy of The Wedge may be downloaded at the following web address: https://ncse.com/creationism/general/wedge-document; last accessed 03/02/2017)

Dobzhansky, T; 1958; Species after Darwin: in S. A Barnett (ed.); *A Century of Darwin:* Heinemann; London; p. 19-55.

Dobzhansky, T; 1973; Nothing in Biology Makes Sense Except in the Light of Evolution: *The American Biology Teacher;* V. 35; p. 125-129.

Doudna, Jennifer A. and Sternberg, Samuel H.; 2017: *A Crack in Creation – Gene*

Editing and the Unthinkable Power to Control Evolution: Houghton Mifflin Harcourt; New York, New York; 281 p.

Edwards, Governor of Louisiana, et al. v. Aguillard et al.; 1987; No. 85-1513; Supreme Court of the United States; 482 U.S. 578; 107 S. Ct. 2573; 1987 U.S. Lexis 2729; 96 L. Ed. 2d 510; 55 U.S.L.W. 4860. (The entire wording of this court case can be found at the following website: http://www.talkorigins.org/faqs/edwards-v-aguillard.html; accessed 02/21/2017; and in Moore, 2002, p. 323-352.)

Eldredge, Niles; 2005; *Darwin - Discovering the Tree of Life*: W. W. Norton & Company; New York and London; 256 p.

Eldredge, Niles and Gould, Stephen Jay; 1972; Punctuated Equilibria: An Alternative to Phyletic Gradualism; in Schopf, Thomas J.M. (ed.); *Models in Paleobiology*: Freeman Cooper and Company; San Francisco; p. 82-115.

Endler, John A; 1978; A Predator's View of Animal Color Patterns: *Evolutionary Biology*; Vol. 11; p. 319-364.

Endler, John A.; 1980; Natural Selection on Color Patterns in *Poecilia reticulata*: *Evolution*; Vol. 34; Issue 1; p. 76-91.

Englbrecht, Claudia; 2003; Nature and Nuture: Genetics Essay, American Museum of Natural History; New York; 3 p. (This essay was for a course on Genetics, Genomics, and Genethics taught by the American Museum of Natural History.)

Erwin, Douglas H. and Valentine, James W.; 2013; *The Cambrian Explosion - The Construction of Animal Biodiversity*: Roberts and Company; Greenwood Village, Colorado; 406 p.

Fairbanks, Daniel J.; 2012; *Evolving - The Human Effect and Why it Matters:* Prometheus Books; Amherst, New York; 352 p.

Fastovsky, David E. and Weishampel, David B.; 2005; *The Evolution and Extinction of the Dinosaurs (Second Edition)*: Cambridge University Press; Cambridge, U. K. and New York, New York, U. S. A.; 485 p.

Finlayson, Clive; 2009; *The Humans Who Went Extinct – Why Neanderthals Died Out and We Survived:* Oxford University Press; Oxford and New York; 273 p.

Flanagan N. S.; Tobler, A.; Davison, A.; Pybus, O. G.; Kapan, D. D.; Planas, S.; Linares, M.; Heckel, D.; and McMillan, W. O.; 2004; Historical Demography of Müllerian Mimicry in the Neo-tropical *Heliconius* Butterflies: Proceedings of the National Academy of Sciences of the United States of America; Vol.101; p. 9704-9709.

Fogden, Michael and Fogden, Patricia; 1997; *Wildlife of the National Parks and Reserves of Costa Rica*: Foundation Neotropica; San Jose, Costa Rica; 166 p.

Forrest, Barbara; 2001; The Wedge at Work: How Intelligent Design Creationism is Wedging Its Way Into the Cultural and Academic Mainstream: in Pennock, Robert T. (ed.); *Intelligent Design Creationism and Its Critics: Philosophical, Theological, and Scientific Perspectives*: The MIT Press; Cambridge, Massachusetts and London, England; p. 5-53 (805 p. in book).

Forrest, Barbara and Gross, Paul R.; 2004; *Creationism's Trojan Horse – The*

Wedge of Intelligent Design: Oxford University Press; Oxford and New York; 401 p.

Foster, John Bellamy; Clark, Brett; and York, Richard; 2008; *Critique of Intelligent Design – Materialism versus Creationism from Antiquity to the Present*: Monthly Review Press; New York; 240 p.

Frank, J. H.; 1990; Bromeliads and Mosquitoes: Entomology Circular No. 331; Fla. Dept. Agric. & Consumer Serv., Division of Plant Industry; 2 p.

Freeman, Scott and Herron, Jon C.; 2004; *Evolutionary Analysis (3rd ed.)*: Pearson/Prentice Hall; Upper Saddle River, New Jersey; 802 p.

Freeman, Scott and Herron, Jon C; 2007; *Evolutionary Analysis (4th ed.)*: Pearson/Benjamin Cummings; San Francisco, California; 834 p.

Friedman, Gerald M. and Sanders, John E.; 1978; *Principles of Sedimentology*: John Wiley & Sons, Inc.; New York; 808 p.

Friedman, Michael J.; 2006; The Evolution of Hominid Bipedalism: *Honors Projects*; Paper 16; at http://digitalcommons.iwu.edu/socanth_honproj/16; accessed 06/02/2014.

Futuyma, Douglas J; 2009; *Evolution* (2nd ed.): Sinauer Associates, Inc.; Sunderland, Massachusetts; 633 p.

Gaines, Robert R. and Droser, Mary L.; 2003; Paleoecology of the Familiar Trilobite *Elrathia kingii*: An Early Exaerobic Zone Inhabitant: *Geology*; Vol. 31; No. 11; p. 941-944.

Gauthier, Jacques A.; 1986; Saurischian Monophyly and the Origin of Birds: in Padian, Kevin (ed.); The Origin of Birds and the Evolution of Flight: *Memoirs of the California Academy of Sciences*; No. 8; p. 1-55.

Genereux, Diane P. and Bergstrom, Carl T.; 2005; Evolution in Action: Understanding Antibiotic Resistance: in Cracraft, Joel and Bybee, Roger W. (eds.): *Evolution Science and Society: Educating a New Generation*: Revised Proceedings of the BSCS, AIBS Symposium, November 2004, Chicago, Il.; BSCS, Colorado Springs, Colorado; Chapter 13, p. 145-153.

Gee, Henry; 2013; *The Accidental Species - Misunderstandings of Human Evolution*: The University of Chicago Press; Chicago and London; 203 p.

Gibbons, Ann; 2006; *The First Human – The Race to Discover Our Earliest Ancestors*: Published by Doubleday (a division of Random House, Inc.); New York, London, Toronto, Sydney, Auckland; 306 p.

Gibbons, Ann; 2010; Close Encounters Of the Prehistoric Kind: *Science*; Vol. 328; Issue 5979; p. 680-684.

Gibbons, Ann; 2015; Revolution in Human Evolution: *Science*; Vol. 349; Issue 6246; p. 362-366.

Gilbert, Lawrence E.; 1982; The Coevolution of a Butterfly and a Vine: *Scientific American*; Vol. 247; Issue 2; p.110-121.

Gillispie, Charles C.; 1951; *Genesis and Geology - A Study in the Relations of Scientific Thought, Natural Theology, and Social Opinion in Great Britain, 1790-1850*: Harper & Brothers, Publishers; New York; 306 p.

Gingerich, Philip D.; 1979; The Stratophenetic Approach to Phylogeny

Reconstruction in Vertebrate Paleontology: In Cracraft, J. and Eldredge, N. (eds.); p. 41-77; *Phylogenetic Analysis and Paleontology*; Columbia University Press; New York.

Gingerich, Philip D.; 1983; Rates of Evolution: Effects of Time and Temporal Scaling: *Science*; Vol. 222; p. 159-161.

Gingerich, Philip D.; 1993; Quantification and Comparison of Evolutionary Rates: *American Journal of Science*; Vol. 293-A; p. 453-478.

Gingerich, Philip D.; 2009; Rates of Evolution: *Annual Review of Ecology, Evolution, and Systematics;* Vol. 40; p. 657-675.

Godin, J. G. and Dugatkin, L. A.; 1996; Female Mating Preference for Bold Males in the Guppy, *Poecilia reticulata*: *Proc. Natl. Acad. Sci. USA*; Vol. 93, pp. 10262-10267.

Gould, Stephen Jay; 1980; This View of Life: Hen's Teeth and Horse's Toes: *Natural History*; Vol. 89; No. 7; p. 24-28.

Gould, Stephen Jay; 1983: *Hen's Teeth and Horse's Toes – Further Reflections in Natural History*: W. W. Norton & Company; New York and London; 414 p.

Gould, Stephen Jay; 1989; *Wonderful Life – The Burgess Shale and the Nature of History*: W. W. Norton & Company; New York and London; 347 p.

Gould, Stephen Jay; 2002; *The Structure of Evolutionary Theory*: The Belknap Press of Harvard University Press; Cambridge, Massachusetts and London, England; 1433 p.

Gould, Stephen Jay and Eldredge, Niles; 1977; Punctuated Equilibria: The Tempo and Mode of Evolution Reconsidered: *Biology*; V. 3, p. 115-151.

Grant, Bruce S.; 1999: Fine Tuning of the Peppered Moth Paradigm: *Evolution*; Vol. 53; No. 3; p. 980-984.

Grant, Bruce S.; 2004: Allelic Melanism in American and British Peppered Moths: *Journal of Heredity*; Vol. 95; Issue 2; p. 97-102.

Grant, B. Rosemary and Grant, Peter R.; 2003; What Darwin's Finches Can Teach Us about the Evolutionary Origin and Regulation of Biodiversity: *BioScience*; Vol. 53, No. 10; p. 965-975.

Grant, Peter R. and Grant, B. Rosemary; 1995; Predicting Microevolutionary Responses to Directional Selection on Heritable Variation: *Evolution*; Vol. 49; No. 2; p. 241-251.

Grant, Peter R. and Grant, B. Rosemary; 2006; Evolution of Character Displacement in Darwin's Finches: *Science*; VOL 313; 14 July 2006; p. 224-226.

Grant, Peter R. and Grant, B. Rosemary; 2008; *How and Why Species Multiply – The Radiation of Darwin's Finches:* Princeton University Press; Princeton and Oxford; 218 p.

Grant, Peter R. and Grant, B. Rosemary; 2014; 40 Years of Evolution – Darwin's Finches on Daphne Major Island: Princeton University Press; Princeton and Oxford; 400 p.

Grant, K. Thalia and Estes, Gregory B.; 2009; *Darwin in Galapagos - Footsteps to a New World*: Princeton University Press; Princeton and Oxford; 362 p.

Grant, Taran.; Frost, Darrel R.; Caldwell, Janalee P.; Gagliardo, Ron; Haddad, Cello F. B.; Kok, Phillip J. R.; Means, D. Bruce; Noonan, Brice P.; Schargel, Walter E. and Wheeler, Ward C.; 2006; Phylogenetic Systematics of Dart-poison

Frogs and Their Relatives (Amphibia: Athesphatanura: Dendrobatidae): Bulletin of the American Museum of Natural History; No. 299; 262 p.; 79 Figs.; 37 Tables; 8 Appendices.

Green, Richard E.; Krause, Johannes; Briggs, Adrian W.; Maricic, Tomislav; Stenzel, Udo; Kircher, Martin; Patterson, Nick; Li, Heng; Zhai, Weiwei; Fritz, Markus Hsi-Yang; Hansen, Nancy F.; Durand, Eric Y.; Malaspinas, Anna-Sapfo; Jensen, Jeffrey D.; Marques-Bonet, Tomas; 7, Alkan, Ca; Prüfer, Kay; Meyer, Matthias; Burbano, Hernán A.; Good, Jeffrey M.; Schultz, Rigo; Aximu-Petri, Ayinuer; Butthof, Anne; Höber, Barbara; Höffner, Barbara; Siegemund, Madlen; Weihmann, Antje; Nusbaum, Chad; Lander, Eric S.; Russ, Carsten; Novod, Nathaniel; Affourtit, Jason; Egholm, Michael; Verna, Christine; Rudan, Pavao; Brajkovic, Dejana; Kucan, Željko; Gušic, Ivan; Doronichev, Vladimir B.; Golovanova, Liubov V.; Lalueza-Fox, Carles; de la Rasilla, Marco; Fortea, Javier; Rosas, Antonio; Schmitz, Ralf W.; Johnson, Philip L. F.; Eichler, Evan E.; Falush, Daniel; Birney, Ewan; Mullikin, James C.; Slatkin, Montgomery; Nielsen, Rasmus; Kelso, Janet; Lachmann, Michael; Reich, David; Pääbo, Svante; 2010; A Draft Sequence of the Neandertal Genome: *Science*; Vol. 328; Issue 5979; p. 7010-7022.

Greenburg, Michael M.; 1983; Constitutional Issues Surrounding the Science-Religion Conflict in Public Schools: The Anti-Evolution Controversy: Pepperdine Law Review; Vol. 10; Issue 2; Article 4; at http://digitalcommons.pepperdine.edu/plr/vol10/iss2/4; accessed 10/31/2016.

Griffiths, Anthony J. F., Gelbart, William M., Lewontin, Richard C., Miller, Jeffrey H.; 2002; *Modern Genetic Analysis - Integrating Genes and Genomes (Second Ed.)*: W. H. Freeman and Company; New York; 736 p.

Grosse, A. V. and Libby, W. F.; 1947; Cosmic Radiocarbon and Natural Radioactivity of Living Matter: *Science*; V. 106; No. 2743; p. 88-89.

Guerrero, Angeles Gavira and Frances, Peter (senior eds.); 2009; *Prehistoric Life – The Definitive Visual History of Life on Earth*: DK Publishing; New York, New York; 512 p.

Guyer, Craig and Donnelly, Maureen A.; 2005; *Amphibians and Reptiles of La Selva, Costa Rica, and the Caribbean Slope: A Comprehensive Guide*: University of California Press; Berkeley, Los Angeles, and London; 299 p.

Haack, Susan; 2007; *Defending Science – Within Reason: Between Scientism and Cynicism*: Prometheus Books; Amherst, New York; 411 p.

Haack, Susan; 2013; Six Signs of Scientism: Part 1: *Skeptical Inquirer*, V. 37, No. 6; p. 40-45.

Haack, Susan; 2014; Six Signs of Scientism: Part 2: *Skeptical Inquirer*, V. 38, No. 1; p. 43-47.

Haldane, J. B. S; 1949; Suggestions as to Quantitative Measurement of Rates of Evolution: Evolution; Vol. 3; No. 1; p. 51-56.

Hanson, James; 2009; *Storms of My Grandchildren*: Bloomsbury USA Publishers; New York; 304 p.

Hannon, Michael; 2010; Scopes Trial (1925): University of Minnesota Law Library; at http://darrow.law.umn.edu/trialpdfs/SCOPES_TRIAL.pdf; accessed 10/11/2016.

Henig, Robin; 2000; *The Monk in the Garden - The Lost and Found Genius of Gregor Mendel, the Father of Genetics*; Houghton Mifflin Company; Boston and New York; 292 p.

Hennig, Willi (as translated by D. Dwight Davis and Rainer Zangerl); 1966, 1979 (Illinois reissue 1999); *Phylogenetic Systematics*; University of Illinois Press; Urbana, Illinois; 280 p.

Hess, Harry H.; 1962; History of Ocean Basins: in Engel, A. E. J.; James, Harold L.; and Leonard, B. F.: Petrologic Studies: A Volume to Honor A. F. Buddington: Geological Society of America; Boulder, CO; p. 599–620. (This paper may be accessed at the following website: http://www.mantleplumes.org/WebDocuments/Hess1962.pdf, accessed 03/17/2015.)

Hester, Jeff; Burstein, David; Blumenthal, George; Greeley, Ronald; Smith, Bradford; and Voss, Howard G.; 2007; 21ST Century Astronomy: W. W. Norton & Company; New York and London; 641 p. + Appendices.

Hicks, David J.; 1995; Epiphytes - Plants Up A Tree: *South American Explorer*; No. 40; p. 17-20.

Hillis, David M.; 2005; Health Applications of the Tree of Life: in Cracraft, Joel and Bybee, Roger W. (eds.): *Evolution Science and Society: Educating a New Generation*; Revised Proceedings of the BSCS, AIBS Symposium, November 2004, Chicago, Il.; BSCS, Colorado Springs, Colorado; Chapter 12, p. 139-144.

Hillis, David M.; 2007; Constraints in Naming Parts of the Tree of Life: *Molecular Phylogenetics and Evolution;* Vol. 42; p. 331-338.

Hillis, David M. and Wilcox, Thomas P.; 2005; Phylogeny of the New World True Frogs (*Rana*): *Molecular Phylogenetics and Evolution*; Vol. 34; p. 299-314.

Hinkle, Gregory; Wetterer, James K; Schultz, Ted R.; Sogin, Mitchell L; 1994; Phylogeny of the Attine Ant Fungi Based on Analysis of Small Subunit Ribosomal RNA Gene Sequences: *Science*; V. 266; 9 December 1994; p. 1695-1697.

Holdrege, Craig; 2010; The Story of an Organism: Common Milkweed: Copyright 2010 by The Nature Institute; at http://scholar.google.com/scholar_url?url=http%3A%2F%2Fnatureinstitute.org%2Ftxt%2Fch%2Fimages%2Fmilkweed%2FMilkweed.pdf&hl=en&sa=T&oi=gga&ct=gga&cd=0&ei=9ODrVfbqF9OIjAG0y5PQCg&scisig=AAGBfm3Qphvx--IXjtE7l8KvATBNJPw1Hg&nossl=1&ws=1600x675 ; accessed 09/06/2015.

Holland, Heinrich D.; 2006; The Oxygenation of the Atmosphere and Oceans: *Philosophical Transactions of The Royal Society, Biological Sciences*; Vol. 361; p. 903-915.

Holldobler, Bert and Wilson, E. O; 2009; *The Super-Organism – The Beauty, Elegance, and Strangeness of Insect Societies*; W. W. Norton and Company; New York, New York; 522 p.

Holldobler, Bert and Wilson, E. O; 2011; *The Leafcutter Ants – Civilization by Instinct*: W. W. Norton & Company; New York and London; 160 p.

Howard, Jonathan C.; 2009; Why Didn't Darwin Discover Mendel's Laws?: *Journal of Biology*; V. 8, No. 2, Article 15; p. 15.1-15.8. (Originally published online. A pdf version of this article can be found at: http://jbiol.com/content/pdf/jbiol123.pdf).

Humes, Edward; 2007; *Monkey Girl – Evolution, Education, Religion, and the Battle for America's Soul*: Harper Collins Publishers; New York, NY; 380 p.

Hutchins, Robert M. (ed.); 1952; Biographical Note Preceding Galileo Galilee's Dialogues Concerning the Two New Sciences: *Great Books of the Western World*, V. 28, Encyclopedia Britannica, Inc.; p. 125-126.

Irmscher, Christoph; 2013; *Louis Agassiz - Creator of American Science*: Houghton Mifflin Harcourt; Boston and New York; 434 p.

Jackman, Todd R. and Wake, David B.; 1994; Evolutionary and Historical Analysis of Variation in the Blotched Forms of Salamanders of the Ensatina Complex (Amphibia: Plethodontidae): *Evolution*; Vol. 48; No. 3; p. 876-897.

Jinek, Martin; Chylinski, Krzysztof; Fonfara, Ines; Hauer, Michael; Doudna, Jennifer A.; Charpentier, Emmanuelle; 2012; A Programmable Dual-RNA-Guided DNA Endonuclease in Adaptive Bacterial Immunity: *Science*; Vol 337; Issue 6096; p. 816-821.

Johanson, Donald and Edey, Maitland; 1981; *Lucy - The Beginnings of Humankind*: Simon and Schuster; 409 p.

Johnson, George B. and Raven, Peter H.; 1998; *Biology – Principles & Explorations*: Holt, Rinehart, and Winston; Orlando, Florida; 1072 p.

Johnson, Norman A.; 2007; *Darwinian Detectives - Revealing the Natural History of Genes and Genomes*: Oxford University Press; Oxford and New York; 220 p.

Johnson, Phillip E.; 1990; *Evolution as Dogma: The Establishment of Naturalism*: Haughton Publishing Company; Dallas, Texas; 37 p.

Johnson, Phillip E.; 1991; *Darwin on Trial*: InterVarsity Press (Originally published by Regnery Publishing; Washington, D.C.); Downers Grove, Illinois; 247 p.

Jones, Steve; 2000; *Darwin's Ghost - The Origin of Species Updated*: Ballantine Publishing Group, a division of Random House, Inc.; New York; 377 p.

Jones, Steve; 2012; *The Serpent's Promise – The Bible Interpreted Through Modern Science*: Pagasus Books LLC; New York, New York; 436 p.

Joron, Mathieu; Wynne, I. R.; Lamas, G; Mallet, James; 1999; Variable Selection and the Coexistence of Multiple Mimetic Forms of the Butterfly *Heliconius numata*: *Evolutionary Ecology*; Vol. 13; p. 721–754.

Joron, Mathieu; Papa, Riccardo; Beltran, Margarita; Chamberlain, Nicola; Mavarez, Jesus; Baxter, Simon; Abanto, Moises; Bermingham, Eldredge; Humphray, Sean J.; Rogers, Jane; Beasley, Helen; Barlow, Karen; ffrench-Constant, Richard H.; Mallet, James; McMillan, W. Owen; Jiggins, Chris D.; 2006; A Conserved Supergene Locus Controls Colour Pattern Diversity in *Heliconius* Butterflies: *PLoS Biology*; Vol. 4; Issue 10; p. 1831-1840.

Kardong, Kenneth; 2008; *An Introduction to Biological Evolution (2nd ed.)*: The McGraw-Hill Companies, Inc.; New York, New York; 352 p.

Kerr, Peter J.; Ghedin, Elodie; DePasse, Jay V.; Fitch, Adam; Cattadori, Isabella M.; Hudson, Peter J.; Tscharke, David C.; Read, Andrew F.; and Holmes, Edward C.; 2012; Evolutionary History and Attenuation of Myxoma Virus on Two Continents: *PLOS Pathogens*; Vol. 8; Issue 10; 9 p. (This article may be

downloaded at the following website:
http://journals.plos.org/plospathogens/article?id=10.1371/journal.ppat.100295
0; last accessed 11/30/2015).

Kettlewell, H. B. D.; 1955; Selection Experiments on Industrial Melanism in the
Lepidoptera: *Heredity*; Vol. 9; p. 323-342.

Kettlewell, H. B. D.; 1958; A Survey of the Frequencies of *Biston betularia* (L.) (Lep.)
and its Melanic Forms in Great Britain: *Heredity*; Vol. 12; No. 1; p. 51-72.

Kettlewell, H. B. D.; 1965; Insect Survival and Selection for Pattern: *Science*; Vol.
148; No. 3675; p. 1290-1296.

Kimbel, William H.; 2005; The Human Species on the Tree of Life: in Cracraft, Joel
and Bybee, Roger W. (eds.); *Evolution Science and Society: Educating a
New Generation:* Revised Proceedings of the BSCS, AIBS Symposium,
November 2004, Chicago, Il.; BSCS, Colorado Springs, Colorado; Chapter 8,
p. 93-98.

Kious, W. Jacquelyne and Tilling Robert I.; 1996; *This Dynamic Earth – The Story of
Plate Tectonics*: U. S. Government Printing; Washington, D. C.; 76 p. (This
book may be ordered as a paper copy, but an online pdf version may be
obtained at the following web address:
http://pubs.usgs.gov/gip/dynamic/dynamic.pdf.)

Kitzmiller, et al. v. Dover Area School District, et al. 400 F. Supp. 2d
707, Docket No. 4cv2688. (M.D. Pa. 2005).

Kolbert, Elizabeth; 2014; *The Sixth Extinction – An Unnatural History*: Henry Holt
and Company; New York, New York; 319 p.

Kovarik, Alois F.; 1929; Biographical Memoir of Bertram Borden Boltwood 1870-
1927: National Academy of Sciences of the United States of America,
Biographical Memoirs, V. XIV – Third Memoir; p. 69-96. (May be accessed at:
http://www.nasonline.org/publications/biographical-memoirs/memoir-
pdfs/boltwood-bertram-b.pdf.)

Kritsky, Gene; 1992; Darwin's *Archaeopteryx* Prophecy: *Archives of Natural History*;
Vol. 19; Issue 3; p. 407-410.

Kritsky, Gene; 2001; Darwin's Madagascan Hawk Moth Prediction: *American
Entomologist*; Vol. 37; p. 206-210.

Kuchta, Shawn R.; 2005; Experimental Support for Aposematic Coloration in the
Salamander *Ensatina eschscholtzii xanthoptica*: Implications for Mimicry of
Pacific Newts: *Copeia*; 2005; No. 2; p. 265–271.

Kuchta, Shawn R.; Krakauer, Alan H.; and Sinervo, Barry; 2008; Why Does the
Yellow-Eyed Ensatina Have Yellow Eyes? Batesian Mimicry of Pacific Newts
(Genus *Taricha*) by the Salamander *Ensatina eschscholtzii xanthoptica*:
Evolution; Vol. 62; No. 4; p. 984-990.

Kuchta, Shawn R.; Parks, Duncan S.; Mueller, Rachel Lockridge; and Wake, David
B.; 2009a; Closing the Ring: Historical Biogeography of the Salamander Ring
Species *Ensatina eschscholtzii*: *Journal of Biogeography*; Vol. 36; p. 982–
995.

Kuchta, Shawn R.; Parks, Duncan S.; and Wake, David B.; 2009b; Pronounced
Phylogeographic Structure On a Small Spatial Scale: Geomorphological
Evolution and Lineage History In the Salamander Ring Species *Ensatina*

eschscholtzii In Central Coastal California: *Molecular Phylogenetics and Evolution*; Vol. 50; p. 240-255.

Lamarck, Jean Baptiste (Translated by Carozzi, Albert V.); 1964; *Hydrogeology (illustrated edition):* University of Illinois Press; 152 p.

Lane, Nick; 2009; *Life Ascending – The Ten Great Inventions of Evolution:* W. W. Norton & Company; New York and London; 344 p.

Lang, J. M. and Benbow, M. E.; 2013; Species Interactions and Competition: *Nature Education Knowledge* (Nature Group Publication); Vol. 4; No. 4; 8 (at the following website: http://www.nature.com/scitable/knowledge/library/species-interactions-and-competition-102131429; accessed 10/27/2015).

Larson, Edward J.; 1997; *Summer of the Gods – The Scopes Trial and America's Continuing Debate Over Science and Religion*: BasicBooks (A member of the Perseus Books Group); New York, New York; 318 p.

Larson, Edward J.; 2003; *Trial and Error – The American Controversy Over Creation and Evolution, Third Edition:* Oxford University Press; Oxford, England and New York, New York; 276 p.

Larson, Edward J.; 2004; *Evolution - The Remarkable History of a Scientific Theory*: Modern Library (a division of Random House, Inc.); New York; 337 p.

Lawrence, Eleanor; 1998; Iguanas Ride the Waves: *Nature* (News); Published Online 15 October 1998; at http://www.nature.com/news/1998/981015/full/news981015-3.html; accessed 05/05/2015.

Lebo, Lauri; 2008; *The Devil in Dover – An Insider's Story of Dogma v. Darwin in Small-Town America*: The New Press; New York and London; 238 p.

Lee, Michael S. Y.; Cau, Andrea; Naish, Darren; and Dyke, Gareth J.; 2014; Sustained Miniaturization and Anatomical Innovation in the Dinosaurian Ancestors of Birds; *Science*; 1 August 2014; Vol. 345; Issue 6196; p. 562-566.

Lee, Michael S. Y.; Soubrier, Julien; and Edgecombe, Gregory D.; 2013; Rates of Phenotypic and Genomic Evolution during the Cambrian Explosion: *Current Biology;* Vol. 23; Issue 19; p. 1889-1895. (an abstract of this article may be found at the following website: http://www.cell.com/current-biology/abstract/S0960-9822(13)00916-0).

Leenders, Twan; 2001; *A Guide to Amphibians and Reptiles of Costa Rica*: Distribuidores Zona Tropical; Miama, Florida; 305 p.

Lemon v. Kurtzman; 1971; 403 US 602, 91 S. Ct. 2105, 29 L. Ed. 2d 745; Supreme Court (1971).

Levin, Harold L.; 1999; *The Earth Through Time* (sixth ed.): Saunders College Publishing – Harcourt Brace College Publishers; Orlando, Florida; 568 p.

Linder, Douglas O.; 2008; State v. John Scopes ("The Monkey Trial") – An Introduction: at http://law2.umkc.edu/faculty/projects/ftrials/scopes/scopes.htm; accessed 10/15/2016.

Losos, Jonathan B.; 2017; *Improbable Destines – Fate, Chance, and the Future of Evolution*: Riverhead Books (An imprint of Penguin Random House LLC); New York, New York; 368 p.

Love, J. David; Reed, Jr., John C.; and Pierce, Kenneth L.; 2007; *Creation of the Teton Landscape:* U. S. Geological Survey; 132 p.

Lucas, Spencer; 2005; *Dinosaurs: The Textbook* (5th ed.): McGraw-Hill Publishers; 304 p.

Maan, Martine E. and Cummings, Molly E.; 2009; Sexual Dimorphism and Directional Sexual Selection On Aposematic Signals in a Poison Frog: Proceedings of the National Academy of Sciences; Vol. 106; No. 45; p.19072–19077. This article contains supporting information online at www.pnas.org/cgi/content/full/0903327106/DCSupplemental.

MacFadden, Bruce J.; 1992; *Fossil Horses: Systematics, Paleobiology, and Evolution of the Family Equidae*: Cambridge University Press; Cambridge and New York; 369 p.

MacFadden, Bruce J.; 2005; Fossil Horses – Evidence of Evolution: *Science*; 18 March 2005; Vol. 307; No. 5716; p. 1728-1730.

Majerus, Michael E. N.; 2009; Industrial Melanism in the Peppered Moth, *Biston betularia*: An Excellent Teaching Example of Darwinian Evolution in Action: *Evolution Education Outreach*; Vol. 2; p. 63-74.

Mallet, James; 2010; Shift Happens! Shifting Balance and the Evolution of Diversity in Warning Colour and Mimicry: *Ecological Entomology*; Vol. 35; Suppl. 1; p. 90–104.

Margulis, Lynn; 1998; *Symbiotic Planet – A New View of Evolution*: Published by Basic Books (A Member of the Perseus Book Group); New York, New York; 147 p.

Markel, Howard; 2013; The Real Story Behind Penicillin: The PBS Newshour; at http://www.pbs.org/newshour/rundown/the-real-story-behind-the-worldsfirstantibiotic/; accessed 07/14/2017

Marshall, Charles R.; 1999; Missing Links in the History of Life: in Scotchmoor, Judy and Springer, Dale (eds.); *Evolution – Investigating the Evidence*; The Paleontological Society: Vol. 9; 1999; Printed by New Image Press; Pittsburgh, PA; p. 119-144.

Martin, Anthony; 2012; Out of One's Depth in the Ediacaran (Tag Archives: *Treptichnus pedum*): at http://www.georgialifetraces.com/tag/treptichnus-pedum/; accessed 12/02/2014.

Martin, Robert A.; 2004; *Missing Links – Evolutionary Concepts & Transitions Through Time*: Jones and Bartlett Publishers; Sudbury, Massachusetts; 303 p.

Matzke, Nick; 2004; Icon of Obfuscation: Jonathan Wells' book Icons of Evolution and Why Most of What it Teaches About Evolution is Wrong: Chapter 5: Haeckel's Embryos: in *The TalkOrigins Archive* at http://www.talkorigins.org/faqs/wells/iconob.html#haeckel-embryo; accessed 08/13/2014.

Mayell, Hillary; 2004; *Fossil Pushes Upright Walking Back 2 Million Years, Study Says*: National Geographic News; at http://news.nationalgeographic.com/news/2004/09/0902_040902_upright_hominid.html, accessed 06/01/2014.

Mayr, Ernst; 2001; *What Evolution Is*: Basic Books (A Member of the Perseus Books

Group); New York, New York; 318 p.

Mayr, Ernst; 1991; *One Long Argument - Charles Darwin and the Genesis of Modern Evolutionary Theory*: Harvard University Press; Cambridge, Massachusetts; 195 p.

McLean v. Arkansas Board of Education; 1982; 529 F. Supp. 1255, 1258-1264 [ED Ark.]. (Judge Overton's entire Memorandum Opinion may be found at the following web site: http://www.talkorigins.org/faqs/mclean-v-arkansas.html, accessed 02/14/2017; and also in Moore, 2002, p. 305-323.)

McGinnis, Helen J.; 1984; *Carnegie's Dinosaurs*: University of Pittsburgh Press: Pittsburgh, PA; 119 p.

McNeil, Donald G., Jr.; 2010; Precursor to H.I.V. Was in Monkeys for Millenniums: *The New York Times*; September 16, 2010.

Menard, H. W.; 1967; Sea Floor Spreading, Topography, and the Second Layer: *Science*; Vol. 157; No. 3791; p. 923-924.

Mendel, Gregor; 1866; Versuche über Pflanzen-Hybriden: Verhandlungen des naturforschenden Vereines, Abhandlungen, Brünn 4, p. 3-47. (See English annotated English translation at http://www.mendelweb.org/Mendel.html.)

Merrill, R. M.; Dasmahapatra, K. K.; Davey, J. W.; Dell'Aglio, D. D.; Hanly, J. J.; Huber, B.; Jiggins, C. D.; Joron, M.; Kozak, K. M.; Llaurens, V.; Martin; S. H.; Montgomery; S. H.; Morris; J.; Nadeau; N. J; Pinharanda, A. L.; Rosser, N.; Thompson, M. J.; Vanjari, S.; Wallbank, R. W. R; and Yu, Q.; 2015; The Diversification of *Heliconius* Butterflies: What Have We Learned in 150 Years?: *Journal of Evolutionary Biology*; Vol. 28; Issue 8; p. 1417–1438.

Meuche, Ivonne; Brusa, Oscar; Linsenmair, K. Eduard; Keller, Alexander; and Pröhl, Heike; 2013; Only Distance Matters – Non-choosy Females in a Poison Frog Population: *Frontiers in Zoology*; Vol. 10; p. 1-16. (The electronic version of this article can be found online at: http://www.frontiersinzoology.com/content/10/1/29.)

Meyer, Axel; 2006; Repeating Patterns of Mimicry: *PLoS Biology*; Vol. 4; Issue 10; p. 1675-1677. (A pdf of this article may be found at the following website: http://www.plosbiology.org/article/fetchObject.action?uri=info:doi/10.1371/journal.pbio.0040341&representation=PDF.)

Meyer, Stephen C.; 2013; *Darwin's Doubt – The Explosive Origin of Animal Life and the Case for Intelligent Design*: Harper Collins Publishers; New York, New York; 498 p.

Meyers, Charles W.; 1987; New Generic Names for Some Neotropical Poison Frogs (Dendrobatidae): *Papeis A vulsos de Zoologia*; Museu de Zoologia da Universida de de Sao Paulo; Vol. 36; No. 25; p. 301-306.

Meyers, Charles W. and Daly, John W.; 1983; Dart-Poison Frogs: *Scientific American*; Vol. 248; Issue 2; p. 120-133.

Meyers, Charles W.; Daly, John W.; and Martinez, Victor; 1984; An Arboreal Poison Frog (*Dendrobates*) from Western Panama: American Museum of Natural History; *American Museum Novitates*; Number 2783, p. 1-20.

Meyers, P. Z.; 2009; Darwin and the Vermiform Appendix: *Pharyngula*: at http://scienceblogs.com/pharyngula/2009/08/22/darwin-and-the-vermiform-appen/; accessed 08/23/2014.

Miller, Kenneth R. and Levine, Joseph S.; 2008; Prentice Hall *Biology*: Pearson Prentice Hall; Upper Saddle River, New Jersey and Boston, Massachusetts; 1144 p.

Miller, Stanley L., 1953, A Production of Amino Acids Under Possible Primitive Earth Conditions: *Science*; Vol. 117; Issue 3046; p. 528-529.

Miller, Stanley L. and Urey, Harold C.; 1959; Organic Compound Synthesis on the Primitive Earth: *Science*, Vol. 130, Issue 3370, p. 245-251.

Miller, Stephen A. and Harley, John P.; 2007; Zoology (7th ed.): The McGraw Hill Companies, Inc.; New York, New York, 588 p.

Miller, Stephen A. and Harley, John P.; 1994; *Zoology (2nd ed.):* Wm. C. Brown Publishers; Dubuque, Iowa; 664 p.

Monroe, James S. and Wicander, Reed; 2011; *The Changing Earth – Exploring Geology and Evolution (6th ed.):* Cengage Learning; Independence, Ky.; 736 p.

Monroe, James S. and Wicander, Reed; 2005; *Physical Geology – Exploring the Earth, 5th Edition*: Thomson – Brooks/Cole; United States (Belmont, CA); 644 p.

Montgomery, David R.; 2012; *The Rocks Don't Lie – A Geologist Investigates Noah's Flood*: W. W. Norton & Company; New York, London; 302 p.

Moore, Randy; 2002; *Evolution in the Courtroom – A Reference Guide*: ABC-CLIO, Inc.; Santa Barbara, California. Denver, Colorado. Oxford, England; 381 p.

Moore, Raymond C.; Lalicker, Cecil G.; and Fischer, Alfred G.; 1952; Invertebrate Fossils: McGraw-Hill Book Company, Inc.; New York, Toronto, London; 766 p.

Moritz C.; Schneider, C. J.; and WAKE David B; 1992; Evolutionary Relationships Within the *Ensatina eschscholtzii* Complex Confirm the Ring Species Interpretation: *Systematic Biology*; Vol. 41; p. 273-291.

Muhs, Daniel R.; 2013; The Contributions of Donald Lee Johnson to Our Understanding of the Quaternary History of the California Channel Islands: *Geological Society of America Abstracts with Programs;* Vol. 45, No. 7; p.632.

Müller, Fritz; 1879; Ituna and Thyridia; A Remarkable Case of Mimicry in Butterflies; *Proceedings of the Entomological Society of London*; 1879; p. xx–xxix.

National Academy of Sciences; 1999; *Science and Creationism – A View from the National Academy of Sciences (Second Edition)*: National Academy Press; Washington, D. C.; 48 p. (A free PDF copy of this title may be downloaded at http://www.nap.edu/catalog/6024.html.)

National Academy of Sciences; 2010; Understanding Climate's Influence on Human Evolution: National Academy Press; Washington, D. C.; 128 p.

National Center for Science Education; 2008; The Wedge Document: at http://ncse.com/creationism/general/wedge-document; accessed 05/16/2014.

National Human Genome Research Institute; 2003; Human Genome Project: at http://www.genome.gov/; accessed 06/05/2014.

Newman, Arnold; 2008; *Tropical Rainforest – Our Most Valuable and Endangered Habitat with a Blueprint for Its Survival into the Third Millennium*: Checkmark Book, An Imprint of Facts and File, Inc.; New York, New York; 260 p.

Nichol, John; 1990; *The Mighty Rainforest*: David and Charles Publishers; London;

200 p.

Nishida, Ritsuo; 2002; Sequestration of Defensive Substances from Plants by Lepidoptera: *Annual Review of Entomolology*; Vol. 47; p. 57–92.

Normile, Dennis; 2017; Genome Editing - China Sprints Ahead in CRISPR Therapy Race: *Science*, News – In Depth; Vol. 358; Issue 6359; p. 20-21.

NOVA; 2009; The DNA of Human Evolution: at http://www.pbs.org/wgbh/nova/evolution/dna-human-evolution.html, accessed 10/06/2010.

Novgorodova, T. A.; 2002; Study of Adaptations of Aphids (Homoptera, Aphidinea) to Ants: Comparative Analysis of Myrmecophilous and Nonmyrmecophilous Species: *Entomological Review*; Vol. 82; No. 5; p. 569–576.

Numbers, Ronald L.; 2006; *The Creationists – From Scientific Creationism to Intelligent Design - Expanded Edition):* Harvard University Press; Cambridge, Massachusetts and London, England; 606 p.

Offenberg, Joachim; 2001; Balancing between Mutualism and Exploitation: the Symbiotic Interaction Between *Lasius* Ants and Aphids: *Behavioral Ecology and Sociobiology;* Vol. 49; Issue 4; p. 304-310.

Ogg, James G.; Ogg, Gabi; and Gradstein, Felix M; 2008; *The Concise Geologic Time Scale*: Cambridge University Press; Cambridge, UK; 177 p.

Omar, Gomaa I. and Steckler, Michael S.; 1995; Fission Track Evidence on the Initial Rifting of the Red Sea: Two Pulses, No Propagation: *Science*; Vol. 270; No. 5240; p. 1341-1344.

Pääbo, Svante; 2014; *Neanderthal Man – In Search of Lost Genomes*: Basic Books, A Member of the Perseus Books Group; New York; 275 p.

Paley, William; 1802 (with an Introduction and Notes by Mathew D. Eddy and David Knight [eds.], 2006, reissued 2008); *Natural Theology or Evidence of the Existence and Attributes of the Deity, Collected from the Appearances of Nature*: Oxford University Press; Oxford and New York; 342 p.

Papagianni, Dimitra and Morse, Michael A.; 2013; *The Neanderthals Rediscovered – How Modern Science is Rewriting Their Story*: Thames & Hudson Ltd. in London, England and Thames & Hudson Inc. in New York, New York; 208 p.

Paul, Gregory S.; 1988; *Predatory Dinosaurs of the World – A Complete Illustrated Guide* (A New York Academy of Sciences Book): Simon and Schuster; New York, New York; 464 p.

Pauley, Thomas K., 2014; West Virginia Wildlife Resources: Toads and Frogs: West Virginia Division of Natural Resources; Online Booklet; 13 p.; at http://www.wvdnr.gov/publications/toadsfrogs/index.html; accessed 09/01/2014.

Pauly, Gregory B.; Hillis, David M.; and Cannatella, David C.; 2009; Taxonomic Freedom and the Role of Official Lists of Species Names: Herpetologica; Vol. 65; No. 2; p. 115-128.

PBS; Penzias and Wilson Discover Cosmic Microwave Radiation1965: http://www.pbs.org/wgbh/aso/databank/entries/dp65co.html: accessed 10/03/2014.

Peng, S.; Babcock, L. E.; and Cooper, R. A.; 2012; The Cambrian Period: in

Gradstein, Felix; Ogg, James; Schmitz, Mark; and Ogg, Gabi (eds.); *The Geologic Time Scale 2012* (2 Volume Set): Elsevier; The Netherlands; 1176 p.

Pennock, Robert T.; 2001; Why Creationism Should Not Be Taught in the Public Schools: in Pennock, Robert T. (ed.): *Intelligent Design Creationism and Its Critics – Philosophical, Theological, and Scientific Perspectives*; The MIT Press; Cambridge, Massachusetts and London, England; Chapter 35; p. 755-777.

Pennock, Robert T.; 2005; On Teaching Evolution and the Nature of Science: in Cracraft, Joel and Bybee, Roger W. (eds.): *Evolution Science and Society: Educating a New Generation*: Revised Proceedings of the BSCS, AIBS Symposium, November 2004, Chicago, Il.; BSCS, Colorado Springs, Colorado; p. 7-12.

Pereira, Ricardo J.; Monahan, William B.; and Wake, David B.; 2011; Predictors for Reproductive Isolation in a Ring Species Complex Following Genetic and Ecological Divergence: *BioMed Central Evolutionary Biology*; Vol. 11: 194. (doi:10.1186/1471-2148-11-194) (The electronic version of this article is the complete one and can be found online at: http://www.biomedcentral.com/1471-2148/11/194).

Pereira, Ricardo J. and Wake, David B.; 2009; Genetic Leakage After Adaptive and Nonadaptive Divergence in the *Ensatina eschscholtizi* Ring Species: *Evolution*; Vol. 63; No. 9; p. 2288–2301.

Perlman, Robert L.; 2013; *Evolution and Medicine*: Oxford University Press; Oxford, United Kingdom; 162 p.

Pinto-Tomás, Adrián A.; Anderson, Mark A.; Suen, Garret; Stevenson, David M.; Chu, Fiona S. T; Cleland, W. Wallace; Weimer, Paul J.; and Currie, Cameron R.; 2009; Symbiotic Nitrogen Fixation in the Fungus Gardens of Leaf-Cutter Ants: *Science*; Vol. 326; No. 5956; p. 1120-1123.

Pond, Caroline M.; Silvertown, Jonathan; Robinson, David; Skelton, Peter; Nettle, Daniel; Grady, Monica; and Slapper, Gary; 2008; *99% Ape – How Evolution Adds Up*: The University of Chicago Press; Chicago and London; 224 p.

Popper, Karl; 2002 (First published 1935; first English edition published 1959); *The Logic of Scientific Discovery*: Routledge Publishers (in Routledge Classics); London and New York; 513 p.

Porter, Susannah M.; 2004; Closing the Phosphatization Window: Testing for the Influence of Taphonomic Megabias on the Pattern of Small Shelly Fossil Decline: *Palaios*, 2004, V. 19, p. 178–183.

Poulsen, Michael and Currie, Cameron R.; 2010; Symbiont Interactions in a Tripartite Mutualism: Exploring the Presence and Impact of Antagonism between Two Fungus-Growing Ant Mutualists: PLoS ONE; Vol. 5; Issue 1; e8748; doi:10.1371/journal.pone.0008748; This article may by viewed at the following website: http://currielab.wisc.edu/pubs/PoulsenCurrie2010.pdf; accessed 08/13/2014.

Powell, R. L., Burton, D. K., Crisp, E. L., and Stone, D. D.; 2001; Description and

Preliminary Taphonomic Implications of the Partial Skeletal Remains of a Diplodocid Sauropod from the Upper Jurassic Morrison Formation of Utah: *Geol. Soc. Amer. Abstracts with Programs*; Vol. 33, No. 2.

Priscu, John C.; accessed 2014; Origin and Evolution of Life on a Frozen Earth: in Evolution of Evolution - 150 Years of Darwin's "On the Origin of Species"; National Science Foundation; at http://www.nsf.gov/news/special_reports/darwin/textonly/polar_essay1.jsp; accessed 09/25/2014.

Pröhl, Heike and Hödl, Walter; 1999; Parental Investment, Potential Reproductive Rates, and Mating System in the Strawberry Dart-Poison Frog, *Dendrobates pumilio*: *Behavioral Ecology and Sociobiology*; Vol. 46; No. 4; p. 215-220.

Prothero, Donald R.; 2004; *Bringing Fossils to Life - An Introduction to Paleobiology (2nd ed.):* McGraw-Hill; New York, New York; 503 p.

Prothero, Donald R.; 2007; *Evolution - What the Fossils Say and Why It Matters*: Columbia University Press; New York; 381 p.

Prothero, Donald R.; 2013; *Reality Check – How Science Deniers Threaten Our Future*: Indiana University Press; Bloomington and Indianapolis; 369 p.

Prothero, Donald R. and Schwab, Fred; 1996; *Sedimentary Geology – An Introduction to Sedimentary Rocks and Stratigraphy*; W. H. Freeman and Company; New York; 575 p.

Prudic, Kathleen L.; Khera, Smriti; Solyom, Aniko; and Timmermann, Barbara N.; 2007; Isolation, Identification, and Quantification of Potential Defensive Compounds in the Viceroy Butterfly and its Larval Host-Plant, Carolina Willow: *Journal of Chemical Ecology;* Vol. 33; p. 1149–1159.

Quammen, David; 1996;*The Song of the Dodo – Island Biogeography in an Age of Extinctions*: Scribner (Scribner and design are trademarks of Macmillan Library Reference USA, Inc., under license by Simon & Schuster; the publisher of this work.); New York, New York; 702 p.

Quammen, David; 2004; Was Darwin Wrong?: *National Geographic*; November 2004 (This article may be read online at: http://ngm.nationalgeographic.com/2004/11/darwin-wrong/quammen-text, accessed 04/06/2014)

Quammen, David (General Editor); 2008; *Charles Darwin – On the Origin of Species – The Illustrated Edition*: Sterling Publishing Co., Inc.; New York, New York; 544 p.

Repcheck, Jack; 2003; *The Man Who Found Time - James Hutton and the Discovery of the Earth's Antiquity*: Perseus Publishing, Perseus Book Group; Cambridge MA; 247 p.

Reznick, David N.; 2010; *The* Origin *Then and Now - An Interpretive Guide to the* Origin of Species: Princeton University Press; Princeton and Oxford; 432 p.

Reznick, David and Endler, John A.; 1982; The Impact of Predation on Life History in Trinidadian Guppies (*Poecilia reticulata*): *Evolution*; Vol. 36; No. 1; p. 160-177.

Richardson, Michael K. and Keuck, Gerhard; 2002; Haeckel's ABC of Evolution and Development: *Biological Review Cambridge Philosophical Society*; Vol. 77; p. 495-528.

Richards-Zawacki, Corinne L. and Cummings, Molly E.; 2010; Intraspecific Reproductive Character Displacement in a Polymorphic Poison Dart Frog, *Dendrobates pumilio*: *Evolution*; Vol. 65; No. 1; p. 259–267.

Richards-Zawacki, Corinne L.; Wang, Ian J.; and Summers, Kyle; 2012; Mate Choice and the Genetic Basis for Colour Variation in a Polymorphic Dart Frog: Inferences from a Wild Pedigree: *Molecular Ecology*; Vol. 21, p. 3879–3892.

Ridley, Mark; 1996; *Evolution, Second Edition*: Blackwell Science, Inc.; Cambridge, Massachusetts; 719 p.

Ridley, Matt; 1993; *The Red Queen - Sex and the Evolution of Human Nature*: Harper-Collins Publishers; New York; 405 p.

Ringler, Eva; Pašukonis, Andrius; Hödl, Walter; and Ringler, Max; 2013; Tadpole Transport Logistics in a Neotropical Poison Frog: Indications for Strategic Planning and Adaptive Plasticity in Anuran Parental Care: *Frontiers in Zoology*; Vol. 10; p. 1-9. (The electronic version of this article is the complete one and can be found online at: http://www.frontiersinzoology.com/content/10/1/67.)

Ritland, David B. and Brower, Lincoln P.; 1991; The Viceroy Butterfly is Not a Batesian Mimic: *Nature*; Vol. 350; Issue 6318; p. 497-498.

Rivas-San Vicente, Mariana and Plasencia, Javier; 2011; Salicylic Acid Beyond Defence: Its Role in Plant Growth and Development: *Journal of Experimental Botany*, Vol. 62, No. 10, p. 3321–3338.

Robinson, Tara Rodden; 2005; *Genetics for Dummies*: Wiley Publishing, Inc.; Indianapolis, Indiana; 364 p.

Rockwood, Larry L.; 2006; *Introduction to Population Ecology*: Blackwell Publishing; Malden, MA, U.S.A., Oxford, U.K., Carlton, Victoria, Australia; 339 p.

Romer, Alfred Sherwood; 1962; *The Vertebrate Body (3rd ed.)*: W. B. Saunders Company; Philadelphia and London; 627 p.

Rowan, Rob; 1998; Review – Diversity and Ecology of Zooxanthellae on Coral Reefs: *Journal of Phycology*; Vol. 34; Issue 3; p. 407–417.

Rudge, David W.; 2005; The Beauty of Kettlewell's Classic Experimental Demonstration of Natural Selection: *Bioscience*; Vol. 55; Issue 4; p. 369-375.

Ruse, Michael; 2005; *The Evolution-Creation Struggle*: Harvard University Press; Cambridge, Massachusetts and London, England; 327 p.

Ruse, Michael; 2009a; The History of Evolutionary Thought: in Ruse, Michael and Travis, Joseph (eds.); *Evolution -The First Four Billion Years*; 979 p.; The Belknap Press of Harvard University Press; Cambridge, Massachusetts and London, England; p. 1-48.

Ruse, Michael; 2009b; Myth 23 – That "Intelligent Design" Represents a Scientific Challenge to Evolution: in Numbers, Ronald L. (ed.); Galileo Goes to Jail – And Other Myths About Science and Religion; 302 p.; Harvard University Press; Cambridge, Massachusetts and London, England; p. 206-223.

Sagan, Carl; 1996; *The Demon-Haunted World – Science as a Candle in the Dark*: Ballantine Books, Published by Random House Publishing Group (a division of Random House, Inc.); New York; 457 p.

Sagan, Lynn; 1967; On the Origin of Mitosing Cells: *Journal of Theoretical Biology*;

Vol. 14; p. 225-274.

Saporito, Ralph A.; Zuercher, Rachel; Roberts, Marcus; Gerow, Kenneth G.; and Donnelly, Maureen A.; 2007; Experimental Evidence for Aposematism in the Dendrobatid Poison Frog *Oophaga pumilio*: *Copeia*; No. 4; p.1006–1011.

Sáyago, Roberto; Lopezaraiza-Mikel, Martha; Quesada, Mauricio; Álvarez-Añorve, Mariana Yolotl; Cascante-Marín, Alfredo; Bastida; Jesus Ma; 2013; Evaluating Factors That Predict the Structure of a Commensalistic Epiphyte–Phorophyte Network: *Proceedings of the Royal Society B: Biological Sciences*; Vol. 280; p. 1-9. (May be downloaded at the following website: http://rspb.royalsocietypublishing.org/content/280/1756/20122821; accessed 10/02/2015.)

Schopf, J. William; 2011; The Paleobiological Record of Photosynthesis: *Photosynthesis Research*; Vol. 107; No. 1; p. 87-101.

Schulz, Sandra S. and Wallace, Robert E.; 2013; The San Andreas Fault: at http://pubs.usgs.gov/gip/earthq3/; accessed 04/22/2015.

Scott, Eugenie C.; 2004; *Evolution Vs. Creationism – An Introduction*: Greenwood Press; Westport, CT; 272 p.

Scott, Eugenie C.; 2016; Chapter 18 (p. 284-299) Creationism and Intelligent Design: in Losos, Jonathan B. and Lenski, Richard E. (eds.); 2016; *How Evolution Shapes Our Lives – Essays on Biology and Society*: Princeton University Press; Princeton and Oxford; 396 p.

Scott, Eugenie C. and Branch, Glen (eds.); 2006; *Not in Our Classrooms – Why Intelligent Design is Wrong for Our Schools*: Beacon Press; Boston, Mass; 184 p.

Secord, James A. (ed.); 1997; *Charles Lyell's Principles of Geology*: Penguin Classics (a division of Penguin Books) (Originally published as *Principles of Geology* by John Murray, 1830-33 in three volumes); London, England; 472 p. (Includes an introduction by James A. Secord, ed.).

Sepkoski, J. John, Jr.; 1993; Chapter One, Foundations - Life in the Oceans: in Gould, Stephen J. (ed.), 1993, *The Book of Life*: W. W. Nortan & Company, New York and London, 256 p.; Chapter One, p. 36-63.

Shaw, Kerry L.; Mendelson, Tamra C.; and Borgia, Gerald; 2005; Evolution By Sexual Selection: in Cracraft, Joel and Bybee, Roger W. (eds.): *Evolution Science and Society: Educating a New Generation:* Revised Proceedings of the BSCS, AIBS Symposium, November 2004, Chicago, Il.; BSCS, Colorado Springs, Colorado; Chapter 9, p. 99-106.

Shipman, Pat; 1998; *Taking Wing – Archaeopteryx and the Evolution of Bird Flight*: Simon & Schuster; New York, New York; 336 p.

Shingleton, Alexander W.; Stern, David L.; and Foster, William A.; 2005; The Origin of a Mutualism: a Morphological Trait Promoting the Evolution of Ant-Aphid Mutualisms: *Evolution*; Vol. 59; No. 4; p. 921–926.

Sherratt, Thomas N.; 2008; The Evolution of Müllerian Mimicry: *Naturwissenschaften*: Vol. 95; p. 681–695.

Singham, Mano; 2009; *God vs. Darwin: The War Between Evolution and Creationism in the Classroom*; Rowan & Littlefield Publishers, Inc.; London, New York, Toronto, Plymouth (UK); 171 p.

Skybreak, Ardea; 2006; *The Science of Evolution and The Myth of Creationism*: Insight Press; Chicago, Illinois; 338 p.

Slack, Gordy; 2007; *The Battle Over the Meaning of Everything – Evolution, Intelligent Design, and a School Board in Dover PA*: John Wiley & Sons; San Francisco, California; 228 p.

Smith, H. F.; Fisher, R. E.; Everett, M. L.; Thomas, A. D.; Bollinger, R. Randall; Parker, W.; 2009; Comparative Anatomy and Phylogenetic Distribution of the Mammalian Cecal Appendix: *Journal of Evolutionary Biology*; Vol. 22; Issue 10; October 2009; p. 1984-1999.

Smith, Thomas Bates; 1990; Natural Selection of Bill Characters in the Two Bill Morphs of the African Finch *Pyrenestes ostrinus*: *Evolution*; Vol. 44; No. 4; July 1990; p. 832-842.

Smith, Thomas Bates; 1993; Disruptive Selection and the Genetic Basis of Bill Size Polymorphism in the African Finch *Pyrenestes*: *Nature*; Vol. 363; June 1993; p. 618-620.

Sobel, Dava; 2000; *Galileo's Daughter:* Penguin; 420 p.

Specter, Michael (Photographs by Girard, Greg); 2016; DNA Revolution: *National Geographic*; Vol. 230; No. 2; p. 30-55.

Spencer, Matt and Crisp, Edward L.; 2002; Description and Preliminary Taphonomy of the Partial Skeletal Remains of a Juvenile Diplodocid Sauropod from the Jurassic Morrison Formation of Utah: *Geol. Soc. Amer. Abstracts with Programs*; Vol. 34; No. 2; p. A-96.

Stanley, Steven M.; 1981; *The New Evolutionary Timetable - Fossils, Genes, and the Origin of Species*: Basic Books, Inc., Publishers; New York; 222 p.

State of Tennessee; 1925; The Butler Act – An Act Prohibiting the Teaching of Evolution Theory in the Universities, Normals and all Other Public Schools of Tennessee…and to Provide Penalties for the Violations Thereof: Public Acts of the State of Tennessee; Chapter No. 27; House Bill No. 185; Passed by the Sixty-Fourth General Assembly; Signed Into Law by Governor Austin Peay, March 21, 1925.

Stebbins, Robert C.; 1949; Speciation in Salamanders of the Plethodontid Genus *Ensatina*: *University of California Publications in Zoology*; Vol. 48; p. 377-526.

Stebbins, Robert C.; 1957; Intraspecific Sympatry in the Lungless Salamander *Ensatina eschscholtzii*: *Evolution*, Vol. 11, No. 3; p. 265–270.

Stone, Irving; 1980; *The Origin - A Biographical Novel of Charles Darwin*: Doubleday and Company, Inc.; Garden City, New York; 743 p.

Stott, Rebecca; 2012; *Darwin's Ghosts -The Secret History of Evolution*: 2013 Spiegel & Grau Trade Paperback Edition; Spiegel & Grau Trade Paperbacks; New York; 402 p.

Stringer, Chris; 2012a; *Lone Survivors – How We Came to be the Only Humans on Earth*: St. Martin's Griffin; New York; 320 p.

Stringer, Chris; 2012b; The status of *Homo heidelbergensis* (Schoetensack 1908): *Evolutionary Anthropology: Issues, News, and Reviews*; Vol. 21; Issue 3; p. 101-107.

Sulloway, Frank J.; 2005; The Evolution of Charles Darwin: *Smithsonian*; Vol. 36; No. 9; p. 58-69.

Summers, Kyle; 1992; Mating Strategies in Two Species of Dart-poison Frogs: A Comparative Study: *Animal Behavior*; Vol. 43; p. 907-919.

Summers, K.; Bermingham, E.; Weigt, L.; McCafferty, S.; and Dahlstrom, L.; 1997; Phenotypic and Genetic Divergence in Three Species of Dart-Poison Frogs with Contrasting Parental Behavior: *Journal of Heredity*; Vol. 88; Issue 1; p. 8-13.

Suttner, Lee J.; 1974; *Laboratory Exercises on Principles of Interpretation of the Evolution of the Earth*: Burgess Publishing Co.; 155 p.

Tarbuck, Edward J., Lutgens, Frederick K., and Tasa, Dennis G.; 2013; *Earth Science (13th ed.):* Prentice Hall; Upper Saddle River, New Jersey; 768 p.

Tattersall, Ian; 2012; *Masters of the Planet – The Search for Our Human Origins*: Palgrave Macmillan; New York, New York; 266 p.

Tattersall, Ian, 2015; *The Strange Case of the Rickety Cossack – And Other Cautionary Tales from Human Evolution*: Palgrave Macmillan; New York, New York; 244 p.

Tenover, Fred C., Biddle, James W., and Lancaster, Michael V.; 2001; Increasing Resistance to Vancomycin and Other Glycopeptides in *Staphylococcus aureus*: *Emerging Infectious Diseases*; Vol. 7, No. 2, March-April 2001 (Special Issue); p. 327-332. (A pdf of this article may be found at the following web site: https://www.ncbi.nlm.nih.gov/pmc/articles/PMC2631729/pdf/11294734.pdf; accessed 07/16/2017.)

Thackray, J. C.; 1976; The Murchison–Sedgwick controversy: *Journal of the Geological Society*; Vol. 132; No. 4; p. 367-372.

Thaxton, Charles B.; Bradley, Walter L.; Olsen, Roger L.; 1984; *The Mystery of Life's Origin – Reassessing Current Theories*: Published by Lewis and Stanley; Dallas, Texas; 228 p. (A pdf of this book may be found at the following web site: https://www.krusch.com/books/evolution/Mystery_of_Lifes_Origin.pdf; accessed 03/21/2017.)

The Chimpanzee Sequencing and Analysis Consortium; 2005; Initial sequence of the chimpanzee genome and comparison with the human genome: *Nature*, Vol. 437, 1 September 2005; p. 69-87. (Here is a link to a pdf file of this paper-http://www.nature.com/nature/journal/v437/n7055/pdf/nature04072.pdf)

Thomas More Law Center; 2017; at https://www.thomasmore.org/about/about-the thomas-more-law-center-1/; last accessed 04/03/2017.

Thompson, John N.; 2013; *Relentless Evolution*: The University of Chicago Press; Chicago and London; 512 p.

Tillery, Bill W.; 2007; *Physical Science (Seventh Edition):* McGraw-Hill Higher Education; New York, New York; 729 p.

Travis, John; 2015; Making the Cut – Genome-editing Technology Shows Its Power: *Science*; Vol. 350; Issue 6267; p. 1456-1457.

Tree of Life Web Project: http://tolweb.org/tree/ ; accessed 04/15/2014.

Tudge, Colin; 2008; *The Bird – A Natural History of Who Birds Are, Where They Came From, and How They Live*: Crown Publishers; New York; 462 p.

University of California at Berkeley; Early Concepts of Evolution: Jean Baptiste

Lamarck: http://evolution.berkeley.edu/evolibrary/article/history_09; accessed
October 30, 2013).

University of California Museum of Paleontology at Berkeley; Adam Sedgwick (1785-1873): http://www.ucmp.berkeley.edu/history/sedgwick.html; accessed
11/13/2013.

University of California Museum of Paleontology at Berkeley; *Archaeopteryx*:
An Early Bird:
http://www.ucmp.berkeley.edu/diapsids/birds/archaeopteryx.html; accessed
07/31/2014.

University of California Museum of Paleontology at Berkeley; Biowarfare:
http://evolution.berkeley.edu/evolibrary/article/biowarfare_01, accessed
05/09/2014.

University of California Museum of Paleontology at Berkeley; Erasmus Darwin:
http://www.ucmp.berkeley.edu/history/Edarwin.html; accessed October 30,
3013.

University of California Museum of Paleontology at Berkeley; Solnhofen Limestone:
http://www.ucmp.berkeley.edu/mesozoic/jurassic/solnhofen.html; accessed
08/02/2014.

Urry, Lisa A.; Cain, Michael L.; Wasserman, Steven A.; Minorsky, Peter V.; Jackson, Robert B.; and Reece, Jane B.; 2014; *Campbell Biology in Focus*: Pearson Education, Inc.; Glenview, Illinois; 905 p.

U. S. Fish & Wildlife Service, Key Deer Fact Sheet:
http://www.fws.gov/southeast/pubs/facts/nkdcon.pdf; accessed 05/01/2014.

U. S. Fish & Wildlife, National Key Deer Refuge:
http://www.fws.gov/nationalkeydeer/; accessed 05/01/2014.

U. S. Geological Survey; 1991; Radiometric Time Scale:
http://pubs.usgs.gov/gip/geotime/radiometric.html; accessed 03/06/2014.

Van Valen, Leigh; 1973; A New Evolutionary Law: *Evolutionary Theory*; Vol. 1; p. 1-30.

van Wyhe, John (ed); 2002-; The Complete Work of Charles Darwin Online;
at http://darwin-online.org.uk/; last accessed 01/09/2018.

van Wyhe, John; 2013; *Dispelling the Darkness - Voyage in the Malay Archipelago and the Discovery of Evolution by Wallace and Darwin*: World Scientific Publishing Co. Pte. Ltd.; Singapore; 420 p.

Vine, F.J.; 1966; Spreading of the Ocean Floor: New Evidence: *Science*; Vol.154 No. 3755; p.1405–1415.

Vine, F. J.; and Matthews, D. H.; 1963; Magnetic Anomalies Over Oceanic Ridges: *Nature*; Vol. 199; No. 4897; p. 947–949.

Vine, F. J.; and Wilson, J. Tuzo; 1965; Magnetic Anomalies Over a Young Oceanic Ridge off Vancouver Island: *Science*; Vol. 150; No. 3695; p. 485–489.

Wainwright, Mark; 2007; The Mammals of Costa Rica – A Natural History and Field Guide: A Zona Tropical Publication from Comstock Publishing Associates; Cornell University Press; Ithaca and London; 529 p.

Wake, David B.; 1997; Incipient Species Formation in Salamanders of the *Ensatina* Complex: *Proc. Natl. Acad. Sci.* USA; Vol. 94, p. 7761–7767.

Wake, David B. and Yanev, Kay P.; 1986; Geographic Variation in Allozymes in a

"Ring Species", the Plethodontid Salamander *Ensatina eschscholtzii* of Western North America: *Evolution*; Vol 40; p. 702-715.

Wake, David B.; Yanev, Kay P; and Brown, Charles W.; 1986; Intraspecific Sympatry in a "Ring Species", the Plethodontid Salamander *Ensatina eschscholtzii* in Southern California: *Evolution*; Vol. 40; No. 4; p. 866-868.

Wake, David B.; Yanev, Kay P.; and Frelow, M. M.; 1989; Sympatry and Hybridization in a "Ring Species": the Plethodontid Salamander *Ensatina eschscholtzii*: in *Speciation and Its Consequences*; D. Otte and J. A. Endler (eds.); Sinauer Associates, Inc., Massachusetts; p.134-157.

Walter, Robert C.; 1997; Potassium-Argon/Argon-Argon Dating Methods: in Taylor, R. E. and Aitken, Martin J. (eds.); *Chronometric Dating in Archaeology, Advances in Archaeological and Museum* Science; Vol. 2; Springer US; p. 97-126.

Ward, Peter D.; 2006; *Dinosaurs, Birds, and Earth's Ancient Atmosphere – Out of Thin Air*: Joseph Henry Press; Washington, D. C.; 282 p.

Warny, Peter R.; Meshaka, Jr., Walter; and Klippel, Amy N.; 2012; Larval Growth and Transformation Size of the Green Frog, *Lithobates clamitans melanota* (Rafinesque, 1820), in South-Eastern New York: *Herpetology Notes*; Vol. 5; p. 59-62.

Watson, J. D.; 1968 (copyright renewed by James D. Watson, 1996); *The Double Helix - A Personal Account of the Discovery of the Structure of DNA*: Simon & Schuster (First Touchstone Edition 2001); New York, New York; 226 p.

Watson, James D. and Crick, Francis H.C.; 1953a; A Structure for Deoxyribose Nucleic Acid: *Nature*; Vol. 171; p. 737-738.

Watson, James D. and Crick, Francis H.C.; 1953b; Genetical Implications of the Structure of Deoxyribonucleic Acid: *Nature*; Vol 171; p. 964-967.

Weiner, Jonathan; 1994; *The Beak of the Finch – A Story of Evolution in Our Time*: Vintage Books, A Division of Random House, Inc.; New York; 332 p.

Wicander, Reed and Monroe, James; 2010; *Historical Geology - Evolution of Earth and Life Through Time (6th ed.)*: Brooks/Cole; Pacific Grove, California; 444 p.

Wicander, Reed and Monroe, James; 2013; *Historical Geology - Evolution of Earth and Life Through Time (7th ed.)*: Brooks/Cole, Cengage Learning; Belmont, California; 432 p.

Wikipedia; Adam Sedgwick; http://en.wikipedia.org/wiki/Adam_Sedgwick; accessed 11/13/2013.

Wikipedia; *Archaeopteryx*; http://en.wikipedia.org/wiki/Archaeopteryx; accessed 07/31/2014.

Wikipedia; Arthur Holmes: http://en.wikipedia.org/wiki/Arthur_Holmes; accessed 03/06/2014.

Wikipedia; Bertram Boltwood: http://en.wikipedia.org/wiki/Bertram_Boltwood; accessed 3/06/2014.

Wikipedia; Charles Doolittle Walcott; http://en.wikipedia.org/wiki/Charles_Doolittle_Walcott; accessed 11/05/2014.

Wikipedia; Discovery of Cosmic Microwave Background Radiation:

http://en.wikipedia.org/wiki/Discovery_of_cosmic_microwave_background_ra diation; accessed 10/03/2014.

Wikipedia; Georges Cuvier: http://en.wikipedia.org/wiki/Georges_Cuvier; accessed 12/08/2013.

Wikipedia; Henri Becquerel: http://en.wikipedia.org/wiki/Henri_Becquerel; accessed 02/09/2014.

Wikipedia; Hugo de Vries: http://en.wikipedia.org/wiki/Hugo_de_Vries; accessed 11/10/2013.

Wikipedia; James Hutton: http://en.wikipedia.org/wiki/James_Hutton; accessed 12/02/2013.

Wikipedia; James Ussher; http://en.wikipedia.org/wiki/James_Ussher; accessed 11/28/2013.

Wikipedia; Key deer: http://en.wikipedia.org/wiki/Key_deer; accessed 05/01/2014.

Wikipedia; Leopard frog: http://en.wikipedia.org/wiki/Leopard_frog; accessed 05/12/2014

Wikipedia; *Lithobates*: http://en.wikipedia.org/wiki/Lithobates; accessed 05/12/2014.

Wikipedia; *Lithobates clamitans*; http://en.wikipedia.org/wiki/Lithobates_clamitans; accessed 09/01/2014.

Wikipedia; Marie Skłodowska-Curie: http://en.wikipedia.org/wiki/Marie_Curie; accessed 02/09/2014.

Wikipedia; Michael Majerus; https://en.wikipedia.org/wiki/Michael_Majerus; accessed 06/24/2015.

Wikipedia; Pierre Curie: http://en.wikipedia.org/wiki/Pierre_Curie; accessed 02/09/2014.

Wikipedia; Solnhofen Plattenkalk: http://en.wikipedia.org/wiki/Solnhofen_Plattenkalk; accessed 08/02/2014.

Wikipedia; White-tailed deer: http://en.wikipedia.org/wiki/White-tailed_deer; accessed 05/01/2014.

Williams, Becky L.; Hanifin, Charles T.; Brodie, Edmund D. Jr; and Brodie III, Edmund D.; 2010; Tetrodotoxin Affects Survival Probability of Rough-Skinned Newts (*Taricha granulosa*) Faced with TTX-Resistant Garter Snake Predators (*Thamnophis sirtalis*): Chemoecology; Vol. 20; p. 285-290.

Wilmot, N. V. and Fallick, A. E.; 1989; Original Mineralogy of Trilobite Exoskeletons: Palaeontology; Vol. 32; Part 2; p. 297-304.

Wilson, J. Tuzo; 1965; Transform Faults, Oceanic Ridges, and Magnetic Anomalies Southwest of Vancouver Island: *Science*; Vol. 150; No. 3695; p. 482-485.

Winchester, Simon; 2001; *The Map That Changed the World - William Smith and the Birth of Modern Geology*: HarperCollins Publishers Inc.; New York, New York; 329 p.

Woese, Karl R; Kandler, Otto; and Wheelis, Mark L.; 1990; Towards a Natural System of Organisms: Proposal for the Domains Archaea, Bacteria, and Eucarya: *Proc. Nati. Acad. Sci. USA*; Vol. 87; p. 4576-4579.

Wood, Bermard; 2014; Welcome to the Family: *Scientific American*; Vol. 311; No. 3; p. 42-47.

Wood, Bernard; 2010; Reconstructing Human Evolution: Achievements,

Challenges, and Opportunities: *Proc. Natl. Acad. Sci. USA*; Vol. 107; Supplement 2; p. 8902–8909. (This article may be downloaded from the following website: www.pnas.org/cgi/doi/10.1073/pnas.1001649107.)

Wood, Bernard; 2002; Palaeoanthropology: Hominid Revelations from Chad: *Nature*; Vol. 418; July 11, 2002; p. 133-135.

Wood, Bernard and Lonergan, Nicholas; 2008; Review - The Hominin Fossil Record: Taxa, Grades and Clades: *Journal of Anatomy* (Printed in the United Kingdom); Vol. 212; p. 354-376.

Wood, Bernard and Richmond, Brian G.; 2000; Review - Human Evolution: Taxonomy and Paleobiology: *Journal of Anatomy* (Printed in the United Kingdom); Vol. 196; p. 19-60.

Wulf, Andrea; 2008 (First Vintage Books Edition, March 2010); *The Brother Gardeners – Botany, Empire, and the Birth of an Obsession*: Vintage Books (A Division of Random House, Inc.); New York; 354 p.

Zimmer, Carl; 1998 (first Touchstone edition, 1999); *At the Water's Edge – Fish with Fingers, Whales with Legs, and How Life Came Ashore but Then Went Back to Sea*: A Touchstone Book, Published by Simon and Schuster; New York, New York; 290 p.

Zimmer, Carl; 2010; *The Tangled Bank – An Introduction to Evolution*: Roberts and Company Publishers; Greenwood Village, Colorado; 385 p.

Zimmer, Carl; 2015; Breakthrough DNA Editor Born of Bacteria: *Quanta Magazine*; online at https://www.quantamagazine.org/crispr-natural-history-in-bacteria-20150206/; accessed 10/10/2017.

Zimmer, Carl and Emlen, Douglas; 2013; *Evolution - Making Sense of Life*: Roberts and Company Publishers; Greenwood Village, Colorado; 680 p.

Zotz, Gerhard and Hietz, Peter; 2001; The Physiological Ecology of Vascular Epiphytes: Current Knowledge, Open Questions: *Journal of Experimental Botany*; Vol. 52; Issue 364; p. 2067-2078.

Zuchowski, Willow; 2007; *Tropical Plants of Costa Rica – A Guide to Native and Exotic Flora*: A Zona Tropical Publication from Comstock Publishing Associates (a division of Cornell University Press); Ithaca and London; 529 p.

Zuykov, Michael A. and Harper, David A. T.; 2007; *Platystrophia* (Orthida) and New Related Ordovician and Early Silurian Brachiopod Genera: *Estonian Journal of Earth Sciences*; Vol. 56; p. 11-34.

INDEX

D

E

G

I

M

N

O

Rodinia, 237

Roman Catholic, 408

Roman Catholic Church, 31

Roman Inquisition, 31

Romer, Alfred S., 250

ropical wet forests in the Caribbean lowlands of northeastern Costa Rica, 337

rossopterygian lobe-fined fishes, 162

rough-skinned newt (*Taricha granulosa*), 318

Rubidium-87:Strontium-87 radiometric dating, 105

S

Sacramento and San Joaquin valleys, 307

Sagan, Lynn, 234

Sahelanthropus tchadensis, 358, 359, 365

Salmonella, 419

saltational evolution, 27

San Andreas Fault, 198

San Diego, California, 389

sandstones, 46

Santa Maria and San Cristóbal Islands, 316

Santorum, Rick, 406

Sarcopterygii, 249

Saturn, 211

savannahs, 364

savannahs with isolated forested areas, 364

scala naturae, 1

scarlet fever, 12

Schopf, William, 229

science classrooms of the public schools, 403

scientific creationism, 386

scientific inquiry, 30, 369, 372

scientific inquiry methodology, 31

scientific materialism, 370

Scientific Methodology, 32, 34

scientific naturalism, 370

scientific theories, 369

Scopes Monkey Trial, 374

Scopes Trial, 374

Scopes, John Thomas, 376

Scotland, 317

Scottish Enlightenment, 1

seafloor spreading, 193, 194

seafloor spreading, hypothesis of, 192

secondary chemical compounds, 350

secondary education (primarily in public schools), 374

secondary sexual characteristics, 281

secondary sexual characteristics of male guppies, 281

Secretary of State, 375

secular legislative purpose, 391

Sedgwick, Adam, 8, 73

sediment, 41

sedimentary facies, 75

sedimentary rocks, 39

seed, 19

selection for the more fit, 292

selection pressure, 260, 266, 295, 366, 367

selective breeding, 261

self-fertilization, 24

self-fertilize, 19

self-pollinate, 19

self-pollination, 19

Semibalanus, 317

semilunate carpal bone, 157

semisynthetic penicillin derivitives, 420

sequester toxins from milkweed plants, 351

Serial Endosymbiosis Hypothesis, 234

Serial Endosymbiosis Theory (SET), 234

Series, 91

seven major lithospheric plates, 189

Seventh Day Adventist, 417

Sex (germ) cells, 111

sex cells, 111, 284

sex cells (gametes)., 21

sex chromosomes, 112

sexual dimorphism, 291, 338

sexual reproduction, 232, 319

sexual selection, 280, 281, 338

sexual selection in guppy populations, 282

sexually dimorphic, 314

sexually dimorphic characteristics, 280

sexually reproducing, 20

shales, 51

shared characters, 131

shared derived homologous characters, 134

Shrewsbury (Boarding) School, 7

Shubin, Neil, 161

Sierra Nevada Mountains, 306

siliciclastic detrital sedimentary rocks, 42

siltstone, 50

siltstones, 50

Silurian System, 74

simian immunodeficiency viruses (SIV), 424

Simpson, George Gaylord, 29

single nucleotide polymorphisms (SNPs), 170

single nucleotide polymorphisms (SNPs),, 124

single supercontinent, 185

Sinsk biota in Russia, 246

Sirius Passet fauna of Greenland, 246

X

Y

Z

ABOUT THE AUTHOR

Edward L. Crisp is an emeritus professor of geology at West Virginia University at Parkersburg. He received a B.S. in geology (minor in biology) from Morehead State University, Morehead, Kentucky; a M.S. in geology from the University of Kentucky, Lexington, Kentucky; and a Ph.D. in geology (major area: paleontology, minor area: biology) from Indiana University, Bloomington, Indiana. Dr. Crisp spent the first fifteen years of his career as a petroleum exploration geologist, primarily working the geology of the subsurface in the Texas Gulf Coast region, except for an eighteen-month stint in Damascus, Syria, exploring for petroleum in Syria. He then spent the remaining twenty-four years of his career as a professor in higher education, the last twenty-two years at WVU at Parkersburg, teaching primarily geology, earth science, astronomy, and physical science; but several years also teaching the two courses, *the paleobiology of dinosaurs* and *the principles of biologic evolution.* Dr. Crisp lives in Vienna, West Virginia with his wife Susan Sowards.